Wavelets in Geophysics

Wavelet Analysis and Its Applications

The subject of wavelet analysis has recently drawn a great deal of attention from mathematical scientists in various disciplines. It is creating a common link between mathematicians, physicists, and electrical engineers. This book series will consist of both monographs and edited volumes on the theory and applications of this rapidly developing subject. Its objective is to meet the needs of academic, industrial, and governmental researchers, as well as to provide instructional material for teaching at both the undergraduate and graduate levels.

This fourth volume of the series is a compilation of twelve papers devoted to wavelet analysis of geophysical processes. In addition to an introductory review article written by the editors, this volume covers such important areas as atmospheric turbulence, seismic data analysis, detection of signals in noisy environments, multifractal analysis, and analysis of long memory geophysical processes. The series editor is grateful to Professor Foufoula-Georgiou and Dr. Kumar for their effort in compiling and editing this excellent volume, and would like to thank the authors for their very fine contributions.

This is a volume in
WAVELET ANALYSIS AND ITS APPLICATIONS

CHARLES K. CHUI, SERIES EDITOR
Texas A&M University, College Station, Texas

A list of titles in this series appears at the end of this volume.

Wavelets in Geophysics

Edited by

Efi Foufoula-Georgiou

*St. Anthony Falls Hydraulic
Laboratory
Department of Civil Engineering
University of Minnesota
Minneapolis, Minnesota*

Praveen Kumar

*Universities Space Research Association
Hydrological Sciences Branch
NASA–Goddard Space Flight Center
Greenbelt, Maryland*

ACADEMIC PRESS

San Diego New York Boston
London Sydney Tokyo Toronto

This book is printed on acid-free paper. ∞

Copyright © 1994 by ACADEMIC PRESS, INC.
All Rights Reserved.
No part of this publication may be reproduced or transmitted in any form or by any
means, electronic or mechanical, including photocopy, recording, or any information
storage and retrieval system, without permission in writing from the publisher.

Academic Press, Inc.
A Division of Harcourt Brace & Company
525 B Street, Suite 1900, San Diego, California 92101-4495

United Kingdom Edition published by
Academic Press Limited
24-28 Oval Road, London NW1 7DX

Library of Congress Cataloging-in-Publication Data

Wavelets in geophysics / edited by Efi Foufoula-Georgiou, Praveen
    Kumar.
         p.   cm. -- (Wavelet analysis and its applications ; v. 4)
      Includes bibliographical references and index.
      ISBN 0-12-262850-0
      I. Foufoula-Georgiou, Efi,   II. Kumar, Praveen.   III. Series.
   QC806.W38   1994
    550'.1'5152433--dc20   94-16953
                                                                   CIP

PRINTED IN THE UNITED STATES OF AMERICA
94   95   96   97   98   99   MM   9   8   7   6   5   4   3   2   1

*To
Katerina, Thomas
and
Ilina*

Contents

Contributors

Numbers in parentheses indicate the pages on which the authors' contributions begin.

JOHN D. ALBERTSON (81), *Hydrologic Science, University of California at Davis, Davis, California 95616*

KEVIN E. BREWER (213), *Department of Geological Sciences/172, University of Nevada, Reno, Reno, Nevada 89557*

YVES BRUNET (129), *Laboratoire de Bioclimatologie, INRA, BP 81, 33883 Villenave d'Ornon, France*

CHIA R. CHU (81), *Department of Civil Engineering, National Central University, Chungli, Taiwan*

SERGE COLLINEAU (129), *Laboratoire de Bioclimatologie, INRA, BP 81, 33883 Villenave d'Ornon, France*

ANTHONY DAVIS (249), *Universities Space Research Association, NASA–Goddard Space Flight Center, Greenbelt, Maryland 20771*

EFI FOUFOULA-GEORGIOU (1), *St. Anthony Falls Hydraulic Laboratory, Department of Civil Engineering, University of Minnesota, Minneapolis, Minnesota 55414*

NIMAL K. K. GAMAGE (45), *Program in Atmospheric and Oceanic Sciences, Astrophysics, Planetary and Atmospheric Sciences Department, University of Colorado, Boulder, Colorado 80309*

PETER GUTTORP (325), *Department of Statistics, University of Washington, Seattle, Washington 98195*

CARL R. HAGELBERG (45), *National Center for Atmospheric Research, Boulder, Colorado 80307*

J. F. HOWELL (107), *Oceanic and Atmospheric Sciences, Oregon State University, Corvallis, Oregon 97331*

GABRIEL G. KATUL (81), *School of the Environment, Duke University, Durham, North Carolina 27708*

PRAVEEN KUMAR (1), *Universities Space Research Association, Hydrological Sciences Branch, NASA-Goddard Space Flight Center, Greenbelt, Maryland 20771*

SARAH A. LITTLE (167), *Department of Geology and Geophysics, Woods Hole Oceanographic Institution, Woods Hole, Massachusetts 02543*

PAUL C. LIU (151), *NOAA Great Lakes Environmental Research Laboratory, Ann Arbor, Michigan 48109*

L. MAHRT (107), *Oceanic and Atmospheric Sciences, Oregon State University, Corvallis, Oregon 97331*

ALEXANDER MARSHAK (249), *Science Systems and Applications, Inc., Lanham, Maryland 20706*

MARC B. PARLANGE (81), *Hydrologic Science, and Department of Agricultural and Biological Engineering, University of California at Davis, Davis, California 95616*

DONALD B. PERCIVAL (325), *Applied Physics Laboratory, University of Washington, Seattle, Washington 98195*

CHRIS J. PIKE (183), *Centre for Cold Ocean Resources Engineering (C-CORE), Memorial University of Newfoundland, St. John's, Newfoundland, Canada A1B 3X5*

NAOKI SAITO (299), *Schlumberger-Doll Research, Old Quarry Road, Ridgefield, Connecticut 06877, and Department of Mathematics, Yale University, New Haven, Connecticut 06520*

STEPHEN W. WHEATCRAFT (213), *Department of Geological Sciences/172, University of Nevada, Reno, Reno, Nevada 89557*

WARREN WISCOMBE (249), *NASA–Goddard Space Flight Center, Climate and Radiation Branch, Greenbelt, Maryland 20771*

Preface

Over the past decade, wavelet transforms have been formalized into a rigorous mathematical framework and have found numerous applications in diverse areas such as harmonic analysis, numerical analysis, signal and image processing, nonlinear dynamics, fractal and multifractal analysis, and others. Although wavelet transforms originated in geophysics (for the analysis of seismic signals), it is only very recently that they are being used again in the geophysical sciences. Properties that make wavelets attractive are time–frequency localization, orthogonality, multirate filtering, and scale-space analysis, to name a few.

This volume is the first collection of papers using wavelet transforms for the understanding, analysis, and description of geophysical processes. It includes applications of wavelets to atmospheric turbulence, ocean wind waves, characterization of hydraulic conductivity, seafloor bathymetry, seismic data, detection of signals from noisy data, multifractal analysis, and analysis of long memory geophysical processes. Most of the papers included in this volume were presented at the American Geophysical Union (AGU) Spring Meeting in Baltimore, May 1993, in a special Union session organized by us entitled "Applications of Wavelet Transforms in Geophysics." We feel that this volume will serve geophysicists as an introduction to the versatile and powerful wavelet analysis tools and will stimulate further applications of wavelets in geophysics as well as mathematical developments dictated by unique demands of applications.

The first chapter in this volume is a review article by Kumar and Foufoula-Georgiou. The purpose of this article is to provide the unfamiliar reader with a basic introduction to wavelets and key references for further study. Wavelet transforms are contrasted with the Fourier transforms and windowed Fourier transforms that are well known to geophysicists; this contrast highlights the important property of time–frequency localization in wavelet transforms, which is essential for the analysis of nonstationary and transient signals. Continuous and discrete, as well as orthogonal, nonorthogonal, and biorthogonal wavelet transforms are then reviewed, and the concept of multiresolution analysis is presented. Several examples of one- and two-dimensional wavelets and information on wavelet construction are given. Finally, some sources of available wavelet analysis software packages are included which may help the interested reader get started in exploring wavelets.

The next four chapters present results from the application of wavelet analysis to atmospheric turbulence. In Chapter 2, Hagelberg and Gamage develop a wavelet-based signal decomposition technique that preserves intermittent coherent structures. Coherent structures in velocity and temperature in the atmospheric boundary layer account for a large portion of flux transport of momentum, heat, trace chemicals, and particulates. The authors' technique partitions signals into two components: one containing coherent structures characterized by sharp transitions and intermittent occurrence, and the other containing the remaining portion of the signal (essentially characterized by smaller length scales and the absence of coherent events). They apply this decomposition to vertical velocity, virtual potential temperature, and buoyancy flux density fields. Chapter 3, by Katul, Albertson, Chu, and Parlange, applies orthonormal wavelets to atmospheric surface layer velocity measurements to describe space–time relations in the inertial subrange. The local nature of the orthonormal wavelet transform in physical space aids the identification of events contributing to inertial subrange intermittency buildup, which can then be suppressed to eliminate intermittency effects on the statistical structure of the inertial subrange.

In Chapter 4, Howell and Mahrt develop an adaptive method for decomposing a time series into orthogonal modes of variation. In contrast to conventional partitioning, the cutoff scales are allowed to vary with record position according to the local physics of the flow by utilizing the Haar wavelet decomposition. For turbulence data, this decomposition is used to distinguish four modes of variation. The two larger modes, determined by spatially constant cutoff scales, are characterized as the mesoscale and large eddy modes. The two smaller scale modes are separated by a scale that depends on the local transport characteristics of the flow. This adaptive cutoff scale separates the transporting eddy mode, responsible for most of the flux, from the nontransporting nearly isotropic motions. Chapter 5, by Brunet and Collineau, applies wavelets to the analysis of turbulent motions above a maize crop. Their results indicate that organized turbulence exhibits the same structure above a forest and a maize crop, apart from a scale factor, and supports the interesting postulate that transfer processes over plant canopies are dominated by populations of canopy-scale eddies, with universal characteristics. The authors also propose a methodology for separating turbulence data into large- and small-scale components using the filtering properties of the wavelet transform.

In Chapter 6, Liu applies wavelet spectrum analysis to ocean wind waves. The results reveal significant new insights on wave grouping parameterizations, phase relations during wind wave growth, and detection of wave breaking characteristics. Chapter 7, by Little, demonstrates the usefulness of wavelet analysis in studying seafloor bathymetry and especially identifying the location and scarp-facing direction of ridge-parallel faulting. In Chapter 8, Pike proposes a wavelet-based methodology for the analysis of high resolution acoustic signals for sub-seabed feature extraction and classification of scatterers. The key idea is that of displaying the energy dissipation in the time–frequency plane, which allows a more distinct description of the signal (compared to the Fourier transform spectra) and thus provides a better means of extracting seabed properties by correlating them to the attenuation of high resolution acoustic signals. In Chapter 9, Brewer and Wheatcraft investigate the wavelet transform as a tool for reconstructing small-scale variability in hydraulic conductivity fields by incorporating the scale and location information of each sample when interpolating to a finer grid. The developed multiscale reconstruction method is compared to traditional interpolation schemes and is used to examine the issue of optimum sample size and density for stationary and fractal random fields.

Chapter 10, by Davis, Marshak, and Wiscombe, explores the use of wavelets for multifractal analysis of geophysical phenomena. They show the applicability of wavelet transforms to compute simple yet dynamically meaningful statistical properties of one dimensional geophysical series. Turbulent velocity and cloud liquid water content are used as examples to demonstrate the need for stochastic models having both additive (nonstationary) and multiplicative (intermittent) features. Merging wavelet and multifractal analysis seems promising for both wavelet and multifractal communities and is especially promising for geophysics, where many signals show structures at all observable scales and are often successfully described within a multifractal framework.

The wavelet transform partitions the frequency axis in a particular way: it iteratively partitions the low-frequency components, leaving the high-frequency components intact at each iteration. For some processes or applications this partition might not achieve the best decomposition, as partition of the high-frequency bands might also be necessary. Wavelet packets provide such a partition and are used in Chapter 11, by Saito, for simultaneous signal compression and noise reduction of geophysical signals. A maximum entropy criterion is used to obtain the best basis out of the many bases that the redundant wavelet packet representation provides. The method is applied to synthetic signals and to some geophysical data, for example, a radioactivity profile of subsurface formation and a migrated seismic section. Finally, in Chapter 12, Percival and Guttorp examine a particular measure of variability for long memory processes (the Allan variance) within the wavelet framework and show that this variance can be interpreted as a Haar wavelet coefficient variance. This suggests an approach to assessing the variability of general wavelet classes which will be useful in the study of power-law processes extensively used for the description of geophysical time series. A fairly extensive bibliography of wavelet analysis in geophysics is included at the end of this volume.

Several individuals provided invaluable help in the completion of this volume. Special thanks go to the reviewers of the book chapters who volunteered their time and expertise and provided timely and thoughtful reviews. The first author thanks Mike Jasinski, Hydrologic Sciences Branch at NASA–Goddard Space Flight Center and the Universities Space Research Association for their support during the completion of this project. We are grateful to Charu Gupta Kumar, who converted most of the chapters to $\LaTeX$ format and typeset and edited the entire volume. Without her expertise and dedication the timely completion of this volume would not have been possible. Finally, we also thank the Academic Press Editor, Peter Renz, for his efficient help during the final stages of this project.

Efi Foufoula-Georgiou
Minneapolis, Minnesota
March, 1994

Praveen Kumar
Greenbelt, Maryland

Wavelet Analysis in Geophysics: An Introduction

Praveen Kumar and Efi Foufoula-Georgiou

Abstract. Wavelet analysis is a rapidly developing area of mathematical and application-oriented research in many disciplines of science and engineering. The wavelet transform is a localized transform in both space (time) and frequency, and this property can be advantageously used to extract information from a signal that is not possible to unravel with a Fourier or even windowed Fourier transform. Wavelet transforms originated in geophysics in early 1980's for the analysis of seismic signal. After a decade of significant mathematical formalism they are now also being exploited for the analysis of several other geophysical processes such as atmospheric turbulence, space-time rainfall, ocean wind waves, seafloor bathymetry, geologic layered structures, climate change, among others. Due to their unique properties, well suited for the analysis of natural phenomena, it is anticipated that there will be an explosion of wavelet applications in geophysics in the next several years. This chapter provides a basic introduction to wavelet transforms and their most important properties. The theory and applications of wavelets is developing very rapidly and we see this chapter only as a limited basic introduction to wavelets which we hope to be of help to the unfamiliar reader and provide motivation and references for further study.

§1. Prologue

The concept of wavelet transforms was formalized in early 1980's in a series of papers by Morlet et al. [42, 43], Grossmann and Morlet [24], and Goupillaud, Grossmann and Morlet [23]. Since this formalism and some significant work by Meyer ([38, 39] and references therein), Mallat [30, 31], Daubechies [13, 15], and Chui [6] among others, the wavelets have become pervasive in several diverse areas such as mathematics, physics, digital signal processing, vision, numerical analysis, and geophysics, to name a few. Wavelet transforms are integral transforms using integration kernels called wavelets. These wavelets are essentially used in two ways when studying

Wavelets in Geophysics
Efi Foufoula-Georgiou and Praveen Kumar (eds.), pp. 1–43.

1

processes or signals: (i) as an integration kernel for analysis to extract information about the process, and (ii) as a basis for representation or characterization of the process. Evidently, in any analysis or representation, the choice of the basis function (or kernel) determines the kind of information that can be extracted about the process. This leads us to the following questions: (1) what kind of information can we extract using wavelets? and (2) how can we obtain a representation or description of the process using wavelets?

The answer to the first question lies on the important property of wavelets called time-frequency localization. The advantage of analyzing a signal with wavelets as the analyzing kernels is that it enables one to study features of the signal *locally* with a detail matched to their scale, i.e., broad features on a large scale and fine features on small scales. This property is especially useful for signals that are either non-stationary, or have short lived transient components, or have features at different scales, or have singularities. The answer to the second question is based on seeing wavelets as the elementary building blocks in a decomposition or series expansion akin to the familiar Fourier series. Thus, a representation of the process using wavelets is provided by an infinite series expansion of dilated and translated versions of a *mother wavelet*, each multiplied by an appropriate coefficient. For processes with finite energy this wavelet series expansion is optimal, i.e., it offers an optimal approximation to the original signal, in the least squares sense.

In what follows we give a brief introduction to the mathematics of wavelet transforms and where possible an intuitive explanation of these results. The intention of this introduction is two fold: (i) to provide the unfamiliar reader with a basic introduction to wavelets, and (ii) to prepare the reader to grasp and appreciate the results of the articles that follow as well as the potential of wavelet analysis for geophysical processes. At times we have sacrificed mathematical rigor for clarity of presentation in an attempt to not obscure the basic idea with too much detail. We also hasten to add that this review is far from complete both in terms of the breadth of topics chosen for exposition and in terms of the treatment of these topics. For example, the important topic of wavelet packets has not been discussed (see for example [53], [8], and the article by Saito [50] in this volume). It is recommended that the interested reader who is not a mathematician and is meeting wavelet analysis for the first time begins with the book by Meyer [39] and continues with the books by Daubechies [15] and Benedetto and Frazier [3]. There are also several nice introductory articles on several aspects of wavelets, as for example, those by Mallat [30, 31], Rioul and Vetterli [47], and the article by Farge [19] on turbulence, among others. Also, considerable insight can be gained by the nice book reviews by Meyer [40] and Daubechies [18].

This article is organized as follows. Section 2 discusses Fourier and windowed Fourier transforms and introduces continuous wavelet transforms and their time-frequency localization properties. In section 3, the wavelet transform is presented as a time-scale transform, and the wavelet variance and covariance (alternatively called wavelet spectrum and cospectrum) are discussed. A link is also made between non-stationary processes and wavelet transforms akin to the link between stationary processes and Fourier transforms. Section 4 presents some examples of commonly used one-dimensional wavelets (Haar wavelet, Mexican hat wavelet, and Morlet wavelet). In section 5, discrete wavelet transforms (orthogonal, non-orthogonal, and biorthogonal) are introduced and the concept of multiresolution analysis presented. In section 6, we present extensions of continuous and discrete wavelets in a two-dimensional space. Finally, in section 7 we present some concluding remarks and give information on obtaining available software packages for wavelet analysis.

§2. Time-Frequency Analysis

The original motive for the development of wavelet transform was (see [23]) "... of devising a method of acquisition, transformation and recording of a seismic trace (i.e., a function of one variable, the time) so as to satisfy the requirements listed below:

1. The contributions of different frequency bands (i.e., of the different intervals of the Fourier conjugate variable) are kept reasonably separated.

2. This separation is achieved without excessive loss of resolution in the time variable (subject, of course, to the limitation of the uncertainty principle).

3. The reconstruction of the original function from its "representation" or "transform" is obtained by a method which is (a) capable of giving arbitrary high precision; (b) is robust, in the sense of being stable under small perturbations. "

The first two conditions essentially characterize the property known as time-frequency localization. Recall that although the Fourier transform of a function $f(t)$, given as

$$\mathcal{F}f \equiv \hat{f}(\omega) = \int_{-\infty}^{\infty} f(t)e^{-i\omega t}\, dt, \qquad (1)$$

gives the information about the frequency content of a process or signal, it gives no information about the location of these frequencies in the time domain. For example figures 1a,b show two signals – the first consisting of

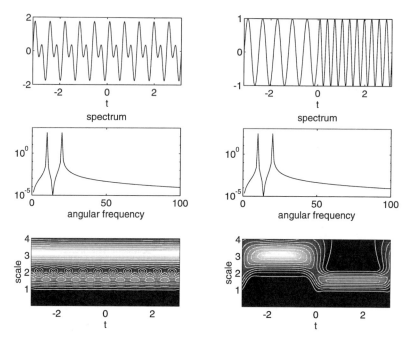

Figure 1. Spectral and wavelet analysis of two signals. The first signal **(a) (upper left)** consists of superposition of two frequencies ($\sin 10t$ and $\sin 20t$), and the second consists of the same two frequencies each applied separately over half of the signal duration **(b) (upper right)**. Figures **(c) (middle left)** and **(d) (middle right)** show the spectra of signals, i.e., $|f(\omega)|^2$ vs ω, in (a) and (b) respectively, and **(e) (lower left)** and **(f) (lower right)** show the magnitude of their wavelet transforms (using Morlet wavelet) respectively.

two frequencies ($\sin 10t$ and $\sin 20t$) superimposed for the entire duration of the signal and the second consisting of the same frequencies, but each one applied separately for half of signal duration. Figures 1c,d show the spectrum, i.e., $|f(\omega)|^2$, of these two signals, respectively. As is clearly evident, the spectrum is quite incapable of distinguishing between the two signals.

Time varying frequencies are quite common in music, speech, seismic signals, non-stationary geophysical processes, etc. To study such processes, one seeks transforms which will enable one to obtain the frequency content of a process locally in time. There are essentially two methods that have been developed to achieve this (within the limits of the uncertainty principle which states that one cannot obtain arbitrary good localization in both time and frequency): (a) windowed Fourier transform, and (b) wavelet transform. These two methods are discussed in the following subsections. Figures 1e,f display the magnitude of the wavelet transform of the signals shown in Figures 1a,b, and clearly show the ability of the wavelet transform to distinguish between the two signals.

2.1. Windowed Fourier transform

2.1.1. Definition

In the Fourier transform framework, time localization can be achieved by windowing the data at various times, say, using a windowing function $g(t)$, and then taking the Fourier transform. That is, the windowed Fourier transform (also called the short-time fourier transform), $Gf(\omega, t)$, is given by

$$Gf(\omega, t) \ = \ \int_{-\infty}^{\infty} f(u)g(u-t)e^{-i\omega u} \, du \tag{2}$$

$$= \ \int_{-\infty}^{\infty} f(u)g_{\omega, t}(u) \, du \tag{3}$$

where the integration kernel is $g_{\omega, t}(u) \equiv g(u-t)e^{-i\omega u}$. This transform measures locally, around the point t, the amplitude of the sinusoidal wave component of frequency ω. The window function $g(t)$ is usually chosen as a real, even function with the maximum concentration of energy in the low frequency components. Notice that the analyzing kernel $g_{\omega, t}(u)$ has the same support[1] for all ω and t, but the number of cycles vary with the frequency ω (see Figure 2).

The representation of the function $f(t)$ on the time-frequency plane, i.e., (ω, t) plane, thus obtained is called the phase-space representation.

The windowed Fourier transform is an energy preserving transformation or isometry, i.e.,

$$\int_{-\infty}^{\infty} |f(t)|^2 \, dx = \frac{1}{2\pi} \int_{-\infty}^{\infty} \int_{-\infty}^{\infty} |Gf(\omega, t)|^2 \, d\omega \, dt \tag{4}$$

provided $\int_{-\infty}^{\infty} |g(t)|^2 \, dx = 1$ (which we assume from here on). It is invertible with the reconstruction formula given as ([31], eq. 15)

$$f(t) = \frac{1}{2\pi} \int_{-\infty}^{\infty} \int_{-\infty}^{\infty} Gf(\omega, u) \, g(u-t)e^{i\omega t} \, d\omega \, du. \tag{5}$$

The parameters t and ω can be assigned discrete values, say $t = nt_0$

---

[1] support is defined as the closure of the set over which the signal/process is non-zero.

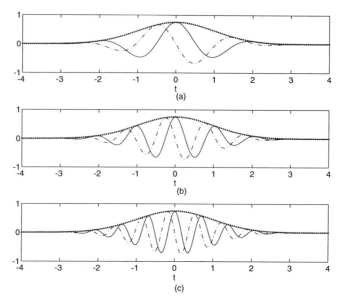

Figure 2. Real (solid lines) and imaginary parts (dot-dashed lines) of the analyzing kernel $g(t)e^{-i\omega t}$ of the windowed Fourier transform at different frequencies: **(a) (top)** $\omega = 3$, **(b) (middle)** $\omega = 6$ and **(c) (bottom)** $\omega = 9$. The dotted line indicates a Gaussian window function $g(t)$.

and $\omega = m\omega_0$, and we obtain the discrete windowed Fourier transform

$$G_d(m, n) = \int_{-\infty}^{\infty} f(u)g(u - nt_0)e^{-im\omega_0 u}\, du. \tag{6}$$

For the discrete windowed Fourier transform to be invertible, the condition $\omega_0 t_0 < 2\pi$ must hold (see [15], sections 3.4 and 4.1).

2.1.2. Time-frequency localization

In order to study the time-frequency localization property of the windowed Fourier transform, we need to study the properties of $|g_{\omega,t}|^2$ and $|\hat{g}_{\omega,t}|^2$ since they determine the features of $f(t)$ that are extracted. Indeed, using Parseval's theorem, equation (3) can be written as

$$Gf(\omega, t) = \frac{1}{2\pi} \int_{-\infty}^{\infty} \hat{f}(\omega')\overline{\hat{g}_{\omega,t}(\omega')}\, d\omega' \tag{7}$$

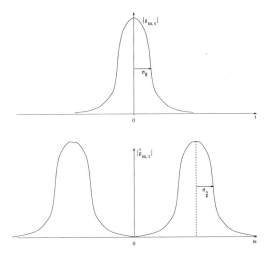

Figure 3. Uncertainties in time (**top**) and frequency (**bottom**) localization in a windowed Fourier Transform for a generic function $g(t)$.

where $\hat{g}_{\omega,t}(\omega')$ is the Fourier transform of $g_{\omega,t}(u)$ and overbar indicates complex conjugate. Let us define the standard deviations of $g_{\omega,t}$ and $\hat{g}_{\omega,t}$ as σ_g and $\sigma_{\hat{g}}$ respectively, i.e.,

$$\sigma_g = \left(\int_{-\infty}^{\infty} (u-t)^2 |g_{\omega,t}(u)|^2 \, du \right)^{1/2} = \left(\int_{-\infty}^{\infty} u^2 g(u)^2 \, du \right)^{1/2} \tag{8}$$

and

$$\sigma_{\hat{g}} = \left(\int_{-\infty}^{\infty} (\omega'-\omega)^2 |\hat{g}_{\omega,t}(\omega')|^2 \, d\omega' \right)^{1/2}. \tag{9}$$

These parameters measure the spread of the function $|g_{\omega,t}|$ and $|\hat{g}_{\omega,t}|$, about t and ω, respectively (see Figure 3). Owing to the uncertainty principle, the products of σ_g^2 and $\sigma_{\hat{g}}^2$ satisfy (see [31])

$$\sigma_g^2 \sigma_{\hat{g}}^2 \geq \frac{\pi}{2}, \tag{10}$$

i.e., arbitrary high precision in both time and frequency cannot be achieved. The equality in the above equation is achieved only when $g(t)$ is the Gaussian, i.e.,

$$g(t) = \pi^{-1/4} e^{-t^2/2}. \tag{11}$$

When the Gaussian function is used as a window, the windowed Fourier transform is called the Gabor transform [22].

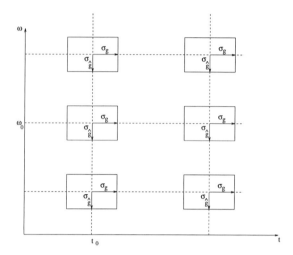

Figure 4. Figure showing the phase-space representation using the windowed Fourier transform.

Once a window function $g(t)$ is chosen both σ_g and $\sigma_{\hat{g}}$ are fixed. Therefore, for any given t_0 and ω_0, the time-frequency resolution can be represented by the fixed size resolution cell $[t_0 \pm \sigma_g \times \omega_0 \pm \sigma_{\hat{g}}]$ (see Figure 4), i.e., the windowed Fourier transform at any point (t_0, ω_0) in the phase-space provides information about $f(t)$ that is localized with an uncertainty of σ_g in the time domain and $\sigma_{\hat{g}}$ in the Fourier domain, and this localization is uniform in the entire phase-space. In other words the entire phase-space is uniformly layered with resolution cells or "bricks" of fixed dimensions. This poses two kinds of limitations. Firstly, if the process has a transient component with a support smaller than σ_g, it is difficult to locate it with precision better than σ_g. Secondly if the process has important features of differing sizes then we can not find an optimal $g(t)$ for analyzing the process. Therefore, window Fourier transform is more suited for analyzing processes where all the features appear approximately at the same scale. The wavelet transform addresses the limitations inherent in the windowed Fourier transform.

2.2. Wavelet transform

In the windowed Fourier transform, the analyzing functions $g_{\omega,t}$ for all ω and t consist of the the same envelope $g(t)$ filled in with sinusoids of frequency ω. Due to the fixed envelope $g(t)$, the resolution cell size in the phase space given by $[\sigma_g \times \sigma_{\hat{g}}]$ is the same for all ω and t. Since higher frequency (or short wavelength) features have smaller support, it would be desirable to have an analyzing function, say $\psi(t)$, such that its standard deviation σ_ψ is small when $\psi(t)$ characterizes high frequency components and vice-versa. This was achieved by decomposing the function $f(t)$ using a two parameter family of functions called *wavelets* (see [42] and [43]). One of the two parameters is the translation parameter as in the windowed Fourier transform case, but the other parameter is a dilation parameter λ instead of the frequency parameter ω.

2.2.1. Definition

The wavelet transform of a function $f(t)$ with finite energy is defined as the integral transform with a family of functions $\psi_{\lambda,t}(u) \equiv \frac{1}{\sqrt{\lambda}}\psi(\frac{u-t}{\lambda})$ and is given as

$$
\begin{aligned}
Wf(\lambda, t) &= \int_{-\infty}^{\infty} f(u)\psi_{\lambda,t}(u)\, du \qquad \lambda > 0 \\
&= \int_{-\infty}^{\infty} f(u)\frac{1}{\sqrt{\lambda}}\psi(\frac{u-t}{\lambda})\, du. \qquad (12)
\end{aligned}
$$

Here λ is a scale parameter, t a location parameter and the functions $\psi_{\lambda,t}(u)$ are called wavelets. In case $\psi_{\lambda,t}(u)$ is complex, we use the complex conjugate $\overline{\psi}_{\lambda,t}(u)$ in the above integration. Changing the value of λ has the effect of dilating ($\lambda > 1$) or contracting ($\lambda < 1$) the function $\psi(t)$ (see Figure 5a), and changing t has the effect of analyzing the function $f(t)$ around the point t. The normalizing constant $\frac{1}{\sqrt{\lambda}}$ is chosen so that

$$
\| \psi_{\lambda,t} \|^2 \equiv \int |\psi_{\lambda,t}(u)|^2\, du = \int |\psi(t)|^2\, dt
$$

for all scales λ (notice the identity $\psi(t) \equiv \psi_{1,0}(t)$). We also choose the normalization $\int |\psi(t)|^2\, dt = 1$. The wavelet transform $Wf(\lambda, t)$ is often denoted as the inner product $\langle f, \psi_{\lambda,t} \rangle$.

Notice that in contrast to the windowed Fourier transform case, the number of cycles in the wavelet $\psi_{\lambda,t}(u)$ does not change with the dilation (scale) parameter λ but the support length does. We will see shortly that when λ is small, which corresponds to small support length, the wavelet transform picks up higher frequency components and vice-versa.

The choice of the wavelet $\psi(t)$ is neither unique nor arbitrary. The function $\psi(t)$ is a function with unit energy chosen so that it has:

1. compact support, or sufficiently fast decay, to obtain localization in space;

2. zero mean, i.e., $\int_{-\infty}^{\infty} \psi(t)\, dt = 0$, although higher order moments

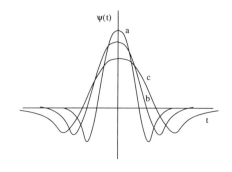

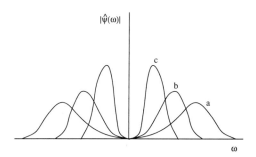

Figure 5. Schematic illustration of the effect of dilation on a "generic" wavelet **(top)** and the corresponding change on its Fourier transform $|\hat{\psi}(\omega)|$ **(bottom)**. When the wavelet dilates, its Fourier transform contracts and vice-versa. (a) $\lambda < 1$, (b) $\lambda = 1$, and (c) $\lambda > 1$.

may also be zero, i.e.,

$$\int_{-\infty}^{\infty} t^k \psi(t)\, dt = 0 \qquad \text{for } k = 0, \ldots, N-1. \tag{13}$$

The requirement of zero mean is called the *admissibility condition* of the wavelet. It is because of the above two properties that the function $\psi(t)$ is called a wavelet. The second property ensures that $\psi(t)$ has a wiggle, i.e., is wave like, and the first ensures that it is not a sustaining wave.

The inverse wavelet transform is given by ([15], eq. 2.4.4)

$$f(t) = \frac{1}{C_\psi} \int_{-\infty}^{\infty} \int_{0}^{\infty} \lambda^{-2} Wf(\lambda, u)\psi_{\lambda,u}(t)\, d\lambda\, du \tag{14}$$

where

$$C_\psi = 2\pi \int_0^\infty \frac{|\hat{\psi}(\omega)|^2}{\omega} \, d\omega < \infty. \tag{15}$$

The wavelet transform is also an energy preserving transformation, i.e., an isometry (up to a proportionality constant), that is,

$$\int_{-\infty}^\infty |f(t)|^2 \, dt = \frac{1}{C_\psi} \int_{-\infty}^\infty \int_0^\infty |Wf(\lambda, u)|^2 \lambda^{-2} \, d\lambda \, du. \tag{16}$$

2.2.2. Time-frequency localization

In order to understand the behavior of the wavelet transform in the frequency domain as well, it is useful to recognize that the wavelet transform $Wf(\lambda, t)$, using Parseval's theorem, can be equivalently written as

$$Wf(\lambda, t) = \frac{1}{2\pi} \int_{-\infty}^\infty \hat{f}(\omega) \overline{\hat{\psi}_{\lambda,t}(\omega)} \, d\omega. \tag{17}$$

Therefore, as in the windowed Fourier transform, we need to study the properties of $|\psi_{\lambda,t}(u)|^2$ and $|\hat{\psi}_{\lambda,t}(\omega)|^2$ to understand the time-frequency localization properties of wavelet transforms. Specifically, we need to understand the behavior of the standard deviations of $|\psi_{\lambda,t}|^2$ and $|\hat{\psi}_{\lambda,t}|^2$, i.e., $\sigma_{\psi_{\lambda,t}}$ and $\sigma_{\hat{\psi}_{\lambda,t}}$. Note that, due to property (13), $\hat{\psi}_{\lambda,t}(\omega = 0) = 0$. Consequently, the center of passing band, $\omega^0_{\hat{\psi}_{\lambda,t}}$, for $\psi_{\lambda,t}(t)$ is located away from the origin $\omega = 0$ (as shown in Figure 5b). It can be obtained as the center of mass (or first moment about the origin) of the right lobe as

$$\omega^0_{\hat{\psi}_{\lambda,t}} = \frac{\int_0^\infty \omega |\hat{\psi}_{\lambda,t}(\omega)|^2 \, d\omega}{\int_0^\infty |\hat{\psi}_{\lambda,t}(\omega)|^2 \, d\omega}. \tag{18}$$

We therefore define the standard deviation (i.e., square root of the second central moment of the right lobe) $\sigma_{\hat{\psi}_{\lambda,t}}$ as

$$\sigma_{\hat{\psi}_{\lambda,t}} = \left(\int_0^\infty (\omega - \omega^0_{\hat{\psi}_{\lambda,t}})^2 |\hat{\psi}_{\lambda,t}(\omega)|^2 \, d\omega \right)^{1/2}. \tag{19}$$

Similarly in the time domain the standard deviation $\sigma_{\psi_{\lambda,t}}$ can be obtained as

$$\sigma_{\psi_{\lambda,t}} = \left(\int_{-\infty}^\infty (u - t_0)^2 |\psi_{\lambda,t}(u)|^2 \, du \right)^{1/2} \tag{20}$$

where t_0 is given as

$$t_0 = \frac{\int_{-\infty}^\infty u |\psi_{\lambda,t}(u)|^2 \, du}{\int_{-\infty}^\infty |\psi_{\lambda,t}(u)|^2 \, du}. \tag{21}$$

It is easy to verify that the following relationships hold:

1. The standard deviation $\sigma_{\psi_{\lambda,t}}$ satisfies

$$\sigma_{\psi_{\lambda,t}} = \lambda\sigma_{\psi_{1,0}}. \qquad (22)$$

2. The standard deviation $\sigma_{\hat{\psi}_{\lambda,t}}$ satisfies

$$\sigma_{\hat{\psi}_{\lambda,t}} = \frac{\sigma_{\hat{\psi}_{1,0}}}{\lambda}. \qquad (23)$$

3. The center of passing band $\omega^0_{\hat{\psi}_{\lambda,t}}$ corresponding to the wavelet $\psi_{\lambda,t}(u)$ satisfies the relationship

$$\omega^0_{\hat{\psi}_{\lambda,t}} = \frac{\omega^0_{\hat{\psi}_{1,0}}}{\lambda}. \qquad (24)$$

From the above relationships one can easily see that as λ increases, i.e., as the function dilates, both $\omega^0_{\hat{\psi}_{\lambda,t}}$ and $\sigma_{\hat{\psi}_{\lambda,t}}$ decrease indicating that the center of passing band shifts towards low frequency components and the uncertainty also decreases, and vice-versa (see also figure 5). In the phase-space, the resolution cell for the wavelet transform around the point $(t_0, \omega^0_{\hat{\psi}_{\lambda,t}})$ is given by $[t_0 \pm \lambda\sigma_{\psi_{1,0}} \times \frac{\omega^0_{\hat{\psi}_{1,0}}}{\lambda} \pm \frac{\sigma_{\hat{\psi}_{1,0}}}{\lambda}]$ (see Figure 6) which has variable dimensions depending on the scale parameter λ. However, the area of the resolution cell $[\sigma_{\psi_{\lambda,t}} \times \sigma_{\hat{\psi}_{\lambda,t}}]$ remains independent of the scale or location parameter. In other words, the phase space is layered with resolution cells of varying dimensions which are functions of scale such that they have a constant area. Therefore, due to the uncertainty principle, an increased resolution in the time domain for the time localization of high frequency components comes with a cost: an increased uncertainty in the frequency localization as measured by $\sigma_{\hat{\psi}_{\lambda,t}}$. One may also interpret the wavelet transform as a mathematical microscope where the magnification is given by $1/\lambda$.

§3. Wavelets and Time-Scale Analysis

3.1. Time-scale transform

Useful information can also be extracted by interpreting the wavelet transform (12) as a time-scale transform. This was well illustrated by Rioul and Vetterli (see [47]) and is sketched below. In the wavelet transform (12) when the scale λ increases, the wavelet becomes more spread out and takes only long time behavior into account, as seen above. However by change of variables, equation (12) can also be written as

$$Wf(\lambda, t) = \int_{-\infty}^{\infty} \sqrt{\lambda} f(\lambda u)\psi(u - \frac{t}{\lambda})\,du. \qquad (25)$$

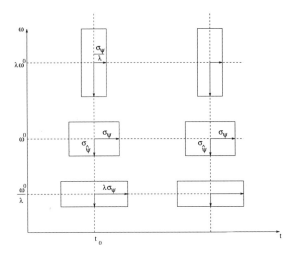

Figure 6. Figure showing the phase-space representation using the wavelet transform.

Since the mapping $f(t) \rightarrow f(\lambda t)$ has the effect of contracting $f(t)$ when $\lambda > 1$ and magnifying it when $\lambda < 1$, the above equation indicates that as the scale grows, a contracted version of the function is seen through a fixed size filter and vice-versa. Thus, the scale factor λ has the interpretation of the scale in maps.

3.2. Scalogram, wavelet variance and covariance

From the isometry of wavelet transform (16) we have

$$\langle f, f \rangle = \int_{-\infty}^{\infty} |f(t)|^2 \, dt = \frac{1}{C_\psi} \int_0^\infty \lambda^{-2} \, d\lambda \int_{-\infty}^{\infty} |Wf(\lambda, t)|^2 \, dt. \qquad (26)$$

In general, for two functions $f(t)$ and $g(t)$ (see [15], equation 2.4.2)

$$\langle f, g \rangle = \frac{1}{C_\psi} \int_0^\infty \lambda^{-2} \, d\lambda \int_{-\infty}^{\infty} Wf(\lambda, t)\overline{Wg(\lambda, t)} \, dt. \qquad (27)$$

By considering the RHS of (26) we see that $|Wf(\lambda, t)|^2/C_\psi \lambda^2$ can be considered as an energy density function on the phase-space or (t, λ) plane, i.e., $|Wf(\lambda, t)|^2 \Delta t \Delta \lambda / C_\psi \lambda^2$ gives the energy on the scale interval $\Delta \lambda$ and time interval Δt centered around scale λ and time t. Flandrin (see [21]) proposed to call the function $|Wf(\lambda, t)|^2$ a *scalogram*. In analogy, the product $Wf(\lambda, t)\overline{Wg(\lambda, t)}$ can be called a *cross scalogram*.

Equation (26) can also be written as

$$\int_{-\infty}^{\infty} |f(t)|^2 \, dt = \frac{1}{C_\psi} \int_0^\infty \lambda^{-2} E(\lambda) \, d\lambda \qquad (28)$$

where

$$E(\lambda) \equiv \int_{-\infty}^{\infty} |Wf(\lambda, t)|^2 \, dt \qquad (29)$$

gives the energy content of a function $f(t)$ at scale λ, i.e., it gives the marginal density function of energy at different scales λ. The function $E(\lambda)$ has been referred to as *wavelet variance* (see [4]) or *wavelet spectrum* (see [25]). In analogy, the function

$$E_{XY}(\lambda) = \int_{-\infty}^{\infty} Wf(\lambda, t)\overline{Wg(\lambda, t)} \, dt \qquad (30)$$

has been referred to as *wavelet covariance* (see [4]) or *wavelet cross-spectrum* (see [25]).

Notice that for a given wavelet $\psi(t)$ the center of passing band $\omega^0_{\hat{\psi}_{\lambda,t}}$ at scale λ is related to that at unit scale through the relation (see equation (24))

$$\omega \equiv \omega^0_{\hat{\psi}_{\lambda,t}} = \frac{\omega^0_{\hat{\psi}_{1,0}}}{\lambda}. \qquad (31)$$

Using this relationship, the scale information can be translated to frequency information. Using $d\omega = -\omega^0_{\hat{\psi}_{1,0}} \lambda^{-2} d\lambda$ and substituting in equation (28) we get

$$\int_{-\infty}^{\infty} |f(t)|^2 \, dt = \frac{1}{C_\psi \omega^0_{\hat{\psi}_{1,0}}} \int_0^\infty E\left(\frac{\omega^0_{\hat{\psi}_{1,0}}}{\omega}\right) d\omega. \qquad (32)$$

By defining

$$E'(\omega) \equiv E\left(\frac{\omega^0_{\hat{\psi}_{1,0}}}{\omega}\right) \qquad (33)$$

the above equation can be written as

$$\int_{-\infty}^{\infty} |f(t)|^2 \, dt = \frac{1}{C_\psi \omega^0_{\hat{\psi}_{1,0}}} \int_0^\infty E'(\omega) \, d\omega. \qquad (34)$$

One would therefore expect that $E'(\omega)$, and thus $E(\lambda)$, is related to the power spectrum $S_f(\omega)$ of $f(t)$. This indeed is the case. It can be shown (see [25]) that

$$E(\lambda) = \frac{1}{2\pi} \int_{-\infty}^{\infty} S_f(\omega) S_{\psi_\lambda}(\omega) \, d\omega \qquad (35)$$

where $S_{\psi_\lambda}(\omega)$ is the spectrum of the wavelet at scale λ. That is, $E(s)$ is the weighted average of the power spectrum of $f(t)$ where the weights are given by the power spectrum of $\psi_\lambda(t)$. This relation is interesting although in characterizing a process through $E(\lambda)$ or $E'(\omega)$, all location information is lost, it does provide certain useful insight ([36], and [27]).

3.3. Non-stationarity and the Wigner-Ville spectrum

One reason for the remarkable success of the Fourier transform in the study of stationary stochastic processes is the relationship between the autocorrelation function and the spectrum as illustrated by the following diagram:

$$
\begin{array}{ccc}
X(t) & \underset{\longleftarrow}{\overset{\mathcal{F}}{\longrightarrow}} & \hat{X}(\omega) \\
\downarrow & & \downarrow \\
R(\tau) = \mathcal{E}[X(t)X(t-\tau)] & \underset{}{\overset{\mathcal{F}}{\longrightarrow}} & S(\omega) = |\hat{X}(\omega)|^2
\end{array}
$$

where $R(\tau)$ and $S(\omega)$ are the auto-covariance function and the power spectrum of the stochastic process $X(t)$, respectively. If an analogous relationship could be developed for non-stationary processes using the wavelet transform, then the properties of the wavelet transform could be harnessed in a more useful way. It turns out that, indeed, such a relationship can be developed.

The wavelet spectrum $E(\lambda)$ or $E'(\omega)$ discussed in the previous section although interesting in its own right, takes us away from the non-stationarity of the process since it is obtained by integrating over t. We, therefore, need something else. This is provided by the Wigner-Ville spectrum. Let us define a general (non-stationary) covariance function $R(t,s)$ as

$$
R(t,s) = \mathcal{E}[X(t)X(s)].
$$

Then the Wigner-Ville spectrum (WVS) is defined as (see [11] for a discussion of WVS and other time-frequency distributions)

$$
WVS_X(t,\omega) = \int_{-\infty}^{\infty} R(t+\frac{\tau}{2}, t-\frac{\tau}{2})e^{-i\omega\tau}\, d\tau. \tag{36}
$$

The $WVS_X(t,\omega)$ is an energy density function as

$$
|X(t)|^2 = \int_{-\infty}^{\infty} WVS_X(t,\omega)\, d\omega \tag{37}
$$

i.e., we get the instantaneous energy by integrating over all frequencies, and the total energy can be obtained as

$$
\int_{-\infty}^{\infty} |X(t)|^2\, dt = \int_{-\infty}^{\infty} \int_{-\infty}^{\infty} WVS_X(t,\omega)\, d\omega\, dt. \tag{38}
$$

The relationship of interest to us is given by the relation between the scalogram and the WVS

$$|WX(\lambda,t)|^2 = \int_{-\infty}^{\infty}\int_{-\infty}^{\infty} WVS_X(u,\omega)\,WVS_\psi\left(\frac{u-t}{\lambda},\lambda\omega\right)\,du\,d\omega \quad (39)$$

i.e., the scalogram can be obtained by affine smoothing (i.e., smoothing at different scales in t and ω directions) of the WVS of X with the WVS of the wavelet. This relationship has been developed by Flandrin (see [21]). As of this writing, we are unaware of any inverse relation to obtain the WVS_X from the scalogram. We can put the key result of this subsection in the following diagrammatic form:

$$
\begin{array}{ccc}
X(t) & \overset{W}{\rightleftharpoons} & WX(\lambda,t) \\
\downarrow & & \downarrow \\
WVS_X(t,\omega) & \overset{\text{equation }(39)}{\longrightarrow} & |WX(\lambda,t)|^2
\end{array}
$$

We, therefore, see that there is an inherent link between the study of non-stationary processes and wavelet transforms akin to the link between stationary processes and Fourier transforms.

§4. Examples of One-Dimensional Wavelets

Due to the flexibility in choosing a wavelet, several functions have been used as wavelets and it would be difficult to provide an exhaustive list. We present here some commonly used wavelets (Haar wavelet, Mexican hat wavelet, and Morlet wavelet) in one-dimensional applications.

4.1. Haar wavelet

The Haar wavelet is the simplest of all wavelets and is given as

$$\psi(t) = \begin{cases} 1 & 0 \le t < \frac{1}{2} \\ -1 & \frac{1}{2} \le t < 1 \\ 0 & \text{otherwise.} \end{cases} \quad (40)$$

In a one-dimensional discretely sampled signal this wavelet can be seen as performing a differencing operation, i.e., as giving differences of non-overlapping averages of observations. In two dimensions an interpretation of the discrete orthogonal Haar wavelet transform has been given in [28].

4.2. Mexican hat wavelet

The Mexican hat wavelet is the second derivative of the Gaussian $e^{-t^2/2}$ given as (see Figure 7)

$$\psi(t) = \frac{2}{\sqrt{3}}\pi^{-1/4}(1-t^2)e^{-t^2/2}. \quad (41)$$

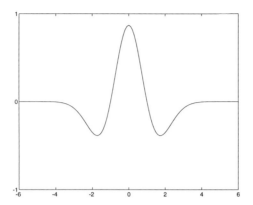

Figure 7. Mexican hat wavelet.

The constant is chosen such that $\parallel \psi \parallel^2 = 1$. This wavelet, being the second derivative of a commonly used smoothing function (the Gaussian), has found application in edge detection (see [34] and [35]).

4.3. Morlet wavelet

The Morlet wavelet is given by

$$\psi(t) = \pi^{-1/4}(e^{-i\omega_0 t - e^{-\omega_0^2/2}})e^{-t^2/2} \tag{42}$$

which is usually approximated as

$$\psi(t) = \pi^{-1/4}e^{-i\omega_0 t}e^{-t^2/2} \qquad \omega_0 \geq 5. \tag{43}$$

Since for $\omega_0 \geq 5$, the second term in (42) is negligible, i.e., $\psi(t) \approx 0$, satisfying the admissibility condition. By Morlet wavelet we now refer to (43). This wavelet is complex, enabling one to extract information about the amplitude and phase of the process being analyzed. The constant is chosen so that $\parallel \psi \parallel^2 = 1$. The Fourier transform of (43) is given by

$$\hat{\psi}(\omega) = \pi^{-1/4}e^{-(\omega-\omega_0)^2/2}. \tag{44}$$

This wavelet has been used quite often in analysis of geophysical processes (for e.g. see [45]) so we shall study it in a little more detail. The Fourier transform of the scaled wavelet $\psi_{\lambda,0}(t)$ is given as

$$\hat{\psi}_{\lambda,0}(\omega) = \lambda\pi^{-1/4}e^{-(\omega_0-\lambda\omega)^2/2} = \lambda\pi^{-1/4}e^{-\frac{\lambda^2}{2}(\frac{\omega_0}{\lambda}-\omega)^2}.$$

This wavelet has the property that its Fourier transform is supported

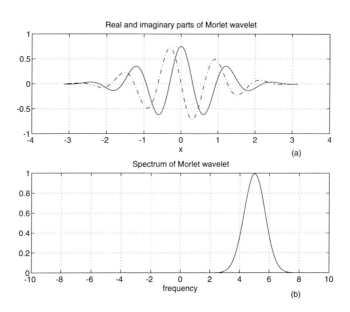

Figure 8. **(a) (top)** Real (solid line) and imaginary part (dot-dashed line) of Morlet wavelet ($\omega_0 = 5$), and **(b) (bottom)** its Fourier transform.

almost[2] entirely on $\omega > 0$, centered at $\omega^0_{\hat{\psi}_{\lambda,t}} = \omega_0/\lambda$ with a spread of $\sigma_{\hat{\psi}_{\lambda,t}} = 1/\lambda$. The wavelet $\psi_{\lambda,t}$ itself is centered at t with a spread of $\sigma_{\psi_{\lambda,t}} = \lambda$.

Figure 8a shows the real and imaginary parts of the Morlet wavelet at unit scale and Figure 8b shows its Fourier transform (with $\omega_0 = 5$). One can interpret the results of analysis of real-valued processes using this wavelet by plotting the square of the modulus and the phase, i.e., $|\langle f, \psi_{\lambda,t} \rangle|^2$ and $\tan^{-1} \frac{Im \langle f, \psi_{\lambda,t} \rangle}{Re \langle f, \psi_{\lambda,t} \rangle}$, on two different plots. Figures 9a,b show these plots for the analysis of a chirp signal (a signal whose frequency changes with t as at^2 where a is some constant). The wavelet transform was obtained using the Fourier transforms of the signal and the wavelet through implementation of equation (17). The scales of analysis are plotted as the ordinate, and the abscissa denotes t. The range of scales of the plots has been decided using the following criteria. If Δt is the sampling interval of $f(t)$ then the center of passing band $\omega_0/\lambda_{\min}$ should be less than or equal to the Nyquist

---

[2] The Fourier transform of (42) is supported entirely on $\omega > 0$ but that of (43) has a negligible mass on $\omega < 0$ for the condition $\omega_0 > 5$.

frequency, i.e., $\omega_0/\lambda_{\min} \leq 2\pi/2\Delta t$, implying

$$\lambda_{\min} \geq \frac{\omega_0 \Delta t}{\pi}. \tag{45}$$

The maximum scale of analysis is obtained by considering the spread of $\psi_{\lambda,t}$. Recognizing that $|\psi_{\lambda,t}|$ decays to 99.9% of its value at $3\sigma_{\psi_{\lambda,t}}$, we impose the condition $3\sigma_{\psi_{\lambda,t}} \leq (t_{\max} - t_{\min})/2$, i.e., the wavelet support should be contained within the data range, giving

$$\lambda_{\max} \leq \frac{t_{\max} - t_{\min}}{6}. \tag{46}$$

The discretization of λ and t for implementation on discrete data is discussed in the following sections.

Large values of $|\langle f, \psi_{\lambda,t}\rangle|^2$ in the phase-space help us identify the scale of the feature and its location on the t axis easily. Figure 9a clearly depicts the decreasing scales in the signal with increasing t. In this figure we notice that large values of the squared modulus appear at relatively large scales on the right hand side of the figure where there are no large scale features. This is due to the apparent periodicity of the Fourier transform of the limited extent signal. The phase plot helps us identify the change of phase of the signal from 0 to 2π or $-\pi$ to π. It is possible to count the number of cycles in a signal. However, this depends upon the scale. As scale decreases, we can see more waves and this gives rise to the bifurcation effect evident in figure 9b. This is helpful in locating singularities and identifying fractal and multifractal nature of processes (for example see [1]). Figure 9b shows some edge effect at small scales due to the periodicity of the Fourier transform of the limited extent signal. This periodicity can be eliminated by taking the discrete Fourier transform of the chirp signal with a sufficient number of zeros appended at the ends. For other methods see [26].

§5. Discrete Wavelet Transforms

When the parameters λ and t in the wavelet transform $\langle f, \psi_{\lambda,t}\rangle$ take on continuous values, it is called continuous wavelet transform. For practical applications the scale parameter λ and location parameter t need to be discretized. One can choose $\lambda = \lambda_0^m$ where m is an integer and λ_0 is a fixed dilation step greater than 1. Since $\sigma_{\psi_{\lambda,t}} = \lambda\sigma_{\psi_{1,t}}$, we can choose $t = nt_0\lambda_0^m$ where $t_0 > 0$ and depends upon $\psi(t)$, and n is an integer. The essential idea of this discretization can be understood by an analogy with a microscope. We choose a magnification, i.e., λ^{-m}, and study the process at a particular location and then move to another location. If the magnification is large, i.e., small scale, we move in small steps and vice-versa. This can be easily accomplished by choosing the incremental step inversely proportional to the magnification (i.e., proportional to the scale

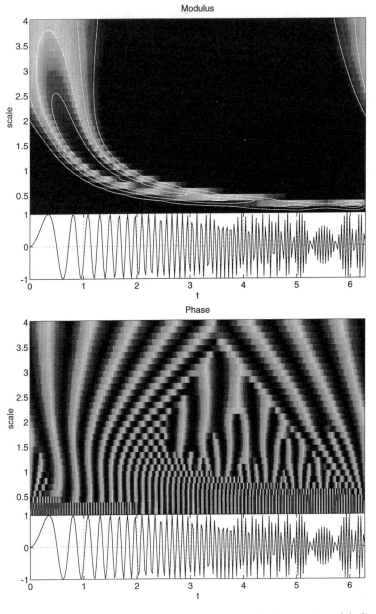

Figure 9.   Analysis of a chirp signal using Morlet wavelet: **(a) (top)** square of the modulus and **(b) (bottom)** phase of $\langle f, \psi_{\lambda,t} \rangle$.

λ_0^m) which the above method of discretization of t accomplishes. We then define

$$\psi_{m,n}(t) = \frac{1}{\sqrt{\lambda_0^m}}\psi(\frac{t - nt_0\lambda_0^m}{\lambda_0^m})$$
$$= \lambda_0^{-m/2}\psi(\lambda_0^{-m}t - nt_0). \qquad (47)$$

The wavelet transform

$$\langle f, \psi_{m,n}\rangle = \lambda_0^{-m/2}\int f(t)\psi(\lambda_0^{-m}t - nt_0)\,dt \qquad (48)$$

is called the discrete wavelet transform.

In the case of the continuous wavelet transform we saw that $\langle f, \psi_{\lambda,t}\rangle$ for $\lambda > 0$ and $t \in (-\infty, \infty)$ completely characterizes the function $f(t)$. In fact, one could reconstruct $f(t)$ using (14). Using the discrete wavelet $\psi_{m,n}$ (with ψ decreasing sufficiently fast) and appropriate choices of λ_0 and t_0, we can also completely characterize $f(t)$. In fact, we can write $f(t)$ as a series expansion, as we shall see in the following subsections. We first study orthogonal wavelets and then the general case.

5.1. Orthogonal wavelet transforms and multiresolution analysis

5.1.1. Orthogonal wavelet transforms

Consider the discrete wavelet transform for $\lambda_0 = 2$ and $t_0 = 1$, i.e.,

$$\psi_{m,n}(t) = 2^{-m/2}\psi(2^{-m}t - n) = \frac{1}{\sqrt{2^m}}\psi(\frac{t - n2^m}{2^m}). \qquad (49)$$

For the purpose of this subsection, let $\psi_{m,n}(t)$ denote the above discretization rather than the general discretization given by equation (47). We will also use the identity $\psi_{00}(t) \equiv \psi(t)$. It is possible to construct a certain class of wavelets $\psi(t)$ such that $\psi_{m,n}(t)$ are orthonormal, i.e.,

$$\int \psi_{m,n}(t)\psi_{m',n'}(t)\,dt = \delta_{mm'}\delta_{nn'} \qquad (50)$$

where δ_{ij} is the Kronecker delta function given as

$$\delta_{ij} = \begin{cases} 1 & \text{if } i = j \\ 0 & \text{otherwise.} \end{cases} \qquad (51)$$

The above condition implies that these wavelets are *orthogonal to their dilates and translates*. One can construct $\psi_{m,n}(t)$ that are not only orthonormal, but such that they form a complete orthonormal basis for all functions that have finite energy [30]. This implies that all such functions $f(t)$ can be approximated, up to arbitrary high precision, by a linear combination of the wavelets $\psi_{m,n}(t)$, i.e.,

$$f(t) = \sum_{m=-\infty}^{\infty} \sum_{n=-\infty}^{\infty} D_{m,n}\psi_{m,n}(t) \tag{52}$$

where the first summation is over scales (from small to large) and at each scale we sum over all translates. The coefficients are obtained as

$$D_{m,n} = \langle f, \psi_{m,n} \rangle \equiv \int f(t)\psi_{m,n}(t)\, dt$$

and, therefore, we can write

$$f(t) = \sum_{m=-\infty}^{\infty} \sum_{n=-\infty}^{\infty} \langle f, \psi_{m,n} \rangle \psi_{m,n}(t). \tag{53}$$

From (53) it is easy to see how wavelets provide a time-scale representation of the process where time location and scale are given by indices n and m, respectively. The equality in equation (53) is in the mean square sense. The above series expansion is akin to a Fourier series with the following differences:

1. The series is double indexed with the indices indicating scale and location;

2. The basis functions have the time-scale (time-frequency) localization property in the sense discussed in section 2.2.

By using an intermediate scale m_0, equation (53) can be broken up as two sums

$$f(t) = \sum_{m=m_0+1}^{\infty} \sum_{n=-\infty}^{\infty} \langle f, \psi_{m,n} \rangle \psi_{m,n}(t) + \sum_{m=-\infty}^{m_0} \sum_{n=-\infty}^{\infty} \langle f, \psi_{m,n} \rangle \psi_{m,n}(t). \tag{54}$$

It turns out that one can find functions $\phi_{m,n}(t)$ defined analogous to $\psi_{m,n}(t)$, i.e.,

$$\phi_{m,n}(t) = 2^{-m/2}\phi(2^{-m}t - n) \tag{55}$$

and satisfying certain properties enumerated in appendix A, such that the first sum on the RHS of equation (54) can be written as a linear combination of $\phi_{m_0,n}$ (see [30]), i.e.,

$$\sum_{n=-\infty}^{\infty} \langle f, \phi_{m_0,n} \rangle \phi_{m_0,n}(t) = \sum_{m=m_0+1}^{\infty} \sum_{n=-\infty}^{\infty} \langle f, \psi_{m,n} \rangle \psi_{m,n}(t). \tag{56}$$

Consequently,

$$f(t) = \sum_{n=-\infty}^{\infty} \langle f, \phi_{m_0,n} \rangle \phi_{m_0,n}(t) + \sum_{m=-\infty}^{m_0} \sum_{n=-\infty}^{\infty} \langle f, \psi_{m,n} \rangle \psi_{m,n}(t) \tag{57}$$

The function $\phi_{m,n}(t)$ is called a *scaling function* and satisfies $\int \phi(t)\,dt = 1$ among its other properties. For example, the scaling function corresponding to the Haar wavelet is the characteristic function of the interval $[0,1)$ given as

$$\phi(t) = \begin{cases} 1 & 0 \leq t < 1 \\ 0 & \text{otherwise.} \end{cases} \qquad (58)$$

The scaling functions and wavelets play a profound role in the analysis of processes using orthogonal wavelets. This analysis framework is known as the wavelet multiresolution analysis framework and is discussed below. Appendix B describes a class of orthogonal wavelets developed by Daubechies [13] and Appendix C briefly discusses the implementation algorithm by Mallat [30].

5.1.2. Multiresolution representation

Equation (56) states that all the features of the process $f(t)$, that are larger than the scale 2^{m_0}, can be approximated by a linear combination of the translates (over n) of the scaling function $\phi(t)$ at the fixed scale 2^{m_0}. Let us represent this approximation by $P_{m_0}f$, i.e.,

$$P_{m_0}f(t) = \sum_{n=-\infty}^{\infty} \langle f, \phi_{m_0,n}\rangle \phi_{m_0,n}(t) \qquad (59)$$

Let us now define

$$Q_m f(t) \equiv \sum_{n=-\infty}^{\infty} \langle f, \psi_{m,n}\rangle \psi_{m,n}(t) \qquad (60)$$

so that equation (57) becomes

$$f(t) = P_{m_0}f(t) + \sum_{m=-\infty}^{m_0} Q_m f(t). \qquad (61)$$

Since m_0 is arbitrary we also have

$$f(t) = P_{m_0-1}f(t) + \sum_{m=-\infty}^{m_0-1} Q_m f(t) \qquad (62)$$

from which we can obtain by subtraction

$$P_{m_0-1}f(t) = P_{m_0}f(t) + Q_{m_0}f(t) \qquad (63)$$

or in general

$$P_{m-1}f(t) = P_m f(t) + Q_m f(t). \qquad (64)$$

This equation characterizes the basic structure of the orthogonal wavelet decomposition (53). As mentioned before $P_m f(t)$ contains all the information about features in $f(t)$ that are larger than the scale 2^m. From equation

(64) it is evident that when we go from the scale 2^m to the next smaller scale 2^{m-1}, we add some detail to $P_m f(t)$ which is given by $Q_m f(t)$. We can, therefore, say that $Q_m f(t)$, or equivalently the wavelet expansion of a function at any scale 2^m, characterizes the difference between the process at two different scales 2^m and 2^{m-1}, or equivalently at two different resolutions.

Representation of a function within the nested structure of equation (64) is called the *wavelet multiresolution representation*. Formally it consists of a sequence of closed subspaces $\{V_m\}_{m \in \mathbf{Z}}$ of $L^2(\mathbf{R})$ where $L^2(\mathbf{R})$ denotes the Hilbert space[3] of all square integrable functions, and $\mathbf{R}$ and $\mathbf{Z}$ denote the set of real numbers and integers, respectively. These subspaces characterize the behavior of a function at different scales or resolutions. For example, V_m characterizes functions at scale 2^m or equivalently at resolution given as 2^{-m} samples per unit length. The subspaces satisfy the following properties:

M1 $V_m \subset V_{m-1}$ for all $m \in \mathbf{Z}$, i.e., a space corresponding to some resolution contains all the information about the space at lower resolution, or equivalently, a space corresponding to some scale contains all the information about the space at larger scale.

M2 $\cup_{m=-\infty}^{\infty} V_m$ is dense in $L^2(\mathbf{R})$ and $\cap_{m=-\infty}^{\infty} V_m = \{0\}$, i.e., as the resolution increases the approximation of the function converges to the original function and as the resolution decreases the approximated function contains less and less information.

M3 $f(t) \in V_m$ if and only if $f(2t) \in V_{m-1}$ for all $m \in \mathbf{Z}$, i.e., all spaces are scaled versions of one space.

M4 $f(t) \in V_m$ implies $f(t - \frac{k}{2^m}) \in V_m \forall k \in \mathbf{Z}$, i.e., the space is invariant with respect to the "integer translations" of a function.

Notice that since $V_m \subset V_{m-1}$ we can write

$$V_{m-1} = V_m \oplus O_m \qquad (65)$$

where O_m is the orthogonal complement of V_m in V_{m-1} (i.e., O_m is the set of all functions in V_{m-1} that are orthogonal to V_m) and $\oplus$ denotes

---

[3]   A Hilbert space **H** is a vector space (possibly infinite dimensional) with an inner product $\langle .,. \rangle$ which is *complete* with respect to the norm $\| f \| = \langle f, f \rangle^{1/2}$ induced by this inner product. A normed space is complete if every Cauchy sequence in that space converges to an element of that space, i.e., for every sequence $\{f_n\} \subset \mathbf{H}$ such that $\| f_m - f_n \| \to 0$ as $m, n \to \infty$, we have $f_n \to f \in \mathbf{H}$ as $n \to \infty$ [44].

the direct sum. Given this structure, representation of a function in V_m is given by $P_m f(t)$ and representation in O_m is given by $Q_m f(t)$ (compare equation (65) with equation (64)). The operators P_m and Q_m are orthogonal projection operators onto the spaces V_m and O_m, respectively. Let $P_m^d f$ and $Q_m^d f$ denote the discrete set of inner products $\{\langle f, \phi_{m,n} \rangle\}$ and $\{\langle f, \psi_{m,n} \rangle\}$, respectively. The set $P_m^d f$ gives the discrete approximation of $f(t)$ at scale 2^m and $Q_m^d f$ gives the *discrete detail approximation* of $f(t)$. Then, in simple words equation (65) says that we need to *add* the information contained in $Q_m^d f$ to $P_m^d f$ to go from one resolution (scale) to the next higher resolution (smaller scale).

The multiresolution analysis framework is not unique. Several multiresolution frameworks can be constructed depending upon the choice of the pair (ϕ, ψ). Recall that the choice of either $\phi(t)$ or $\psi(t)$ determines the other. The simplest of all multiresolution frameworks is the one where V_m is composed of piecewise constant functions. In this case the scaling function is given by equation (58) and the wavelet is the Haar wavelet given by equation (40). For examples of other pairs of (ϕ, ψ) that give rise to the multiresolution framework, see Appendix B and [13, 15] and [31]. For algorithms to construct the pairs (ϕ, ψ), see [52, 51].

5.2. Non-orthogonal wavelet transforms

5.2.1. Frames

We saw in section 5.1 that it is possible to find λ_0, t_0 and $\psi_{m,n}(t)$ as defined in equation (47) such that $\psi_{m,n}(t)$ are orthogonal. This allows a function $f(t)$ to be written as a series expansion as given in equation (53). However, even if $\psi_{m,n}(t)$ are not orthogonal, the function $f(t)$ can be represented completely as a series expansion under certain broad conditions on the wavelet $\psi(t)$, t_0 and λ_0. These discrete wavelets which provide complete representation of the function $f(t)$ are called *wavelet frames* and will be the subject of the next sub-section. We will see that orthogonal wavelets are a special case of this general framework. Let us first define frames.

A sequence of functions $\{\varphi_n\}_{n \in \mathbf{Z}}$ in a Hilbert space $\mathbf{H}$ (see footnote on page 24 for definition of Hilbert Space) is called a *frame* if there exist two constants $A > 0$, $B < \infty$, called *frame bounds*, so that for all functions $f(t)$ in the Hilbert space $\mathbf{H}$ the following holds:

$$A \parallel f \parallel^2 \leq \sum_n |\langle f, \varphi_n \rangle|^2 \leq B \parallel f \parallel^2 . \tag{66}$$

The constant $B < \infty$ guarantees that the transformation $f \rightarrow \{\langle f, \varphi_n \rangle\}$ is continuous and the constant $A > 0$ guarantees that this transformation is invertible and has continuous a inverse. This enables one to: (1) completely characterize the function, and (2) reconstruct the function from its decomposition.

In general, a frame is not an orthonormal basis. It provides a redundant representation of the function $f(t)$. This is analogous, for example, to representing a vector in the Euclidean plane using more than two basis vectors. The ratio A/B is called the redundancy ratio or redundancy factor. Redundant representations are more robust to noise and therefore useful when noise reduction is an issue.

When $A = B$, the frame is called a *tight frame*. In this case there is a simple expansion formula given as

$$f(t) = \frac{1}{A} \sum_n \langle f, \varphi_n \rangle \varphi_n(t). \tag{67}$$

Notice that this formula is very similar to the one obtained for an orthonormal set $\{\varphi_n\}$. In this case, however, $\{\varphi_n\}$ may not even be linearly independent, i.e., there is a large degree of redundancy in the representation. Orthonormal bases arise as a special case. For a tight frame, if $A = B = 1$ and if $\| \varphi_n \| = 1$, then $\{\varphi_n\}$ form an orthonormal basis and we get the usual expansion formula. When $\{\psi_{m,n}\}$ constitute a tight frame then $A = B = C_\psi/t_0 \log \lambda_0$ where C_ψ is defined in equation (15) (see [15], equation 3.3.8). However, in practice it is difficult to get A exactly equal to B, but easier to get A close to B, i.e., $\epsilon = \frac{B}{A} - 1 \ll 1$. Daubechies (see [14], pg. 971) calls such frames *snug frames*. The expansion formula in this case is given as

$$f(t) = \frac{2}{A+B} \sum_n \langle f, \varphi_n \rangle \varphi_n + \gamma \tag{68}$$

where the error γ is of the order of $\frac{\epsilon}{2+\epsilon} \| f \|$. The general case of $A \not\approx B$ is more involved and beyond the scope of this introduction (see [15], for details).

5.2.2. Wavelet frames

Now let L denote the transformation $L : f(t) \rightarrow \{\langle f, \psi_{m,n} \rangle\}$, where $\psi_{m,n}(t)$ is defined by equation (47). We can characterize the function $f(t)$ through the wavelet coefficients $\{\langle f, \psi_{m,n} \rangle\}$ provided the transform L satisfies the condition (66), i.e.,

$$A \| f \|^2 \leq \sum_m \sum_n |\langle f, \psi_{m,n} \rangle|^2 \leq B \| f \|^2 . \tag{69}$$

Given discrete wavelets, we can obtain simple expansions such as in (67) and (68), provided $\psi_{m,n}$ constitute a frame, i.e.,

$$f(t) = \frac{1}{A} \sum_m \sum_n \langle f, \psi_{m,n} \rangle \psi_{m,n}(t). \tag{70}$$

if $\{\psi_{m,n}\}$ is a tight frame, and

$$f(t) = \frac{2}{A+B} \sum_m \sum_n \langle f, \psi_{m,n} \rangle \psi_{m,n} + \gamma \tag{71}$$

when $\{\psi_{m,n}\}$ is a snug frame. Such frames can be constructed for certain choices of λ_0 and t_0, provided $\psi(t)$ satisfies the admissibility condition, i.e., $\int \psi(t)\,dt = 0$, and has compact support or sufficiently fast decay. The conditions for the choice of λ_0 and t_0 are described in Daubechies (see [15], chapter 3). Here it suffices to say that these conditions are fairly broad and admit a very flexible range. For example, for the Mexican hat wavelet (as given in equation (41)), for $\lambda_0 = 2$ and $t_0 = 1$, the frame bounds are $A = 3.223$ and $B = 3.596$ giving $B/A = 1.116$.

One can obtain B/A closer to 1 by choosing $\lambda_0 < 2$. Grossmann et al. [24] suggested decomposing each *octave* into several *voices* (as in music) by choosing $\lambda_0 = 2^{1/M}$ where M indicates the number of voices per octave. With such a decomposition we get

$$\psi_{m,n}^M(t) = 2^{-m/2M} \psi(2^{-m/M} t - n t_0). \tag{72}$$

For the Mexican hat wavelet, by choosing $M = 4$ and $t_0 = 1$ we can obtain $A = 13.586$ and $B = 13.690$ giving $B/A = 1.007$. Such a decomposition using such a *multivoice frame* enables us to cover the range of scales in smaller steps giving a more "continuous" picture. For example, with $M = 4$ we get discrete scales at $\{\lambda = \ldots, 1, 2^{1/4}, 2^{1/2}, 2^{3/4}, 2, 2^{5/4}, 2^{3/2}, 2^{7/4}, 4, \ldots\}$ as against $\{\lambda = \ldots, 1, 2, 4, \ldots\}$ for usual $M = 1$. Figure 9 was created using Morlet wavelet with $M = 4$ and $t_0 = 1$. For this decomposition $A = 6.918$, $B = 6.923$ giving $B/A = 1.0008$. It should be noted that Morlet wavelet, which is not orthogonal, gives a good reconstruction under the framework of equation (71). Multivoice frames are discussed extensively in Daubechies ([15], chapter 3) where more details on the values of A and B for different choices of M and t_0 are given for the mexican hat and the Morlet wavelet.

Redundant representations such as the one presented above, in addition to their noise reduction capability, are useful when representations that are close to the continuous case are sought (see for example [3, 32, 35, 5, 33] and [49]).

5.3. Biorthogonal wavelets

Under the wavelet multiresolution framework, the decomposition and reconstruction of a function is done using the same wavelet, i.e.,

$$f(t) = \sum_m \sum_n \langle f, \psi_{m,n} \rangle \psi_{m,n}(t) \tag{73}$$

where $\{\langle f, \psi_{m,n} \rangle\}$ are the decomposition coefficients. This however, can severely limit the choice of wavelet $\psi(t)$. For example, it has been shown (see [15], theorem 8.1.4) that the only real and compactly supported symmetric or antisymmetric wavelet under a multiresolution framework is the Haar wavelet. In certain applications however, real symmetric wavelets which are smoother and have better frequency localization than the Haar wavelet may be needed. In such situations, biorthogonal wavelets come to the rescue. It is possible to construct two sets of wavelets $\{\psi_{m,n}\}$ and $\{\tilde{\psi}_{m,n}\}$ such that

$$f(t) \quad = \quad \sum_m \sum_n \langle f, \psi_{m,n} \rangle \tilde{\psi}_{m,n}(t) \tag{74}$$

$$= \quad \sum_m \sum_n \langle f, \tilde{\psi}_{m,n} \rangle \psi_{m,n}(t). \tag{75}$$

That is, one can accomplish decomposition using one set of wavelets and reconstruction using another. The wavelets $\psi_{m,n}(t) \equiv \frac{1}{2^{m/2}} \psi(\frac{t}{2^m} - n)$ and $\tilde{\psi}_{m,n}(t) \equiv \frac{1}{2^{m/2}} \tilde{\psi}(\frac{t}{2^m} - n)$ need to satisfy

$$\sum_m \sum_n |\langle f, \psi_{m,n} \rangle|^2 \quad \leq \quad B \parallel f \parallel^2 \tag{76}$$

$$\sum_m \sum_n |\langle f, \tilde{\psi}_{m,n} \rangle|^2 \quad \leq \quad \tilde{B} \parallel f \parallel^2 \tag{77}$$

$$\langle \psi_{m,n}, \tilde{\psi}_{m',n'} \rangle \quad = \quad \delta_{mm'} \delta_{nn'} \tag{78}$$

where B and $\tilde{B}$ are some constants and condition (78) is the condition of biorthonormality. Given such a biorthonormal set, it is possible to construct corresponding scaling functions $\{\phi_{m,n}\}$ and $\{\tilde{\phi}_{m,n}\}$ such that

$$\langle \phi_{m,n}, \tilde{\phi}_{m,n'} \rangle = \delta_{nn'}. \tag{79}$$

Notice that nothing is said about the orthogonality of $\{\psi_{m,n}\}$, $\{\tilde{\psi}_{m,n}\}$, $\{\phi_{m,n}\}$ and $\{\tilde{\phi}_{m,n}\}$ themselves. In general they form a linearly independent basis. Also, there is no condition of orthogonality between the wavelets $\psi(t)$ and $\tilde{\psi}(t)$, and the corresponding scaling functions $\phi(t)$ and $\tilde{\phi}(t)$, respectively. Given these wavelets and scaling functions, one can construct a multiresolution nest, as in the orthonormal case, i.e.,

$$\cdots \subset V_2 \subset V_1 \subset V_0 \subset V_{-1} \subset V_{-2} \subset \cdots$$

$$\cdots \subset \tilde{V}_2 \subset \tilde{V}_1 \subset \tilde{V}_0 \subset \tilde{V}_{-1} \subset \tilde{V}_{-2} \subset \cdots$$

with $V_m = \text{span}\{\phi_{m,n}\}$ and $\tilde{V}_m = \text{span}\{\tilde{\phi}_{m,n}\}$ and the complementary spaces $O_m = \text{span}\{\psi_{m,n}\}$ and $\tilde{O}_m = \text{span}\{\tilde{\psi}_{m,n}\}$. The spaces V_m and O_m ($\tilde{V}_m$ and $\tilde{O}_m$, respectively) are not orthogonal complements in general. Equation (78), however, implies that

$$V_m \perp \tilde{O}_m \quad \text{and} \quad \tilde{V}_m \perp O_m. \tag{80}$$

Another advantage of biorthogonal wavelets is (see [15], section 8.3) that one can have $\psi(t)$ and $\tilde{\psi}(t)$ with different vanishing moments. For example, if $\psi(t)$ has more vanishing moments than $\tilde{\psi}(t)$, one can obtain higher data compression using $\langle f, \psi_{m,n} \rangle$ and a good reconstruction using

$$f(t) = \sum_m \sum_n \langle f, \psi_{m,n} \rangle \tilde{\psi}_{m,n}(t),$$

the sum being restricted to some finite values.

§6. Two-Dimensional Wavelets

6.1. Continuous wavelets

The continuous analogue of wavelet transform (12) is obtained by treating $\boldsymbol{u} = (u_1, u_2)$ and $\boldsymbol{t} = (t_1, t_2)$ as vectors. Therefore for the two dimensional case

$$
\begin{aligned}
\langle f, \psi_{\lambda, \boldsymbol{t}} \rangle \equiv Wf(\lambda, \boldsymbol{t}) &= \int_{-\infty}^{\infty} \int_{-\infty}^{\infty} f(\boldsymbol{u}) \psi_{\lambda, \boldsymbol{t}}(\boldsymbol{u}) \, d\boldsymbol{u} \qquad \lambda > 0 \\
&= \int_{-\infty}^{\infty} \int_{-\infty}^{\infty} f(\boldsymbol{u}) \frac{1}{\lambda} \psi(\frac{\boldsymbol{u} - \boldsymbol{t}}{\lambda}) \, d\boldsymbol{u}.
\end{aligned}
\tag{81}
$$

An analogous inversion formula also holds, i.e.,

$$f(\boldsymbol{t}) = \frac{1}{C_\psi} \int_{-\infty}^{\infty} \int_{-\infty}^{\infty} \int_{\lambda=0}^{\infty} \lambda^{-3} Wf(\lambda, \boldsymbol{u}) \psi_{\lambda, \boldsymbol{u}}(\boldsymbol{t}) \, d\lambda \, d\boldsymbol{u}. \tag{82}$$

The condition of admissibility of a wavelet remains the same, i.e.,

1. compact support or sufficiently fast decay; and

2. $\iint \psi(\boldsymbol{t}) \, d\boldsymbol{t} = 0$.

Two examples of two-dimensional wavelets are discussed in the following subsection.

6.1.1. Two-dimensional Morlet wavelet

Define the vector $\boldsymbol{t} = (t_1, t_2)$ on the two-dimensional plane with $|\boldsymbol{t}| = \sqrt{t_1^2 + t_2^2}$. Then the two dimensional Morlet wavelet is defined as

$$\psi^\theta(\boldsymbol{t}) = \frac{1}{\sqrt{\pi}} e^{-i\Omega^0 \cdot \boldsymbol{t}} e^{|\boldsymbol{t}|^2 / 2} \quad \text{for } |\Omega^0| \geq 5 \tag{83}$$

with Fourier transform

$$\hat{\psi}^\theta(\Omega) = \frac{1}{\sqrt{\pi}} e^{-|\Omega - \Omega^0|^2 / 2} \tag{84}$$

where $\Omega = (\omega_1, \omega_2)$ is an arbitrary point on the two-dimensional frequency plane, and $\Omega^0 = (\omega_1^0, \omega_2^0)$ is a constant. The superscript θ indicates the

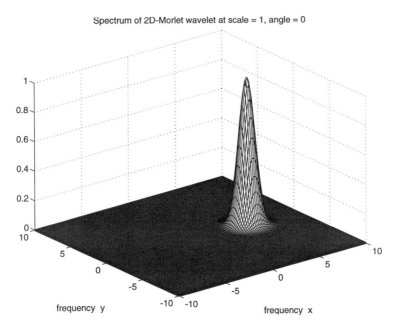

Figure 10. Frequency support of two dimensional Morlet wavelet.

direction of the wavelet, i.e,

$$\theta = \tan^{-1} \frac{\omega_2^0}{\omega_1^0}. \tag{85}$$

The properties of this wavelet are best understood from its spectrum. Figure 10 shows the spectrum of the two-dimensional Morlet wavelet for $\theta = 0$ and $\lambda = 1$. This wavelet is no longer progressive as in the one-dimensional case, i.e., its spectrum is not entirely supported on the positive quadrant. Manipulating Ω^0 by changing θ allows us to change the directional selectivity of the wavelet. For example, by choosing $\Omega^0 = (\omega_1^0, \omega_2^0) = \omega^0(\cos\theta, \sin\theta)$, $\omega^0 \geq 5$, $0 \leq \theta \leq 2\pi$ we get the wavelet transform

$$
\begin{aligned}
\langle f, \psi_{0,t}^\theta \rangle &= \int_{-\infty}^{\infty} \int_{-\infty}^{\infty} f(t_1, t_2) \frac{1}{\sqrt{\pi}} e^{-i\Omega^0 \cdot t} e^{-|t|^2/2} \, dt_1 \, dt_2 \\
&= \frac{1}{\sqrt{\pi}} \int_{-\infty}^{\infty} \int_{-\infty}^{\infty} f(t_1, t_2) e^{-i(\omega_1^0 t_1 + \omega_2^0 t_2)} e^{-(t_1^2 + t_2^2)/2} \, dt_1 \, dt_2 \\
&= \frac{1}{\sqrt{\pi}} \int_{-\infty}^{\infty} \int_{-\infty}^{\infty} \hat{f}(\omega_1, \omega_2) e^{-\{(\omega_1 - \omega_1^0)^2 + (\omega_2 - \omega_2^0)^2\}/2}
\end{aligned}
$$

$$\times e^{i(\omega_1 t_1 + \omega_2 t_2)} \, d\omega_1 \, d\omega_2. \tag{86}$$

The last equation is obtained by using Parseval's theorem. At any arbitrary scale λ, equation (86) can be written as

$$\langle f, \psi_{\lambda,t}^{\theta} \rangle = \frac{\lambda}{\sqrt{\pi}} \int_{-\infty}^{\infty} \int_{-\infty}^{\infty} \hat{f}(\omega_1, \omega_2) e^{-\lambda^2 \{(\omega_1 - \frac{\omega_1^0}{\lambda})^2 + (\omega_2 - \frac{\omega_2^0}{\lambda})^2\}/2}$$

$$\times e^{i(\omega_1 t_1 + \omega_2 t_2)} \, d\omega_1 \, d\omega_2. \tag{87}$$

The above equation indicates that the wavelet transform $\langle f, \psi_{\lambda,t}^{\theta} \rangle$ extracts the frequency contents of the function $f(t)$ around the frequency coordinates $(\omega_1^0/\lambda, \omega_2^0/\lambda) \equiv (\omega^0 \cos\theta/\lambda, \omega^0 \sin\theta/\lambda)$ with a radial uncertainty of $\sigma_{\hat{\psi}_{\lambda,t}} = 1/\lambda$, at the location t. Therefore, by fixing λ and traversing along θ, directional information at a fixed scale λ can be extracted, and by fixing θ and traversing along λ, scale information in a fixed direction can be obtained.

6.1.2. Halo wavelet

Often the directional selectivity offered by Morlet wavelet is not desired and one wishes to pick frequencies with no preferential direction. Dallard and Spedding [12] defined a wavelet by modifying the Morlet wavelet and called it the Halo wavelet because of its shape in the Fourier space. The wavelet itself is defined through its Fourier transform

$$\hat{\psi}(\Omega) = \kappa e^{-(|\Omega| - |\Omega^0|)^2/2} \tag{88}$$

where κ is a normalizing constant. As can be seen from the above expression this wavelet has no directional specificity.

6.2. Orthogonal wavelets

For two-dimensional multiresolution representation, consider the function $f(t_1, t_2) \in L^2(\mathbf{R}^2)$. A multiresolution approximation of $L^2(\mathbf{R}^2)$ is a sequence of subspaces that satisfy the two-dimensional extension of properties M1 through M4 enumerated in the definition of the one-dimensional multiresolution approximation. We denote such a sequence of subspaces of $L^2(\mathbf{R}^2)$ by $(V_m)_{m \in \mathbf{Z}}$. The approximation of the function $f(t_1, t_2)$ at the resolution m, i.e., 2^{2m} samples per unit area, is the orthogonal projection on the vector space V_m.

A two-dimensional multiresolution approximation is called separable if each vector space V_m can be decomposed as a tensor product of two identical subspaces V_m^1 of $L^2(\mathbf{R})$, i.e., the representation is computed by filtering the signal with a low pass filter of the form $\Phi(t_1, t_2) = \phi(t_1)\phi(t_2)$.

For a separable multiresolution approximation of $L^2(\mathbf{R}^2)$,

$$V_m \;=\; V_m^1 \otimes V_m^1 \tag{89}$$

where $\otimes$ represents a tensor product. It, therefore, follows (by expanding V_{m+1} as in (89) and using property M1) that the orthogonal complement O_m of V_m in V_{m+1} consists of the direct sum of three subspaces, i.e.,

$$O_m = (V_m^1 \otimes O_m^1) \oplus (O_m^1 \otimes V_m^1) \oplus (O_m^1 \otimes O_m^1). \tag{90}$$

The orthonormal basis for V_m is given by

$$(2^m \Phi(2^m t_1 - n, 2^m t_2 - k))_{(n,k) \in \mathbf{Z}^2} = (2^m \phi_m(2^m t_1 - n)\phi_m(2^m t_2 - k))_{(n,k) \in \mathbf{Z}^2}. \tag{91}$$

Analogous to the one-dimensional case, the detail function at the resolution m is equal to the orthogonal projection of the function on to the space O_m which is the orthogonal complement of V_m in V_{m+1}. An orthonormal basis for O_m can be built based on Theorem 4 in Mallat (1989a, pg. 683) who shows that if $\psi(t_1)$ is the one dimensional wavelet associated with the scaling function $\phi(t_1)$, then the three "wavelets" $\Psi^1(t_1, t_2) = \phi(t_1)\psi(t_2)$, $\Psi^2(t_1, t_2) = \psi(t_1)\phi(t_2)$ and $\Psi^3(t_1, t_2) = \psi(t_1)\psi(t_2)$ are such that

$$\{(\Psi_{mnk}^1, \Psi_{mnk}^2, \Psi_{mnk}^3)_{(n,k) \in \mathbf{Z}^2}\}$$

is an orthonormal basis for O_m.

The discrete approximation of the function $f(t_1, t_2)$ at a resolution m is obtained through the inner products

$$P_m^d f = \{(f, \Phi_{mnk})_{(n,k) \in \mathbf{Z}^2}\} = \{(f, \phi_{mn}\phi_{mk})_{(n,k) \in \mathbf{Z}^2}\} \tag{92}$$

The discrete detail approximation of the function is obtained by the inner product of $f(t_1, t_2)$ with each of the vectors of the orthonormal basis of O_m. This is, thus, given by

$$Q_m^{d1} f = \{(f, \Psi_{mnk}^1)_{(n,k) \in \mathbf{Z}^2}\}, \tag{93}$$

$$Q_m^{d2} f = \{(f, \Psi_{mnk}^2)_{(n,k) \in \mathbf{Z}^2}\}, \tag{94}$$

and

$$Q_m^{d3} f = \{(f, \Psi_{mnk}^3)_{(n,k) \in \mathbf{Z}^2}\}. \tag{95}$$

The corresponding continuous approximation will be denoted by $Q_m^1 f(t)$, $Q_m^2 f(t)$ and $Q_m^3 f(t)$ respectively. For implementation to discrete data see [31].

The decomposition of O_m into the sum of three subspaces (see equation (90)) acts like spatially oriented frequency channels. Assume that we have a discrete process at some resolution $m + 1$ whose frequency domain is shown in Figure 11 as the domain of $P_{m+1}^d f$. When the same process

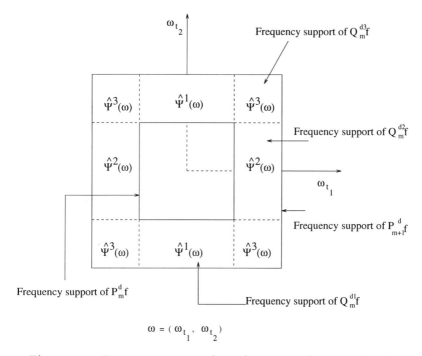

$$\omega = (\, \omega_{t_1}, \ \omega_{t_2}\,)$$

Figure 11. Frequency support of wavelets in two dimensional multiresolution decomposition.

is reduced to resolution m, its frequency domain shrinks to that of $P_m^d f$. The information lost can be divided into three components as shown in Figure 11: vertical high frequencies (high horizontal correlation), horizontal high frequencies (high vertical correlation) and high frequencies in both direction (high vertical and horizontal correlations, for example, features like corners). They are captured as $Q_m^{d1} f$, $Q_m^{d2} f$ and $Q_m^{d3} f$ respectively. This property has been used in [29] to characterize the directional behavior of rainfall. Wavelets with more than three frequency channels can be constructed (see, for example, [10]) but are not discussed herein.

§7. Conclusions

Wavelet theory involves representing general functions in terms of simpler building blocks at different scales and positions. Wavelets offer a versatile and sophisticated tool yet simple to implement, and have already found several applications in a wide range of scientific fields. The list of

Table I

Anonymous ftp site information for wavelet related software.

FTP site	Directory	References/Notes
cs.nyu.edu	pub/wave	[35, 33], C programs
playfair.stanford.edu	pub/software/wavelets	MATLAB Scripts
simplicity.stanford.edu	/pub/taswell	MATLAB scripts
gdr.bath.ac.uk	/pub/masgpn	S software
pascal.math.yale.edu	/pub/software/xwpl	Wavelet Packet Laboratory
bimi@iss.nus.sg	e-mail request	for KHOROS
cml.rice.edu	/pub/dsp/software	MATLAB scripts
wuarchive.wustl.edu	edu/math/msdos/modelling	C programs

wavelet applications increases at a fast pace and includes to date signal processing, coding, fractals, statistics, image processing, astrophysics, physics, turbulence, mathematics, numerical analysis, economics, medical research, target detection, industrial applications, quantum mechanics, geophysics etc. (see, for example, the edited volumes by Chui [7], Ruskai et al. [48], Benedetto and Frazier [3], Farge et al. [20], Meyer and Roques [41], Combes et al. [9], and Beylkin et al. [2], among others). A general literature survey on wavelets can be found in [46]. In geophysics, significant progress has already been made in studying and unraveling structure of several geophysical processes using wavelets (see the extended bibliography of wavelets in geophysics at the end of this volume). We have no doubt that the study of geophysical phenomena, which are by nature complex, and take place and interact at a range of scales of interest, will continue to benefit from the use of the powerful and versatile tools that wavelet analysis has to offer.

There are a number of useful sources of information about wavelet publications and computer software for the implementation of wavelet analysis. An electronic information service called "wavelet digest" exists (at the time of this writing) on the Internet with the address *wavelet@math.scarolina.edu*. Several anonymous ftp sites exist on the Internet from where software for wavelet analysis can be obtained. In Table I we provide a brief list (known to the authors at the time this was written) solely for the purpose of information to readers, without any recommendations, or reference to suitability or correctness of these codes.

A. Properties of Scaling Function

The scaling function satisfies the following properties:

1. $\int \phi(t)\,dt = 1$, i.e., the scaling function is an *averaging function*; compare this with the wavelet that satisfies $\int \psi(t)\,dt = 0$.

2. $\| \phi(t) \| = 1$, i.e., the scaling function is normalized to have unit norm.

3. $\int \phi_{m,n}(t)\psi_{m',n'}(t)\,dt = 0$, i.e., the scaling function is *orthogonal to all the wavelets*.

4. $\int \phi_{m,n}(t)\phi_{m,n'}(t)\,dt = \delta_{nn'}$, i.e., the scaling function is *orthogonal to all its translates at any fixed scale*. Note that unlike the wavelets, the scaling function is not orthogonal to its dilates. In fact,

5. $\phi(t) = \sum_n h(n)\phi(2t - n)$, i.e., the scaling function at some scale can be obtained as a linear combination of itself at the next scale ($h(n)$ are some coefficients called the scaling coefficients). This is a two-scale difference equation (see [16] and [17] for a detailed treatment of such equations).

6. The scaling function and wavelet are related to each other. In fact, one can show that

$$\psi(t) = \sum_n g(n)\phi(2t - n) \tag{A.1}$$

where $g(n)$ are coefficients derived from $h(n)$. That is, the wavelets can be obtained as a linear combination of dilates and translates of the scaling function.

For the particular case of Haar wavelet (see equation (40)) and the corresponding scaling function (equation (58)) $h(0) = h(1) = 1$ and $h(n) = 0$ for all other n, and $g(1) = -g(0) = 1$ and $g(n) = 0$ for all other n.

B. Daubechies' Wavelets

Daubechies [13] developed a class of compactly supported scaling functions and wavelets denoted as ($_N\phi$, $_N\psi$). They were obtained through the solution of the following two-scale difference equations:

$$\phi(t) = \sqrt{2} \sum_{n=0}^{2N-1} h(n)\phi(2t - n) \tag{B.1}$$

$$\psi(t) = \sqrt{2} \sum_{n=0}^{2N-1} g(n)\phi(2t - n) \tag{B.2}$$

where
$$g(n) = (-1)^n h(2N - n + 1) \text{ for } n = 0, 1, \ldots, 2N - 1. \tag{B.3}$$

For techniques to solve the above equations see [52]. The scaling coefficients $h(n)$ are obtained from solutions of high order polynomials ([15], chapter 6, and [51]) and satisfy the following constraints:

1. A necessary and sufficient condition for the existence of a solution to the above two-scale difference equations is

$$\sum_{n=0}^{2N-1} h(n) = \sqrt{2}. \tag{B.4}$$

2. Integer translations and dilations of $\phi(t)$ and $\psi(t)$ form an orthogonal family if the scaling coefficients satisfy

$$\sum_{n=0}^{2N-1} h(n-2k)h(n-2l) = \delta_{kl} \text{ for all } k \text{ and } l. \tag{B.5}$$

3. The constraints

$$\sum_{n=0}^{2N-1} (-1)^{n-1} n^k h(n) = 0 \text{ for } k = 0, 1, \ldots, N-1 \tag{B.6}$$

yield the result that $\psi(t)$ has N vanishing moments, i.e.,

$$\int t^k \psi(t) \, dt = 0 \text{ for } k = 0, 1, \ldots, N-1. \tag{B.7}$$

Daubechies' wavelets of this class have the following properties:

1. They are compactly supported with support length $2N - 1$. The scaling function also has the same support length.

2. As N increases, the regularity of $_N\phi$ and $_N\psi$ also increases. In fact $_N\phi, _N\psi \in C^{\alpha_N}$ (the set of continuous functions that are α_N^{th} order differential) where (N, α_N) pairs for some N are given as (see [13])

$$\{(2, 0.5 - \epsilon), (3, 0.915), (4, 1.275), (5, 1.596), (6, 1.888), (7, 2.158)\}.$$

3. Condition (B.7) of vanishing moments of $\psi(t)$ implies that up to N^{th} order derivatives of the Fourier transform of $\psi(t)$ at the origin are zero, i.e.,

$$\frac{d^k}{d\omega^k} \hat{\psi}(\omega = 0) = 0 \text{ for } k = 0, 1, \ldots, N-1. \tag{B.8}$$

This property essentially implies a form of localization of the Fourier transform $\hat{\psi}(\omega)$.

Figures B1 and B2 show the scale function and the corresponding wavelets for $N = 2$ and $N = 5$ respectively. The figures also depict the magnitude of their Fourier transforms. As can be seen the regularity of the wavelets and the scale functions increase as N increases.

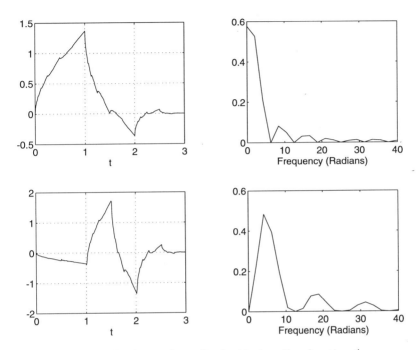

Figure B1. The figure shows Daubechies' scaling function (**upper left**) and the wavelet (**lower left**) of order $N = 2$. The magnitude of their Fourier transforms is shown, respectively, on the right hand column plots.

C. Implementation Algorithm for Orthogonal Wavelets

The implementation algorithm for wavelet multiresolution transforms is simple. From a data sequence $\{c_n^0\}$ (say at resolution level $m = 0$) corresponding to a function $f(t)$ we construct

$$P_0 f(t) = \sum_n c_n^0 \phi(t - n) \tag{C.1}$$

for a chosen $\phi(t)$, i.e., assume $\{c_n^0\}_{n \in \mathbf{Z}} = \{(f, \phi_{0n})\}_{n \in \mathbf{Z}}$. The data sequence at lower resolution can be obtained by

$$c_k^1 = \sum_n h(n - 2k) c_n^0. \tag{C.2}$$

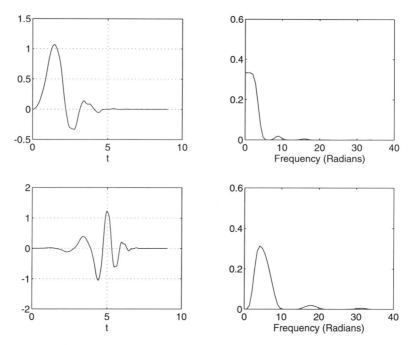

Figure B2. The figure shows Daubechies' scaling function (**upper left**) and the wavelet (**lower left**) of order $N = 5$. The magnitude of their Fourier transforms is shown, respectively, on the right hand column plots.

The detail sequence $\{d_k^1\}_{k \in \mathbf{Z}} = \{(f, \psi_{1k})\}_{k \in \mathbf{Z}}$ is obtained as

$$d_k^1 = \sum_n g(n - 2k)c_n^0. \tag{C.3}$$

Equivalently the above two equations can be written in the matrix notation

$$\{c^1\} = H\{c^0\} \quad \text{and} \quad \{d^{-1}\} = G\{c^0\}. \tag{C.4}$$

The matrices H and G are such that

$$HH^* = I, \qquad GG^* = I \tag{C.5}$$

and H^*H and G^*G are mutually orthogonal projections with

$$H^*H + G^*G = I \tag{C.6}$$

where H^* and G^* are adjoints of H and G respectively and I is the identity matrix. Also,

$$GH^* = 0 \quad \text{and} \quad HG^* = 0. \tag{C.7}$$

The algorithm given in equation (C.2) and (C.3) can be recursively implemented and the data and details at lower and lower resolutions can be obtained. This also highlights another important feature. Once we have the coefficients $h(n)$, we never need to construct the scale function $\phi(t)$ or the wavelet $\psi(t)$ for implementation on a discrete data set. The assumption involved, however, is that $c_n^0 = \int f(t)\phi(t-n)\,dt$, i.e., the integration kernel is $\phi(t)$ corresponding to the chosen $h(n)$'s. The reconstruction of the original sequence can be achieved by using

$$c_n^{m-1} = 2\sum_k h(n-2k)c_k^m + 2\sum_k g(n-2k)d_k^m \tag{C.8}$$

and is exact.

References

1. Arneodo, A., G. Grasseau and M. Holschneider, Wavelet transform of multifractals, *Physical Rev. Let.*, 61(20), 2281-2284, 1988.

2. Beylkin, G., R. R. Coifman, I. Daubechies, S. G. Mallat, Y. Meyer, L. A. Raphael, M. B. Ruskai (eds.), *Wavelets and Their Applications*, Jones and Bartlett Publishers, Boston, 1991.

3. Benedetto J. J., and M. W. Frazier (eds.), *Wavelets: Mathematics and Applications*, CRC Press, Boca Raton, Florida, 1993.

4. Brunet, Y. and S. Collineau, Wavelet analysis of diurnal and nocturnal turbulence above a maize crop, *This Volume*, 1994.

5. Burt, P. J., Fast filter transforms for image processing, *Computer Graphics and Image Processing*, Vol. 16, 20-51, 1981.

6. Chui, C. K., *An Introduction to Wavelets, Wavelet Analysis and its Applications*, Vol. 1, Academic Press, New York, 1992.

7. Chui, C. K., *Wavelets – a Tutorial in Theory and Applications, Wavelet Analysis and its Applications*, Vol. 2, Academic Press, New York, 1992.

8. Coifman, R. R., Y. Meyer and V. Wickerhauser, Size properties of wavelet-packets, in Ruskai et al., 453-470, 1992.

9. Combes, J. M., A. Grossman, Ph. Tchamitchian (eds.), *Wavelets: Time-Frequency Methods and Phase-Space*, Proc. of the Int. Conf., Marseille, Frace, December14–18, 1987, Springer-Verlag, 1989.

10. Cohen, A., Non-separable bidimensional wavelet bases, *Revista Mathemática Iberoamericana*, 9(1), 51-137, 1993.

11. Cohen, L., Time-frequency distributions–a review, *Proc. of the IEEE*, 77(7), 941-981, 1989.

12. Dallard, T. and G. R. Spedding, 2-D wavelet transforms, *Preprint*, 1990.

13. Daubechies, I., Orthonormal bases of compactly supported wavelets, *Commun. on Pure and Appl. Math.*, Vol. XLI, 901-996, 1988.

14. Daubechies, I., The wavelet transform, time-frequency localization and signal analysis, *IEEE Trans. Info. Theory*, 36(5), 1990.

15. Daubechies, I., *Ten Lectures on Wavelets*, SIAM, 1992.

16. Daubechies, I. and J. C. Lagarias, Two-scale difference equations I. existence and global regularity of solutions, *SIAM J. Math. Anal.*, 22(5), 1388-1410, 1991.

17. Daubechies, I. and J. C. Lagarias, Two-scale difference equations II. local regularity, infinite products of matrices and fractals, *SIAM J. Math. Anal.*, 23(4), 1031-1079, 1992.

18. Daubechies, I., Review of the Books: *Wavelets and Operators*, Y. Meyer, (Cambridge University Press, New York, 1993, Cambridge Studies in Advanced Mathematics, 37, Translation from the french edition, Paris, 1990, by D.H. Salinger) and *Wavelets: Algorithms and Applications*, Y. Meyer, (SIAM, Philadelphia, 1993, Translation form the french edition, by R. D. Ryan), Science, 262, 1589-1591, 1993.

19. Farge, M., Wavelet transforms and their applications to turbulence, *Annu. Rev. Fluid Mech.*, 24, 395-457, 1992.

20. Farge, M., J. R. Hunt, J. C. Vassilicos (eds.), *Wavelets, Fractals and Fourier Transforms: New Developments and New Applications*, Oxford University Press, Oxford, 1991.

21. Flandrin, P., Time-frequency and time-scale, *IEEE Fourth Annual ASSP Workshop on Spectrum Estimation and Modeling*, Minneapolis, Minnesota, 77-80, 1988.

22. Gabor, D., Theory of communications, *J. Inst. Elec. Eng.*, Vol. 93, 429-457, 1946.

23. Goupillaud, P., A. Grossmann and J. Morlet, Cycle-octave and related transforms in seismic signal analysis, *Geoexploration*, 23, 85-102, 1984.

24. Grossmann, A. and J. Morlet, Decomposition of Hardy functions into square integrable wavelets of constant shape, *SIAM J. math. Anal.* 15(4), 723-736, 1984.

25. Hudgins L. H., M. E. Mayer and C. A. Friehe, Fourier and wavelet analysis of atmospheric turbulence, in *Progress in Wavelet Analysis and Applications*, Y. Meyer and S. Roques, eds., 491-498, Editions Frontiers, 1993.

26. Jawerth, B, and W. Sweldens, An overview of wavelet based multiresolution analyses, *to appear, SIAM review*, 1994.

27. Katul, G. G., J. D. Albertson, C. R. Chu, and M. B. Parlange, Intermittency in atmospheric turbulence using orthonormal wavelet, *This Volume*, 1994.

28. Kumar, P. and E. Foufoula-Georgiou, A multicomponent decomposition of spatial rainfall fields: 1. segregation of large and small-scale features using wavelet transforms, *Water Resour. Res.*, 29(8), 2515-2532, 1993.

29. Kumar, P. and E. Foufoula-Georgiou, A multicomponent decomposition of spatial rainfall fields: 2. self-similarity in fluctuations, *Water Resour. Res.*, 29(8), 2533-2544, 1993.

30. Mallat, S., A theory for multiresolution signal decomposition: the wavelet representation, *IEEE Tran. on Pattern Anal. and Mach. Intel.*, 11(7), 674-693, 1989.

31. Mallat, S, Multifrequency channel decomposition of images and wavelet models, *IEEE Trans. on Acoustics, Speech and Signal Anal.* 37(12), 2091-2110. Dec. 1989.

32. Mallat, S., Zero-crossings of a wavelet transform, *IEEE Trans. Inform. Theory*, 37(4), 1019-1033, 1991.

33. Mallat, S. and L. Hwang, Singularity detection and processing with wavelets, *IEEE Trans. Inform. Theory*, 38(2), 617-643, 1992.

34. Mallat, S. and S. Zhong, Wavelet transform maxima and multiscale edges, in *Wavelets and Their Applications*, M. B. Ruskai, G. Beylkin, R. Coifman, I. Daubechies, S. Mallat, Y. Meyer, L. Raphael, eds., Jones and Bartlett, Boston, 1992.

35. Mallat, S. and S. Zhong, Complete signal representation from multi-scale edges, *IEEE Trans. Pattern Anal. Machine Intell*, 14(7),710-732, 1992.

36. Meneveau, C., Analysis of turbulence in the orthonormal wavelet representation, *J. Fluid Mech.*, 332, 469-520, 1991.

37. Meyer, Y., Ondolettes et applications, *J. Annu. Soc. Math.*, Soc. Francaise Math., Paris, France, pp. 1-15., 1990.

38. Meyer, Y., *Wavelets and Operators*, (Translation from the French Edition, Paris, 1990, by D. H. Salinger) Cambridge University Press, New York, 1993, Cambridge Studies in Advanced Mathematics, 37.

39. Meyer, Y., *Wavelets: Algorithms and Applications*, SIAM, Philadelphia, 1993.

40. Meyer, Y., Review of Books *An introduction to wavelets* by C. K. Chui, and *Ten Lectures on Wavelets* by I. Daubechies, *Bulletin (New Series) of the American Mathematical Society*, 28(2), 350-360, 1993.

41. Meyer, Y. and S. Roques (eds.), *Progress in Wavelet Analysis and Applications*, Editions Frontieres, France, 1993.

42. Morlet, J., G. Arens, E. Fourgeau, and D. Giard, Wave propogation and sampling theory- part 1: complex signal and scattering in multilayered media, *Geophysics*, 47(2), 203-221, 1982.

43. Morlet, J., G. Arens, E. Fourgeau, and D. Giard, Wave propogation and sampling theory- part 2: sampling theory and complex waves,

Geophysics, 47(2), 222-236, 1982.

44. Naylor, A. W. and G. R. Sell, Linear Operator Theory in Engineering and Science, Springer-Verlag, New York, 1982.
45. Pike, C. J., Analysis of high resolution marine seismic data using the wavelet transform, This Volume, 1994.
46. Pittner, S., J. Schneid, and C. W. Ueberhuber, Wavelet Literature Survey, Institute for Applied and Numerical Mathematics, Technical University Vienna, Wien, Austria, 1993.
47. Rioul, O. and M. Vetterli, Wavelets and signal processing, IEEE Signal Processing Magazine, 14-38, Oct. 1991.
48. Ruskai, M. B., G. Beylkin, R. Coifman, I. Daubechies, S. Mallat, Y. Meyers, and L. Raphael (eds.), Wavelets and Their Applications, Jones and Bartlett Publishers, 1992.
49. Saito, N. and G. Beylkin, Multiresolution representations using the auto-correlation functions of compactly supported wavelets, IEEE Trans. on Signal Processing, 41(12), 3584–3590, 1993.
50. Saito, N., Simultaneous noise suppression and signal compression using a library of orthonormal bases and the minimum description length criterion, This volume, 1994.
51. Strichartz, R. S., How to make wavelets, The Amer. Mathematical Monthly, 100(6), 539-556, 1993.
52. Strang, G., Wavelets and dilation equations: a brief introduction, SIAM Rev., 31(4), 614-627, 1989.
53. Wickerhauser, V., Lectures on wavelet packets algorithms, Preprint, Dept. of Math., Washington Univ., 1991.

The first author would like to thank Mike Jasinski and the Hydrologic Sciences Branch at NASA-Goddard Space Flight Center for the support during the course of this work. The second author would like to acknowledge the support of National Science Foundation grants BSC-8957469 and EAR-9117866, and NASA grant NAG 5-2108. We would also like to acknowledge the thoughtful comments of Don Percival, Naoki Saito and Paul Liu on this chapter.

Praveen Kumar
Universities Space Research Association/
Hydrological Sciences Branch (Code 974)
NASA-Goddard Space Flight Center
Greenbelt, MD 20771
e-mail:praveen@hydromet.gsfc.nasa.gov

Efi Foufoula-Georgiou
St. Anthony Falls Hydraulic Laboratory
Department of Civil and Mineral Engineering
University of Minnesota
Minneapolis, Minnesota 55414
e-mail:*efi@mykonos.safhl.umn.edu*

Applications of Structure Preserving Wavelet Decompositions to Intermittent Turbulence: A Case Study

Carl R. Hagelberg and Nimal K. K. Gamage

Abstract. Coherent structures in velocity and temperature in the atmospheric boundary layer account for a large portion of flux transport of scalars such as momentum, heat, trace chemicals and particulates. Generally, coherent structures are bounded by zones of concentrated shear in velocity or temperature fields.

We develop techniques based on the wavelet transform to provide a signal decomposition which preserves coherent structures. The decomposition is used to separate the signal into two components, one of which contains the important structures as defined by the characteristics of the transform. Embedded within the technique are a coherent structure detection mechanism, an analysis of intermittency resulting in an intermittency index, and filtering techniques. We illustrate the dependence of the coherent structure detection mechanism on the choice of analyzing wavelet demonstrating that anti-symmetric wavelets are appropriate for detecting zones of concentrated shear, while symmetric wavelets are appropriate for detection of zones of concentrated curvature.

We apply these techniques to the vertical velocity component, virtual potential temperature, and buoyancy flux density records from two portions of data gathered during the First International Satellite Land Surface Climatology Project (ISLSCP) Field Experiment 1987 (FIFE 87). We analyze the buoyancy flux corresponding to velocity and temperature structures in a convective, unstable atmospheric boundary layer, and the same boundary layer in a state of declining convection in late afternoon.

Spectral estimates of each of the partitions for the velocity signals are compared to the non–partitioned velocity signals. The characteristics of the partition reveal that the structure containing components of the velocity records follow a classical spectral description having a $-5/3$ power law behavior in the inertial range, while the non–structure components follow a -1 power law. We suggest that classification of turbulence as weak or strong can be based on the intermittency index and the scale distribution of coherent structures.

Wavelets in Geophysics
Efi Foufoula-Georgiou and Praveen Kumar (eds.), pp. 45–80.
Copyright 1994 by Academic Press, Inc.
All rights of reproduction in any form reserved.
ISBN 0-12-262850-0

§1. Introduction

Various physical processes in the atmospheric boundary layer such as the vertical transport of momentum and heat are associated with intermittent, coherent events, such as convective updrafts and plumes occurring due to localized heating and the resulting convection [18, 20]. Characterizing the physics of these events through observations and their behavior under varying conditions is still a topic of considerable continued study [1, 3]. There is a need to verify existing models (conceptual, mathematical or numerical) with observational data, and the need to provide models with parameter estimates for realistic simulation and continued insight into the physics to further develop conceptual models. Additionally, there is a need to determine from observation a set of parameters which best describe the processes associated with intermittent events. Identification of such parameters will suggest new approaches to modeling.

Conditional sampling techniques have been extensively reported in the literature as a means of extracting coherent events from data records [1, 3, 17, 20, 24]. Typically, conditional sampling techniques extract coherent events from a record to form an ensemble. The ensemble is then analyzed to characterize the events. This process loses important intermittency information since the global placement of the events is neglected. Various approaches to the sampling method have been used including "direct" thresholding on the values of scalars [18], thresholding on computed variance over short record patches [24, 25], visual identification of time domain characteristics [3], and thresholding time domain gradients [17]. In the latter case, an indicator record is constructed applying a threshold on, for example, the absolute value of the time derivative of velocity. This allows the definition of intermittency as a function of threshold which is 0 if no events are selected and 1 if the whole record is selected. This technique does not attempt to address the issue of coherence. More recently, conditional sampling methods have been developed based on variants of the wavelet transform [4, 9, 20]. The techniques proposed and demonstrated here, based on the wavelet transform, provide a general and objective means for achieving the desirable characteristics of conditional sampling methods and also provide a measure of intermittency which is directly related to coherent events.

The wavelet transform has evolved in various disciplines which require analysis of signals exhibiting intermittency in various forms. We attempt to illustrate some properties of the wavelet transform and propose certain new ideas useful to the study of atmospheric turbulence and the planetary boundary layer.

Sharp edges contain significant energy at high wave number. A signal consisting of intermittent sharp edges can have varying spectral characteristics depending on the spacing of the events. This is significant since the physical processes of transport of momentum and heat often occur in concentrated events (in space or time) in the atmospheric boundary layer. Consequently, a spectral description of such processes is not adequate to charcterize the intermittency of events.

In this paper we partition signals into two components. One containing coherent structures characterized by sharp transitions and intermittent occurrence, the other component containing the remaining portion of the signal essentially characterized by smaller length or time scales and the absence of coherent events. We envision an intermittency index, a structure type, and perhaps a small-scale spectral characterization as typical parameters useful for characterizing the atmospheric boundary layer. The structure type can be based on the analyzing wavelet. The intermittency index is defined in Section 3 from the local extremes of the wavelet transform. The spectrum may be computed directly from the wavelet transform phase plane and is discussed in Section 2. We note that the partitioning we develop is different from a Fourier based low-pass/high-pass partitioning since a typical structure defined by a large gradient will have frequency components in both bands.

It is possible to show that any symmetric wavelet is the second derivative of a smoothing function, whereas an anti–symmetric wavelet is the first derivative of a smoothing function [4, 22]. The anti–symmetric wavelets are optimally suited for shear–zone or micro–front detection [10]. These anti–symmetric events contribute to the inertial range, dissipation range, and to transport processes. In contrast, the symmetric wavelets are optimal for finding maximum curvature. Hence, if a signal contains a sharp transition, the anti–symmetric wavelet will be highly coherent at the center of the transition while the symmetric wavelet will be coherent at the edges of the transition effectively defining the boundaries of the transition zone. Equivalently, a zero–crossing of the symmetric wavelet transform will coincide with the center of the transition. However, distinguishing between large gradients and small gradients (over a given length scale) is more difficult using a symmetric wavelet [22]. It requires an estimation of the angle of the zero crossing in the transform [4], while the magnitude of the local extremes of the anti–symmetric transform is a direct (relative) measure of the sharpness [10]. Some of these properties are demonstrated using a simulated data set in Section 3.

Combining the optimal properties of anti–symmetry in space, and concentrated support in frequency leads to the natural choice of using the spline wavelets for analyses where strong shear zones characterize the important structures.

In Section 4 we apply the techniques presented to data sampled during FIFE 87 (described in Section 4). The vertical velocity, virtual potential temperature, and (derived) buoyancy flux density fields obtained from the atmospheric boundary layer are studied for two cases corresponding to late morning and late afternoon on the same day. We compute the intermittency index for each of the records. We then construct partition of the signals and compute total buoyancy flux for various combinations of the components. Additionally, spectra for the velocities and their decompositions are studied. A summary is provided in Section 5.

§2. The Wavelet Transform

2.1. The continuous case

We follow [22], where the wavelet transform is considered to be a type of multi–channel tuned filter. This allows the implementation of a fast algorithm and facilitates the interpretation of the wavelet transform as a coherent structure detection mechanism. We review the necessary information for the decomposition and reconstruction algorithms, though further discussion may be found in [22] and [23]. The books [5] and [7] give comprehensive treatment of wavelet transforms and Multi–Resolution Analysis. We limit our discussion to the properties most pertinent to our results.

To establish the notation, we begin with two functions, f and g, in $L^2(\mathbf{R})$ and define their inner–product by

$$(f, g) := \int_{\mathbf{R}} f(t)\, g(t)\, dt. \tag{1}$$

Hence, the $L^2(\mathbf{R})$ norm of f may be written

$$\|f\|_2 = (f, f)^{1/2}\ .$$

Similarly, the convolution of f and g is

$$f * g\,(b) := \int_{\mathbf{R}} f(t)\, g(b - t)\, dt\ . \tag{2}$$

The Fourier Transform of $f(t)$ is

$$f(t)^\wedge = \widehat{f}(\omega) := \int_{\mathbf{R}} e^{-i\omega t} f(t)\, dt\ , \tag{3}$$

and the inverse Fourier transform is denoted $f(t) = \widehat{f}(\omega)^\vee$.

Some further definitions specific to developing the wavelet transform are the scaled version of a function,

$$f_s(t) := \frac{1}{s} f(\frac{t}{s})\quad (s \neq 0); \tag{4}$$

the scaled and translated version of a function,

$$f_{sb}(t) := \frac{1}{s} f(\frac{t-b}{s}) \quad (s \neq 0); \tag{5}$$

and the reflection through the origin of a function,

$$f^-(t) := f(-t) . \tag{6}$$

A **basic wavelet** is an $L^2(\mathbf{R})$ function, $\psi(t)$, which satisfies the following admissibility criteria [5, 7]:

(i) zero mean: $\int dt\, \psi(t) = 0$, and

(ii) regularity: $C_\psi := \int d\omega\, |\hat{\psi}(\omega)|^2 / |\omega| < \infty$.

Additionally we normalize the wavelet to have norm unity, that is, $\|\psi\|_2 = 1$, and since we are interested in time–frequency (equivalently, space–frequency) localization, we specify that ψ be a windowing function [5]. That is, the first moment, $t\psi(t)$, should be integrable. Requiring first and higher moments to be zero imposes higher regularity on the wavelet.

Definition 2.1. The *continuous wavelet transform* of a function $f(t) \in L^2(\mathbf{R})$ at a location b, relative to the basic wavelet ψ at scale s, is defined by

$$W_s f(b) := (f, \psi_{sb}) = \int_{\mathbf{R}} f(t) \frac{1}{s} \psi(\frac{t-b}{s})\, dt . \tag{7}$$

Equivalently,

$$W_s f(b) := f * \psi_s^-(b) . \tag{8}$$

The continuous wavelet transform of an $L^2(\mathbf{R})$ function is a function of scale, $s \neq 0$, and location, b, and as such forms a surface over the (s, b)–plane called the *phase plane* of the wavelet transform.

Note that we have chosen a particular scaling for our definition of the wavelet transform in equation (7) through equation (4). Other choices are possible [2, 10], however, it has been argued that the choice provided by equation (7) facilitates the interpretation of the wavelet transform as a measure of coherence of structures with respect to scale [10]. We note that [10] attempted to distinguish between scaling choices of $\sqrt{s}$ and s in the definition of the wavelet transform by calling scaling by $\sqrt{s}$ the wavelet transform and scaling by s the covariance transform. It appears that the emerging convention in the literature is to disregard this distinction by using the name "wavelet transform" for any choice of scaling, and we adopt this point of view, making distinctions where necessary. Using the scaling $\sqrt{s}$ leads to an energy preserving transform, that is, a type of Parseval's relation for wavelets. While this is attractive for interpreting the wavelet transform as a decomposition of variance, it obscures the interpretation for detecting coherent events.

In the notation of equation (7) the wavelet transform may be interpreted as a covariance between the wavelet at a given dilation, ψ_s and the signal, f, at a specified location in the signal [10]. Using the notation of equation (8) the wavelet transform may also be interpreted as the signal, f, filtered by a particular (non–uniform, band pass) filter, ψ_s [6, 21]. We will discuss the application of the wavelet transform to the construction of filters and the filtering properties of some wavelets in Section 3.

The continuous wavelet transform may be inverted to recover the signal as follows.

$$f(t) = \frac{1}{C_\psi} \int\limits_{\mathbf{R}} db \int\limits_{\mathbf{R}\backslash 0} \frac{ds}{s^2}\, W_s f(b)\psi_{sb}(t) \,, \tag{9}$$

where C_ψ is the constant defined in the regularity condition (ii).

2.2. Dyadic subsampling in scale

We restrict the scale parameter, s to powers of 2. This allows the implementation of the fast wavelet transform algorithm given by [22].

Thus, let $s = 2^j$ for j any integer. For this restriction to yield a usefully invertible transform, the regularity condition (ii) must be modified as follows. The function $\psi \in L^2(\mathbf{R})$ (which satisfies the zero mean condition (i)) is called a *dyadic wavelet* if there exist positive constants $0 < A \leq B < \infty$ such that

$$A \leq \sum_{j=-\infty}^{j=\infty} |\hat{\psi}(2^j\omega)|^2 \leq B \,, \tag{10}$$

which is called the stability condition for the dyadic wavelet [5, 22]. It has been shown that a dyadic wavelet also satisfies the regularity condition (ii) [5].

Definition 2.2. The (continuous) *dyadic wavelet transform* of the signal f at the scale 2^j at location b is defined by

$$W_j f(b) := (f, \psi_{jb}) = \frac{1}{2^j} \int_{\mathbf{R}} f(t)\, \psi(\frac{t-b}{2^j})\, dt \,, \tag{11}$$

where we denote ψ_{2^j} by ψ_j. Equivalently,

$$W_j f(b) := f * \psi_j^-(b) \,. \tag{12}$$

By restricting the scale parameter to powers of 2, we have in effect subsampled the continuous wavelet phase plane. This is sometimes referred to as the semi–discrete wavelet transform. It has been shown that a complete reconstruction of the signal may be accomplished using another wavelet which is *dual* to the analyzing wavelet [5, 7, 22]. That is, the forward wavelet transform is performed using the basic analyzing wavelet, ψ, while the inverse wavelet transform must be accomplished using the dual wavelet, χ, defined as any function capable of performing the inverse transform in the following way.

Definition 2.3. Any function, $\chi \in L^2(\mathbf{R})$, is a *dyadic dual* of a dyadic wavelet, ψ, provided every function, $f \in L^2(\mathbf{R})$, can be written as

$$f(t) = \sum_{j=-\infty}^{\infty} 2^{-3j/2} \int_{\mathbf{R}} W_j f(b) \chi(2^j(t-b)) \, db \ . \tag{13}$$

In order to find the dual wavelet, conditions have been established which relate the dual to the analyzing wavelet. The following result is proven in [5].

Theorem 2.1. *Given a dyadic wavelet, ψ, and any $L^2(\mathbf{R})$ function, χ, such that*

$$\sup_{\omega \in \mathbf{R}} \sum_{j=-\infty}^{\infty} |\hat{\chi}(2^j \omega)|^2 < \infty \ , \tag{14}$$

then χ is a dyadic dual of ψ if and only if

$$\sum_{j=-\infty}^{\infty} \hat{\psi}^*(2^j \omega) \hat{\chi}(2^j \omega) = 1 \ . \tag{15}$$

In equation (15), $\widehat{\psi}^*$ denotes the complex conjugate of $\widehat{\psi}$. There may be many functions, χ, which will satisfy equation 15 for a given ψ. The condition given in equation (14) is a smoothness condition similar to the stability condition imposed on the basic wavelet in equation (10). Equation (15) expresses the restriction to an energy preserving forward–inverse transform relation.

The set of functions $\{\psi_j(t-b)\}$ for $-\infty < j < \infty$ and $b \in \mathbf{R}$, used to define the continuous dyadic wavelet transform, forms a non–orthogonal basis for $L^2(\mathbf{R})$. An orthogonal basis for $L^2(\mathbf{R})$ may be obtained from this set by subsampling the translation parameter b using $b_k = 2^j k$ to obtain the set $\{\psi_{jk}(t) = \psi_j(t - 2^j k)\}$ for j and k integers. By normalizing the orthogonal set we have an orthonormal basis of $L^2(\mathbf{R})$. In fact the proper normalization is given by the following.

$$\eta_{jk}(t) = \frac{1}{2^{j/2}} \psi\Big(\frac{t - 2^j k}{2^j}\Big)$$

$$= \frac{1}{\sqrt{s}}\psi(\frac{t-b}{s}) \quad \text{for} \quad s = 2^j, \ b = 2^j k \ . \tag{16}$$

Thus, by choosing the normalization in equation (4) to be $\psi_s(t) = 1/\sqrt{s}\psi(t/s)$ and subsampling in the orthogonal fashion, we arrive at the orthonormal basis $\{\eta_{jk}\}$. We denote the orthonormal wavelet coefficients obtained from the signal f by $W_{jk}f$ to distinguish them from the dyadic wavelet coefficients given by $W_j f$ in equation (12).

Orthonormal wavelets provide certain advantages. For example, they are self–dual in the sense that $\chi = \psi$ in equation (13), with the integral over b replaced by a sum over k. The orthonormal wavelet coefficients satisfy preservation of energy as a type of Parseval's relation. That is,

$$\|f\|^2 = \sum_{j,k=-\infty}^{\infty} |W_{jk}f|^2 \ . \tag{17}$$

Using the orthonormal wavelet transform it is possible to define the *wavelet spectrum* in the following way [14, 27, 28]. Rewriting equation (17) as

$$\|f\|^2 = \sum_{j=-\infty}^{\infty} E_j$$
$$= \sum_{j=-\infty}^{\infty} \sum_{k=-\infty}^{\infty} |W_{jk}f|^2 \ , \tag{18}$$

we can define the wavelet spectrum of an L_2 function, f, as

$$E_j = \sum_{k=-\infty}^{\infty} |W_{jk}|^2 \ . \tag{19}$$

It is possible to assign a correspondence between the Fourier wavenumber and the scale of the wavelet transform. This can be done formally by assigning the wavenumber at the center of mass of the frequency content of the wavelet at a particular scale [27]. However, for comparing power law trends of wavelet spectra with trends in Fourier spectra the assignment is essentially arbitrary as long as it remains consistent through the dilation of the wavelet.

If the energy spectrum, $|\hat{f}(\omega)|^2/\omega$, obeys a scaling law, say $|\hat{f}(\omega)|^2/\omega \sim \omega^{-p}$, then the wavelet spectrum is expected to possess a scaling law through the correspondence $E_j \leftrightarrow \hat{f}(\omega)|^2 \sim \text{constant} \cdot 2^{j(p-1)}$. (In our notation, $j \to -\infty$ implies $\omega \to \infty$. That is, large positive j corresponds to increasing scale; large negative j corresponds to decreasing scale). The similarity symbol, $\sim$, is used to indicate that for small scales (ω large, j negative and large) the spectrum has a trend corresponding to exponential decay.

2.3. Finite resolution in discrete applications

Since discrete data are limited to finite resolution (smallest scale) determined by a sampling rate it is not possible to perform an analysis at a resolution finer than some fixed scale. Choosing the smallest scale to be unity ($s = 2^j$, $j = 0$) means we cannot represent data on scales between grid points. Actual data are also limited to some finite record length, say $s = 2^j$ for $j = J > 1$. Furthermore, any definitions of functions on the finite grid may fail to follow the continuous properties in the conditions given by equations (10), (14), and (15). This is discussed in detail in [22]. The solution is to introduce a smoothing function, ϕ, whose frequency content represents the fractional loss of energy when performing a wavelet transform followed by the inverse transform (under the restrictions imposed by finite resolution). That is,

$$\hat{\phi}(\omega) = \sum_{j=1}^{\infty} \hat{\psi}(2^j\omega)\hat{\chi}(2^j\omega) \ .$$

The relation in equation (15) then can be written

$$|\hat{\phi}(\omega)|^2 - |\hat{\phi}(2^J\omega)|^2 = \sum_{j=1}^{J} \hat{\psi}^*(2^j\omega)\hat{\chi}(2^j\omega) \ , \tag{20}$$

for $\hat{\psi}^*(2^j\omega)\hat{\chi}(2^j\omega)$ a positive, real, even function.

Denote by S_j the smoothing operator at scale j given by

$$S_j f(t) := f * \phi_j(t) \ , \tag{21}$$

where $\phi_j(t) = 2^{-j}\phi(t2^{-j})$. The information content of the signal smoothed at some coarse scale, $S_J f(t)$, and the wavelet coefficients for all scales up to the coarse scale, $\{W_j f(t)\}_{1 \le j \le J}$, is sufficient to recover the original discrete signal. We accomplish this using the fast wavelet transform algorithm described in [22].

As noted, the orthonormal wavelet coefficients may be obtained from the continuous dyadic wavelet coefficients through appropriate subsampling and scaling. For real applications the data record length is finite and an orthogonal subsampling depends on which data point is assigned as the end of record. Since the wavelet transform is not shift invariant, a shift in the location at which the wavelet coefficients are computed results in a change in the wavelet coefficients, and hence a change in the estimate of the spectrum resulting from the coefficients. By averaging over all possible orthogonal subsamplings of the translation parameter we approach a shift invariant estimate of the spectrum. An equivalent approach would be to average over several globally translated orthogonal wavelet analyses.

When computing spectra of finite resolution data using traditional applications of the fast Fourier transform, it is well known that the resulting spectral estimates are highly dependent on windowing the data ([13, 16]). The wavelet transform creates its own window (the wavelet) and the characteristics of the spectral estimate are then dependent on the regularity of the analyzing wavelet. For an application of these ideas to atmospheric data see [9].

§3. Filtering Properties of Wavelets

3.1. Partitions as filters

We wish to partition a signal into two components, one containing significant structures and the other containing the remaining portion of the signal. Once a partition is obtained, further analysis of each component can proceed. For example, an intermittency index can be assigned to the structure–containing component, and spectral characteristics of the components can be analyzed. In certain communication theory applications the significant structures may correspond to a signal prior to being transmitted, while the remaining portion of the signal corresponds to the noise introduced during transmission. In the case of atmospheric turbulence, structures may be defined as regions of sharp transition of significant amplitude [10, 20], while the remaining portion corresponds to a different physical flow characteristic (not necessarily regarded as "noise").

To create a partition of a signal one must begin with assumptions regarding the nature of the signal. For example, Fourier (band–pass) filtering partitions a signal into components based on frequency content. Some criterion is used to determine the components, such as the location of a frequency band or the location of multiple bands determined by a frequency amplitude cutoff. The partition then consists of the portion of the signal with frequency content inside the bands and the portion with frequency content outside the bands. A filter consists of retaining only one element of the partition. If the significant structures are band limited and the remaining portion of the signal is outside the structure containing band, then the signal will exhibit spectral gaps. The band–pass filters can then be prescribed according to the location of the spectral gaps.

However, if the structure of interest in the signal is a sharp transition it contains significant high frequency content. The high frequency content can not be distinguished from the frequency contribution of several smaller amplitude sharp transitions, or from colored or white noise. In this case the component of interest in the signal is not usefully band limited and the spectrum will not exhibit a clear spectral gap. A Fourier filtering technique is not well suited to creating a partition of the signal in such a case. The wavelet transform can help in this situation since it effectively limits the spectral content under consideration to specific time or space locations. The wavelet phase plane contains information on the coherence between the signal and the wavelet in both position and scale. This can be condensed into a global statistic providing information about the scale at which there are the largest number of coherent events of significant amplitudes. The global statistic for a signal, f, is called the *wavelet variance* [10, 20] and is defined in the continuous case as

$$D^2(s) = \int_{-\infty}^{\infty} [W_s f(b)]^2 \, db .$$ (22)

This is similar to the wavelet spectrum of equation (19), but is not a variance decomposition of the signal with the choice of scaling given in equation (4). (It is possible to interpret the wavelet variance as proportional to the energy (variance) of the signal using appropriate logarithmic axes [4]). However, the quantity D^2 is a variance since it is the squared error about zero of the wavelet transform (which removes the mean from the signal). Note that there are two situations which will contribute to producing a large value of the wavelet variance at a given scale – one large amplitude coherent event or several coherent events of lesser amplitude. For analysis of atmospheric turbulence signals, both situations may be considered significant.

We note that when information about dominant physical length scales of important structures are deduced from the wavelet variance, the continuous scale transform, rather than a dyadic subsampling, should be considered. The subsampling, while providing complete information for inversion, does not allow for a complete physical analysis since important physical scales may be skipped.

While the wavelet variance is a reasonable means of characterizing signals, a characterization by the type of events they contain, and a measure of their intermittency would provide further description. We address this issue in Section 3c.

Locating local maxima in the wavelet variance provides a means of determining the scales at which the structures in the signal are coherent with the wavelet shape, and of significant amplitude and/or number. It is possible to imagine situations where there is more than one local maxima in the wavelet variance. For example, if there is significant white noise in the signal, the wavelet variance will have a large peak at the smallest scale, and a somewhat smaller peak at the scale of the coherent events. Another example is a signal consisting of short pulses. One peak in the wavelet variance corresponds to the scale of the length of the pulse and another peak corresponds to the spacing of the pulses, assuming the spacing and width are different scales [8]. Finally, a superposition of two sinusoidal waves at different frequencies will produce two local maxima in the wavelet variance [4]. In such a situation, each local maximum corresponds to a wavelength and some choice as to their significance must be based on knowledge of the nature of the signal.

Once the dominant event scale or scales are found, one of the obvious methods leading to a partition of the signal is to limit the reconstruction from the phase plane to the scales above and below some scale near the dominant scale, or between two dominant scales. One approach to determining a partition if only one dominant scale is present is to reconstruct from the phase plane information using the large scales down to *some* small scale which is determined from the dominant scale by taking an appropriate fraction of the dominant scale. The partition then consists of the large scale reconstruction and the small scale reconstruction (that is, the remaining portion of the signal). If there is a clear "scale gap" between peaks in the wavelet variance, the scale for the partition could be chosen to be in the gap [4]. We refer to this as *scale threshold partitioning*. Scale threshold partitioning has the problem of neglecting the fact that structures of interest are multiscaled and may have significant information content even at small scales. That is, portions of the information content of a single structure may fall into both elements of the partition. This is an inherent problem in partitioning signals in which multiscale structures are important but are found in the presence of broad band noise. Another inherent problem is determining the appropriate fraction of the dominant scale for the threshold. Examples are given later in this section and in Section 4. Even with these difficulties, a scale threshold partition works very well for separating out white noise at the smallest resolvable scale.

In [23] the rate of growth or decay of significant "ridges" in the phase plane is used as a means of quantifying what is meant by noise. This is a measure of the regularity of a multiscale feature. For highly singular features the wavelet transform will undergo a rapid increase in value at small scales as the scale is decreased. Assuming that the "noise" is less regular than the structures of interest and limited to small scales, a ridge reconstruction ([23, 22]) is performed using only the portions of ridges which do not exhibit an increase at small scales, or by extending ridges to smaller scales keeping the amplitude fixed.

In [23] the use of ridge threshold filtering described above is motivated by the interest in signal compression. For this study we have no need for signal compression and therefore have the information of the entire phase plane at our disposal. We can therefore employ a simpler approach to partitioning which still allows for the multiscale nature of coherent events. Furthermore, for applications to atmospheric turbulence data, we are interested in partitioning the signal rather than an actual filtering (that is, we retain both components of the partition). The method we examine is the reconstruction of the portion of the signal whose wavelet coefficients are larger or smaller than a particular threshold. Reconstruction of the signal using coefficients larger than some fixed value, $|W_j f(n)| > w_c$, preserves the structural features. The remaining coefficients, $|W_j f(n)| < w_c$, are used to reconstruct the portion of the signal not associated with the structures. We call this type of partitioning of a signal *phase plane threshold partitioning*.

Note that phase plane threshold partitioning requires the scaling chosen in equation (4) for the definition of the wavelet transform. If some other scaling is chosen then an appropriately scaled cutoff must be used. For the orthonormal decomposition, in which the scaling is chosen as $1/\sqrt{s}$ (equation 16), the phase plane cutoff is scale dependent and given by $w_c(j) = w_c/\sqrt{2^j}$.

The partitioning of the signal into a component containing structures and a component containing the remainder allows the comparison of the spectra of the respective components. We will demonstrate that the spectral characteristics of each component of a velocity record are quite different. The partitioning can be accomplished using the wavelet partitioning methods described above. The partition is dependent upon the choice of analyzing wavelet and the method of thresholding used. We utilize these characteristics to analyze buoyancy flux in the atmospheric boundary layer in Section 4.

3.2. Detection characteristics of anti–symmetric and symmetric wavelets

In this section we demonstrate certain general characteristics of the differences between using symmetric versus anti–symmetric wavelets for signal analysis. Figure 1a is the vertical velocity record from an aircraft measurement during the FIFE 87 field experiment (described in Section 4). The signal contains structures resembling square or ramped pulses on the order of 10 seconds wide (40 meters). There are approximately eight or nine such events. In order to illustrate the response of the wavelet transform to such structures, consider the simulated signal of Figure 2. This consists of four "pulses" of similar shape, unevenly spaced, with a resolution of 256 points. The shape was chosen to have some characteristics typical of the structures observed in the FIFE 87 vertical velocity record.

The subsequent analyses are performed using the anti–symmetric quadratic spline wavelet and the symmetric cubic spline wavelet shown in Figure 3. The wavelet transform of the simulated signal based on a symmetric wavelet and an anti–symmetric wavelet at a single scale are shown in Figure 4. Note the magnitude of the symmetric transform is large at the boundaries of the transitions, while the magnitude of the anti–symmetric

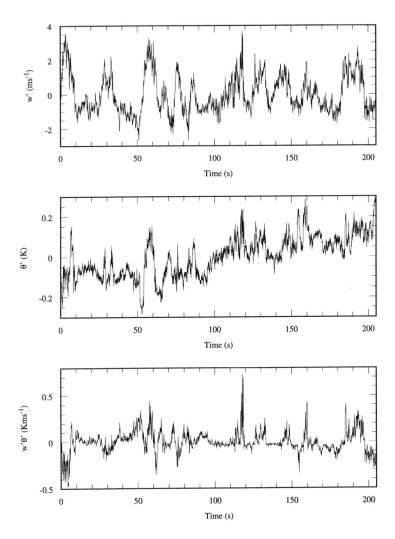

Figure 1. Morning case: **(a)** **(top)** Vertical velocity record shown as the fluctuation about the mean, w' (m/s) $= w - \overline{w}$, where $\overline{w} = -0.0045$ (m/s). **(b)** **(middle)** Virtual potential temperature fluctuation, θ_v' (K). $\overline{\theta_v} = 301.46$ (K). **(c)** **(bottom)** Buoyancy flux density, $w'\theta_v'$ (ms^{-1}K).

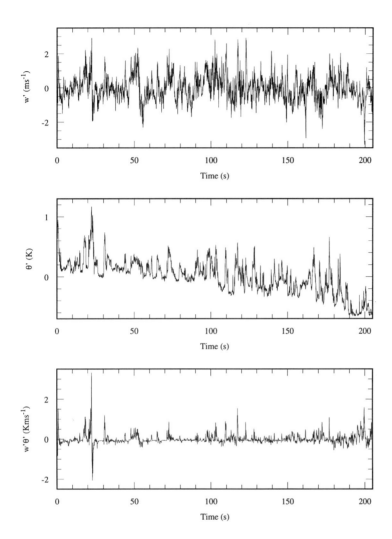

Figure 1. Afternoon case **(d) (top)** Vertical velocity fluctuation, w' (m/s). **(e) (middle)** Virtual potential temperature fluctuation, θ'_v (K). **(f) (bottom)** Buoyancy flux density, $w'\theta'_v$ (ms^{-1}K).

transform is large at the center of the transitions. This characteristic is scale dependent since at scales larger than a particular event size the location of maximum coherence between the wavelet and the signal may be

Figure 2. Simulated signal having features qualitatively similar to features in the vertical velocity record (Figure 1a).

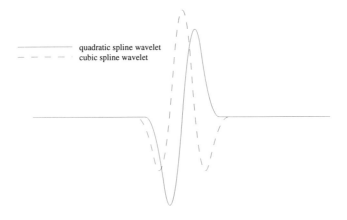

Figure 3. Anti–symmetric quadratic spline (solid line) and symmetric cubic spline (dashed line) used for wavelet analysis.

influenced by characteristics of the signal outside of the event. This is illustrated in Figure 5, where the anti–symmetric transform is shown at three scales. Note that the location of maximal gradient, corresponding to extremes of the transform, migrate as larger scales are considered [22].

We now turn to the notion of creating a partition of the signal. Con-

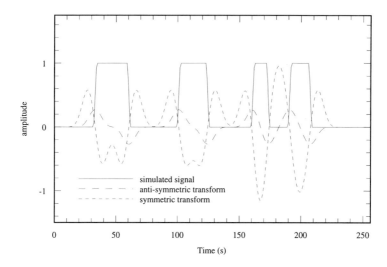

Figure 4. Anti–symmetric (long dashed line) and symmetric (short dashed line) wavelet transforms of the simulated signal (solid line) at the scale 2^3.

sider the signal shown in Figure 6 consisting of the simulated signal given in Figure 2 plus small amplitude sine wave with random phase. This is intended to represent a situation where the desired features are the larger scale structures which contain sharp transitions but are found in the presence of significant smaller scale information.

The dominant scale of structures must first be found using the wavelet variance defined in equation (22). The dyadic equivalent of equation (22) was used to compute an estimate of the wavelet variance of the simulated signals of Figure 2 and Figure 6 using both the symmetric and anti–symmetric wavelets. The result is shown in Figure 7. Four curves are shown corresponding to computing the wavelet variance of the simulated signal without noise and the simulated signal with noise using either a symmetric or anti–symmetric wavelet. The maximum wavelet variance for scales larger than the smallest occurs at the scale 2^5 and is apparently independent of the choice of wavelet. This independence would suggest that identifying structures in a signal is independent of the symmetry of the wavelet used. However, further test cases using a variety of length scales in the structures show that is not always true.

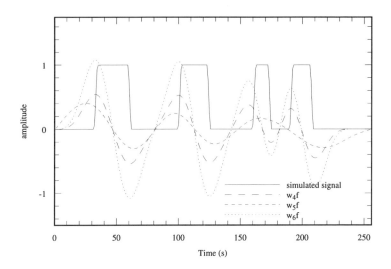

Figure 5. Anti–symmetric wavelet transform (dashed lines) of the simulated signal (solid line) at three scales. Note the migration of the location of the extremes. The location and interpretation of a coherent event is necessarily scale dependent.

Figure 6. Simulated signal which includes small scale information consisting of a mix of white noise and coherent small scale structure.

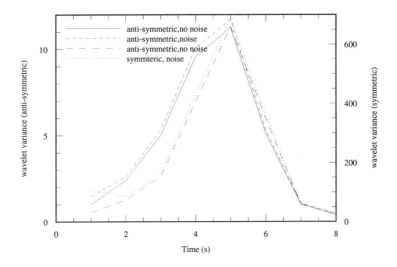

Figure 7. Wavelet variance computed using the discrete form of equation (22)

Once the dominant scale of structures is selected from a (local) maximum of the wavelet variance, the local extremes (either maxima or minima) of the wavelet transform at that scale provides the location of the structures. Moreover, a phase plane partition can then be formed based on a fraction of the largest (in magnitude) of the local extrema at the scale of maximum wavelet variance. This process is illustrated in Figure 8 where the anti–symmetric quadratic spline wavelet has been used.

Once a threshold value has been determined by the method in the previous paragraph, a partition of the signal may be formed. A reconstruction based on the wavelet transform values above the threshold forms one component and the remaining portion of the signal forms the other component. This is shown in Figure 9 where a factor of 0.3 of the largest of the local extrema was used (determined empirically based on the influence of changes in the factor on the variance of the non–structure component).

The result in Figure 9 may be compared to a simple scale threshold partitioning shown in Figure 10, where the scale threshold is 2^1. That is, the smallest scale information forms one element of the partition and all scales above that form the other. This is the best way in which to separate

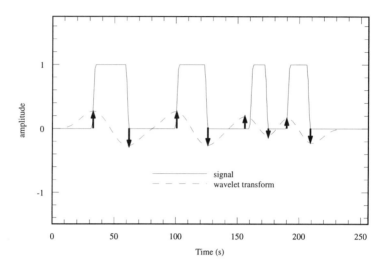

Figure 8. Anti–symmetric wavelet transform (dashed line) of simulated signal (solid line) at the scale of maximum variance (from Figure 7). The arrows indicate the location of the local extremes. A factor times the largest of the extremes is used to determine the phase plane threshold value.

out white noise, since it has a decorrelation length scale of two points. Scale partitioning has been suggested as a method of filtering in [4].

Finally, the phase plane threshold and the scale threshold may be combined to create a partition such as that shown in Figure 11. We have combined a phase plane threshold of a factor of 0.2 (also emperically determined as above) of the largest of the local extremes with a scale threshold of 2^1.

3.3. Quantifying intermittency

Intermittency has many manifestations depending on physical context. One classic example is the laboratory flow constructed by ejecting fluid of a higher than ambient temperature from a jet. The time history of temperature at a point downstream in the turbulent flow will exhibit intermittent spikes as small cores of fluid of high temperature from the jet pass by the sensor. As the sensor is moved from the center of the jet flow to the edge of the domain, the intermittency effect increases. That is, the fraction of the

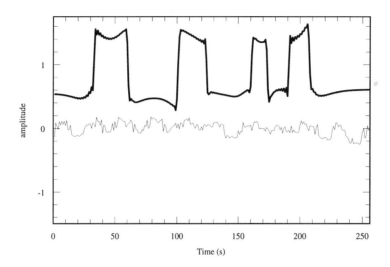

Figure 9. Phase plane threshold partition of the simulated signal shown in Figure 6. A constant was added to the large scale portion to offset the plots.

record which is high temperature decreases as the sensor moves away from the center of the domain. An intermittency index defined as the fraction of the record occupied by high temperature fluid provides information about the degree of entrainment of ambient fluid into the high temperature core. See [26] for a more detailed discussion. In Figure 1c the regions of sustained positive buoyancy flux density are the structures of interest. These structures could be identified by applying a threshold to the data, but the intermittency (fractional record length where data is large) is sensitive to the choice of the threshold. Furthermore, thresholding does not take into account the coherency of the structures. For the vertical velocity record shown in Figure 1a, intermittency depends on the type of event one wishes to define. For example, we are interested in identifying regions of sharp gradient of sufficient magnitude (which can not be identified using a threshold on the data). The wavelet transform provides a means of locating such events. Once their location is known some measure of the intermittency of these occurrences with respect to the total signal length can be devised.

The main purpose of this section is to propose a new measure of in-

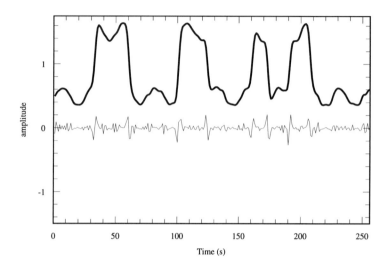

Figure 10. Scale partition (at scale 2^1) of the simulated signal shown in Figure 6. A constant was added to the large scale portion to offset the plots.

termittency computed directly from the wavelet transform information. This measure of intermittency combined with the fraction of buoyancy flux contributed by the structure component of buoyancy record provides an important insight into the nature of the boundary layer convective process. In Section 3b we have discussed finding the dominant scale using the wavelet variance, and locating the local extremes in the wavelet transform at that scale. Using this information the signal is partitioned into a structure component and a non–strucuture component. As a measure of the fraction of the record containing structures, an intermittency index can be formed as

$$\text{intermittency} = \frac{\text{support of structures}}{\text{total record length}}. \tag{23}$$

For example, if the structure component of the buoyancy flux accounts for most of the total flux and the intermittency index is small (very intermittent coherent structures) then there are a few strong convective plumes in the flow. In contrast, if the intermittency index is large (most of the record occupied by coherent structures) then there is a more uniform transport (such as in regions of subsidence or secondary flow).

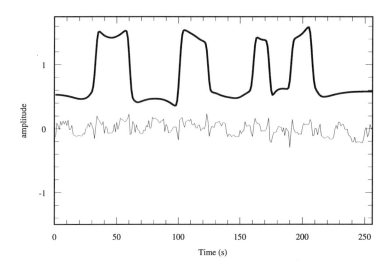

Figure 11. Combination of a scale (at scale 2^1) and phase plane threshold (at 0.25 times the largest extreme) to create a combination partition of the simulated signal of Figure 6. A constant was added to the large scale portion to offset the plots.

This definition of intermittency is dependent on the thresholding used to locate local extremes in the wavelet transform, but does not depend on thresholding the data.

§4. Applications to Atmospheric Data

4.1. Description of the data

We now apply the techniques described in Sections 2 and 3 to two portions of data collected during the First International Satellite Land Surface Climatology Project (ISLSCP) Field Experiment 87 (FIFE87). A description of the experiment and analysis of data using other techniques including conditional sampling may be found in the FIFE special issue of JGR [15]. The purpose of this section is to demonstrate the utility of the techniques presented in the previous sections by providing new insight into the interpretation of specific data. This is an extension of the results shown in [12] where a different segment of the data was analyzed.

An objective of FIFE87 was to gather in situ boundary layer measure-

ments in conjunction with satellite remote measurements to obtain suitable
parameterization schemes. In part, the objectives are to determine the accuracy of, and to calibrate, satellite data and to relate the remotely sensed
data to flux quantities. The intermittent nature of momentum, heat, and
buoyancy fluxes makes this task difficult and may lead to deceptive estimates from remotely sensed data. The in situ data used in this study were
sampled using the National Center for Atmospheric Research (NCAR) King
Air research aircraft in fair weather conditions over a 50 kilometer square
of uniform tallgrass over the Konza National Prairie in northeast Kansas—
the designated FIFE site. A case study for June 6, 1987 from the Intensive
Field Campaign 1 (IFC1) is reported in [11]. June 6th was a clear day designated as the "golden day" for FIFE87 IFC1. We study two portions of
the data from June 6th corresponding to late morning and late afternoon.
These two times present differing physical situations; late morning being
a strongly convective, developing mixed layer, and late afternoon having
subsiding thermal forcing and the resulting weak or decaying convection
combined with a well mixed and sheared boundary layer.

The measurements were obtained on June 6th at 650 meters above
ground in the morning and 1300 meters above ground in the afternoon
with 4 meter resolution along an east to west transect at the northern
edge of the site. The flights remained at a fixed position with respect to
the atmospheric boundary layer top (about 90% of ABL) to ensure consistency in the sampled data. The data were collected over the time intervals
11:26:46 L (Local time (L) = -6 UTC) to 11:30:11 L for the morning leg,
and 15:55:56 L to 15:59:21 L for the afternoon leg. Figure 1a–c shows
the vertical velocity component, the virtual potential temperature, and the
buoyancy flux density derived from the virtual potential temperature and
the vertical velocity for the morning leg. Figure 1d–f shows the corresponding fields for the afternoon. Note the intermittently occurring zones
of sharp transition in the vertical velocity (Figure 1a and d), and the zones
of positive buoyancy flux (Figure 1c and f). Note also the the number and
size of ramp and plume structures (in both velocity and temperature) is
different between the morning and afternoon legs.

4.2. Analysis of the data

Table 1 collects some length scale statistics computed using an anti–
symmetric wavelet transform for vertical velocity components and the virtual potential temperature for the morning and afternoon legs. Listed are
the dominant scale lengths (as derived from the largest local maximum in
the wavelet variance), the number of occurrences of events of this size, and
the intermittency index as defined in Section 3c. Due to large amplitude
low frequency components in the temperature records, the largest variance
for an anti–symmetric wavelet will occur at the largest scale. We therefore

chose to use the next smallest scale at which a local maximum occurred. The determination of dominant scales based on the dyadic wavelet transform subsamples the continuous scale information that would be possible using a continuous scale. However, the dominant scales are in the smaller scales where there is a better sampling density, and the scale distinctions at higher resolution are not particularly relevant to the present analysis.

Table 2 collects the same statistics based on using a symmetric wavelet. The dominant length scales tend to be the same or slightly longer, and the number of events the same or slightly fewer. Correspondingly, the intermittency index is the same or slightly larger (less intermittent).

Both Table 1 and Table 2 show that coherent structures are of smaller scale in the afternoon than the morning and are much more (spatially) frequent. This can be seen by comparing, for example, the vertical velocity record from the morning (Figure 1a) with the vertical velocity record from the afternoon (Figure 1d).

Table 1: Length scale analysis using an anti–symmetric wavelet			
Field	Dominant Length Scale (km)	Number of events	Intermittency Index
Morning case:			
w	1	10	0.59
θ_v	0.5	18	0.53
$w'\theta_v'$	0.5	20	0.61
Afternoon case:			
w	0.125	76	0.57
θ_v	0.25	36	0.54
$w'\theta_v'$	0.06	142	0.54

Table 2. Length scale analysis using a symmetric wavelet			
Field	Dominant Length Scale (km)	Number of events	Intermittency Index
Morning case:			
w	1	10	0.61
θ_v	0.5	19	0.57
$w'\theta_v'$	1	8	0.49
Afternoon case:			
w	0.125	77	0.59
θ_v	.25	36	0.55
$w'\theta_v'$	0.125	83	0.62

A combination phase plane threshold and scale partition has been created for the velocity components (Figure 12a and c) and the buoyancy flux densities (Figure 12b and d) for the morning and afternoon cases. Combination partitions were also created for the temperature records but are not shown. An anti–symmetric wavelet analysis was used in each case. The sharp vertical edges have been maintained in the larger scale partition in each case.

Total buoyancy flux over the signal was computed by summing over the flux density for each of five possibilities – from the original data, $w'\theta_v'$, which will be the reference flux; from the structure partition of the derived buoyancy flux density record, $(w'\theta_v')_{\text{str}}$; from the structure component of the velocity with the original temperature data, $w'_{\text{str}}\theta_v'$; from the structure component of the temperature with the original velocity data, $w'(\theta_{v\text{str}})'$; and from the structure component of velocity with the structure component of temperature, $(w_{\text{str}})'(\theta_{v\ \text{str}})'$. This was done for the morning case and the afternoon case resulting in the ten numbers shown in Table 3 where the total fluxes are shown as a percentage of the reference flux. The morning reference flux was about 40% of the afternoon reference flux.

The flux transport computed using the structure component of velocity, $(w_{\text{str}})'\theta_v'$, accounts for 69% of the reference flux in the morning and 58% of the reference flux in the afternoon. The flux transport computed using the structure component of temperature, $w'(\theta_{v\text{str}})'$, is relatively higher in both the morning (82%) and the afternoon (75%). Additionally, the total flux computed using the structure component of the velocity with the structure component of the temperature, $(w_{\text{str}})'(\theta_{v\text{str}})'$, is 58% in the morning and 42% in the afternoon indicating that the temperature structures are not highly correlated with the velocity structures.

Table 3. Total flux analysis using structure partitions		
Flux density		Total Flux
Morning case:		
$w'\theta_v'$	Buoyancy flux	100%
$(w'\theta_v')_{\text{str}}$	Structure part of flux	100%
$w'_{\text{str}}\theta_v'$	Flux due to velocity structures	69%
$w'(\theta_{v\text{str}})'$	Flux due to temperature structures	82%
$(w_{\text{str}})'(\theta_{v\ \text{str}})'$	Flux of temp. str. due to vel. str.	58%
Afternoon case:		
$w'\theta_v'$	Buoyancy flux	100%
$(w'\theta_v')_{\text{str}}$	Structure part of flux	100%
$w'_{\text{str}}\theta_v'$	Flux due to velocity structures	58%
$w'(\theta_{v\text{str}})'$	Flux due to temperature structures	75%
$(w_{\text{str}})'(\theta_{v\ \text{str}})'$	Flux of temp. str. due to vel. str.	42%

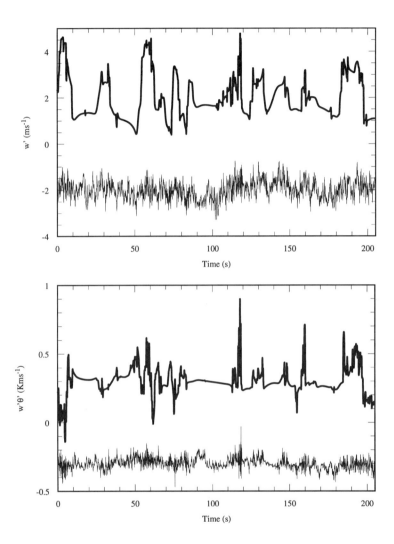

Figure 12. Morning case: Combination phase plane and scale partitions for **(a)** **(top)** vertical velocity and **(b)** **(bottom)** buoyancy flux density, Vertical shifts have been added to offset the components.

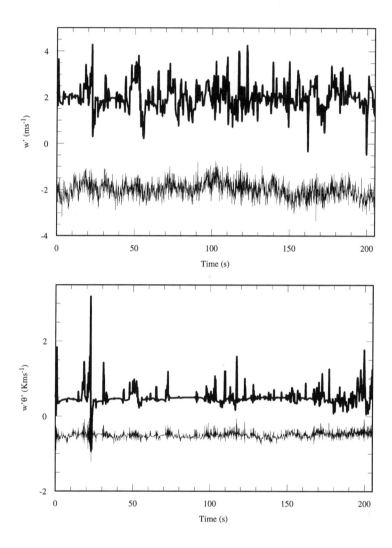

Figure 12. Afternoon case: Combination phase plane and scale partitions for **(c) (top)** vertical velocity and **(d) (bottom)** buoyancy flux density, Vertical shifts have been added to offset the components.

In the lower portion of the convective boundary layer (CBL) the buoyancy generation is due to heating resulting in coherent temperature structures, $(\theta_{vstr})'$, accounting for a large fraction of the buoyancy flux. Near the top of the CBL the buoyancy is driven by accumulated momentum, $(w_{str})'$, of the plume structures [18]. The data were gathered in a middle portion of the CBL (650 meters in the morning; 1300 meters in the afternoon), and the percentage of buoyancy flux due to virtual potential temperature structures being larger than buoyancy flux due to velocity structures reflects this in both the morning and afternoon cases. Comparing the velocity partition of the morning case (Figure 12a) to the velocity partition of the afternoon case (Figure 12c) it is apparent that the morning case contains relatively larger, well–defined structures. This is reflected in a comparison of the buoyancy flux densities as well (Figures 12b and d). This is quantified in the length scale analysis summarized in Table 2.

The late morning boundary layer is characterized by vigorous convective activity which results in large plume structures penetrating from the bottom to the top of the CBL. By afternoon, vertical soundings taken simultaneously with the aircraft data indicate that significant vertical shear of horizontal winds has developed and that the CBL is well mixed. This leads to entrainment of upper air into the CBL and breaks up the plume structures into smaller sized structures. Combined with the weakening heating at the ground, this accounts for the observed smaller and more numerous structures in the afternoon data compared to the morning data. This conceptual model is supported by the statistics gathered in Table 1. The larger vertical velocity structures occur in the morning data, but are fewer in number, while the intermittency remains approximately the same. That is, the total area covered by vertical velocity structures is relatively the same, but the size decreases by the afternoon. The virtual potential temperature structures follow a similar, but less pronounced, pattern.

The Fourier spectra of the morning and afternoon vertical velocity records are shown in Figure 13a and b, respectively. It is apparent that the size of the predominant structures decreases in the afternoon, since most of the energy migrates to short wavelengths. Figure 14 shows the spectral estimates for the structure and non–structure elements of the morning record (Figure 14a) and the afternoon record (Figure 14b). The spectra are plotted as $log(k^{-5/3} f(k))$ vs. $log(k)$ to visualize a $k^{-5/3}$ power law behavior as a flat region. In the inertial range (200m and smaller) both morning and afternoon spectra indicate a $-5/3$ power law behavior (seen as a flat region in Figure 13). The morning spectra contains more scatter due to the smaller number of large coherent structures contained in the record (see Table 1,2). In the afternoon when the record contained a large number of small sized coherent structures, the spectra has less scatter. Many studies have computed energy density spectra of observed turbulence records and either established the presence or absence of the −5/3 power law region. In [19] it was observed using data from several experiments that the −5/3 region is seen in strong turbulence but not in weak turbulence. It is suggested in [12] that the universal −5/3 slope spectra is found in the spectra of the structure partition of velocity, while the non–structure partition is flat or has a −1 power law behavior. In Figure 14 we show the spectra of the velocity structure component (thick line) and the non–structure component (thin line). The structure component of the record contains a −5/3 power law region while the non–structure component has a −1 power law behavior. These results are consistent with [12]. The similarity of the spectra of the structure component to spectra of strong turbulence and those of the non–structure to weak turbulence suggest that a better characterization of turbulence may be achieved by the use of the intermittency index and the size distribution of the coherent structure component of the velocity.

§5. Summary and Conclusions

In this paper we have utilized the wavelet transform as a time (or space) series analysis tool for the analysis of data containing intermittent coherent events. We use a non–orthogonal formulation of the wavelet transform, and choose scaling to emphasize the edge and singularity detection capabilities of the transform. The fast algorithm of [22] is implemented for this purpose.

We use local maxima of the wavelet variance (defined in Section 2) as a means for identifying important scales in the data. This approach has been used in other studies as well [10, 4]. It is important to note that this is a reduction of the wavelet phase plane and is subject to the bias of non–characteristic events. For example, a single event with a large amplitude may determine a local maximum in the wavelet variance when several events of a differing scale may contain the desired structural information.

Using the wavelet transform for detecting coherent events is considered in Section 3. The locations of local extrema of an anti–symmetric wavelet transform correspond to locations of centers of large gradients in the signal. Similarly, the locations of local extrema of a symmetric wavelet transform correspond to locations of large curvature in the signal. Alternatively, the zero crossings of the symmetric wavelet transform correspond to locations of centers of gradients in the signal, but the magnitude (sharpness of the gradient) is then only determined by the slope of the zero crossing. Each type of detection is necessarily scale dependent which must be considered when attempting to identify structures. A crucial choice for structure detection algorithms is between anti–symmetric and symmetric analyzing wavelets. The specific type of wavelet, and its regularity are less critical in applications to real data. This is due in part to the limitations imposed by

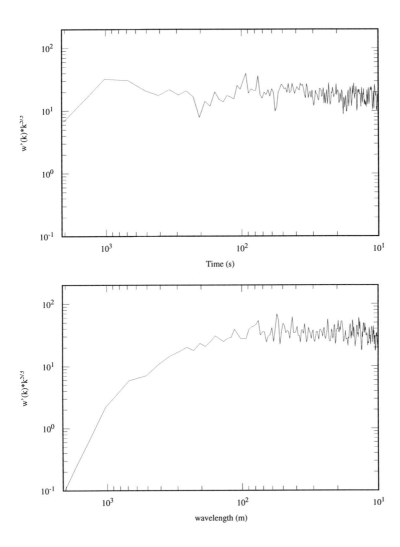

Figure 13. Spectral variance density derived from the vertical velocity component for **(a) (top)** the morning case, and **(b) (bottom)** the afternoon case. The spectra have been multiplied by $k^{5/3}$ so that a $-5/3$ region will appear horizontal in the figure.

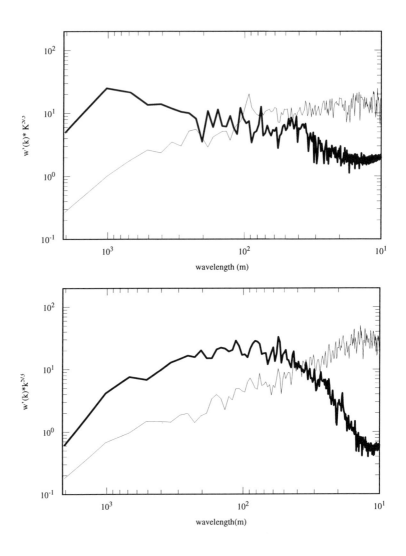

Figure 14. Spectral variance density of structure (thick line) and non-structure (thin line) components of w-velocity shown in Figure 12a and c. **(a) (top)** Morning case, and **(b) (bottom)** afternoon case.

finite resolution and finite record length.

Having determined a dominant length scale, the wavelet transform at the dominant scale of the signal can be used to identify the location of coherent events of that size (by locating local extremes in the transform). We then define an intermittency index based on the ratio of the support of wavelets at these locations to the total record length.

In order to study the properties of the structure containing component of a signal separately from the remaining component we utilize three types of signal partitioning (Section 3). A scale threshold partition, a wavelet phase plane partition, and a combination of both. The partitions may be visualized as taking the wavelet phase plane information derived from the signal and partitioning it into two phase planes based on scale, magnitude of the wavelet transform, or both. Each of the two new phase planes is then used to reconstruct a signal, whose sum is the original signal. The localization in time and frequency of the wavelet transform allows a partition which retains sharp edge characteristics in the structure component which not possible with traditional Fourier band pass techniques.

Finally, these tools are applied to atmospheric data from the FIFE87 experiment. The analysis performed here complements a previous study which used data from the same field experiment [12]. The vertical velocity components and the buoyancy flux density are partitioned using the anti–symmetric wavelet transform for fields obtained during late morning and late afternoon. An intermittency index and dominant scale length are computed for each signal. Additionally, total fluxes were computed and analyzed in terms of the structure components of the signals. We characterize the evolution of the convective boundary layer by comparing the characteristic scale lengths of velocity and temperature structures for the morning and afternoon cases. Additionally, we summarize the contribution of velocity and temperature structures to buoyancy flux by comparing the contributions of various partitions of the signals. The findings are consistent with the conceptual model of a convective boundary layer subject to increasing shear and mixing during the afternoon. The more uniform mixing results in the temperature structures being stronger and well defined in the afternoon.

The spectral energy density of each of the elements of the velocity partitions are compared to the spectral density of the original velocity signals. We have found that the structure component has the classical characteristics of a strong turbulence signal, containing an energetic region at long wavelengths and a transition region having a $-5/3$ behavior. The non–structure component exhibits -1 spectrum usually associated with weak turbulence. Thus, the structure element of the partition behaves (spectrally) like that of strong turbulence while the non–structure element behaves more like that of weak turbulence. We suggest that the spectral characteristics of turbulent velocities are determined by the intermittency and scale lengths of structures, the combination yielding the $-5/3$ spectral slope, and provide a better classification than weak and strong turbulence based on the spectral characteristics.

In this study we have applied a wavelet analysis to signals derived from the atmospheric boundary layer. We chose certain aspects of the types of wavelets used (symmetric and anti–symmetric), and the way in which they are used (non–orthogonal translates) based on known signal processing characteristics of wavelets [22]. The technique encompasses many desirable features of traditional conditional sampling techniques. An objective means for defining and quantifying intermittency of coherent events, and for creating partitions of signals which preserve the sharp gradients characteristic of coherent events has been presented. The objective (i.e. computable) nature of these characteristics will allow large data bases to be processed in a consistent fashion. This should allow further study of the nature of the evolution of structures in differing planetary boundary layer scenarios.

References

1. Antonia, R. A., A. J. Chambers, C. A. Friehe, and C. W. Van Atta, 1979: Temperature ramps in the atmospheric surface layer. *Journal of the Atmospheric Sciences*, **36**, No.1, 99–108.

2. Arneodo, A., G. Grasseau, and M. Holschneider, 1989: Wavelet Transform Analysis of Invariant Measures of Some Dynamical Systems. In *Wavelets* (Combes, J. M., A. Grossman, and Ph. Tchamitchian, Eds.), Springer–Verlag, 315 pg.

3. Bergstrom, H. and U. Hogstrom, 1989: Turbulent exchange above a pine forest II. Organized structures. *Boundary–Layer Meteorology* **49** 231–263.

4. Collineau, S., and Y. Brunet, 1993: Detection of turbulent coherent motions in a forest canopy; Part I: wavelet analysis. *Boundary–Layer Meteorology* **65** 357–379.

5. Chui, C. K., 1992: *An introduction to wavelets*. Academic Press, Inc. 266 pg.

6. Daubechies, I., 1988: Othonormal bases of compactly supported wavelets. *Communications in Pure and Applied Mathematics*, **41**, 909–996.

7. Daubechies, I., 1992: *Ten lectures on wavelets*. CBMS–NSF Regional Conference Series in Applied Mathematics, Soc. for Industrial and Appl. Math. , 357 pg.

8. Gamage, N. K. K., 1989: *Modeling and analysis of geophysical turbulence: use of optimal transforms and basis sets*. Ph.D. Thesis, Oregon

State University, 135 pg. [Available from University Microfilm, 305 N. Zeeb Rd., Ann Arbor, MI 48106.]

9. Gamage, N. and W. Blumen, 1993: Comparative analysis of low–level cold fronts: wavelet, Fourier and empirical orthogonal function decompositions. *Monthly Weather Review* (submitted).

10. Gamage, N. and C. Hagelberg, 1993: Detection and Analysis of Microfronts and Associated Coherent Events using localized transforms. *Jour. of Atm. Sci.* , **50**, No. 5, 750–756.

11. Grossman, R. L. 1992: Convective boundary layer budgets of moisture and sensible heat over an unstressed prairie. *Journal of Geophysical Research*, **97**, no. D17, 18,425–18,438.

12. Hagelberg, C. R. and N. K. K. Gamage, 1993 (submitted): Structure preserving wavelet decompositions of intermittent turbulence. *Boundary–Layer Meteorology*.

13. Harris, F. J., 1978: On the use of windows for harmonic analysis with the discrete Fourier transform. *Proc. of the IEEE*, **66**, 51–83.

14. Hudgins, L. C. A. Friehe, and M. E. Mayer, 1993: Wavelet transforms and atmospheric turbulence. *Physical Review Letters*, **71**, No. 20, 3279–3282.

15. JGR special issue, November, 1992: First ISLSCP Field Experiment. *Journal of Geophysical Research*, **97**, November 1992. 18343–19110.

16. Kristensen, L., P. Kirkegaard, C. W. Fairall, J. C. Kaimal, and D. H. Lenschow, 1992: Advantages of tapering finite data records for spectral analysis. NOAA Technical Memorandum ERL WPL–266, National Technical Information Service, Springfield, VA.

17. Kuznetsov, V. R., A. A. Praskovksy, and V. A. Sablenikov, 1992: Fine-scale turbulence structure of intermittent shear flows. *Journ. of Fluid Mechanics*, **243**, 595–622.

18. Lenschow, D. H. and P. L. Stephens, 1980: The role of thermals in the convective boundary layer. *Boundary Layer Meteorology*, **19**, 509–532.

19. Mahrt, L. and N. Gamage, 1987: Observations of turbulence in stratified flow. *Jour. Atm. Sci.*, **44**, 1106–1121.

20. Mahrt, L., 1991: Eddy asymmetry in the sheared heated boundary layer. *Jour. Atm. Sci.*, **48**, 472–492.

21. Mallat, S., 1989: Multifrequency channel decompositions of images and wavelet models. *IEEE Trans. on Acoustics, Speech, and Signal Processing*, **37**, No. 12, 2091–2110.

22. Mallat, S. and S. Zhong, 1992: Characterization of Signals from Multiscale Edges. *IEEE Trans. on Pattern Anal. and Mach. Intel.*, **14**, no.7, 710–732.

23. Mallat, S. and W. L. Hwang, 1992: Singularity detection and processing with wavelets. *IEEE Trans. on Inform. Theory*, **38**, no.2, 617–643.

24. Schols, J. L. J., 1984: The detection and measurement of turbulent

structures in the atmospheric boundary layer. *Boundary–Layer Meteorology*, **29**, 39–58.

25. Taubenheim, J. 1989: An easy procedure for detecting a discontinuity in a digital time series. *Zeitschrift Meteorol.*, **39**, 344–347.

26. Tennekes, H. and J. L. Lumley, 1972: *A first course in turbulence*, The MIT Press, Cambridge, Mass., 300 pp.

27. Yamada, M. and K. Ohkitani, 1990: Othonormal wavelet expansion and its application to turbulence. *Progress in Theoretical Physics*, **83**, No. 5, 819–823.

28. Yamada, M. and K. Ohkitani, 1991: Orthonormal wavelet analysis of turbulence. *Fluid Dynamics Research*, **8**, 101–115.

This research was supported by the Advanced Study Program of the National Center for Atmospheric Research sponsored by the National Science Foundation (CH), the Airforce Office of Scientific Research, Directorate of Chemical and Atmospheric Sciences grant F49620–92–J–0137 (NG), and NASA grant NAG5–904 (NG).

Carl R. Hagelberg
National Center for Atmospheric Research
P.O. Box 3000
Boulder, Colorado 80307
e-mail: *hagelber@ncar.ucar.edu*

Nimal K. K. Gamage
Program in Atmospheric and Oceanic Sciences
Astrophysics, Planetary and Atmospheric Sciences Department
University of Colorado
Campus Box 391
Boulder, Colorado 80309.
e-mail: *gamage@boulder.colorado.edu*

Intermittency in Atmospheric Surface Layer Turbulence: The Orthonormal Wavelet Representation

Gabriel G. Katul, John D. Albertson, Chia R. Chu, and
Marc B. Parlange

Abstract. Orthonormal wavelet expansions are applied to atmospheric surface layer velocity measurements that exhibited about three decades of inertial subrange energy spectrum. A direct relation between the nth order structure function and the wavelet coefficients is derived for intermittency investigations. This relation is used to analyze power-law deviations from the classical Kolmogrov theory in the inertial subrange. The local nature of the orthonormal wavelet transform in physical space aided the identification of events contributing to inertial subrange intermittency buildup. By suppressing these events, intermittency effects on the statistical stucture of the inertial subrange is eliminated. The suppression of intermittency on the nth order structure function is carried out via a conditional wavelet sampling scheme. The conditional sampling scheme relies on an indicator function that identifies the contribution of large dissipation events in the wavelet space-scale domain. The conditioned wavelet statistics reproduce the Kolmogrov scaling in the inertial subrange and resulted in a zero intermittency factor. A relation between Kolmogrov's theory and Gaussian statistics is also investigated. Intermittency resulted in non-Gaussian statistics for the inertial subrange scales.

§1. Introduction

According to the Kolmogorov theory [25](hereafter referred to as K41), the ensemble average of the nth order velocity difference (Δu_i) between two points separated by spatial distance (r), in the inertial subrange, for high Reynolds number flow is given by

$$\langle |\Delta u_i|^n \rangle = K_n (\langle \epsilon \rangle)^{\frac{n}{3}} r^{\frac{n}{3}} \tag{1}$$

Wavelets in Geophysics   81
Efi Foufoula-Georgiou and Praveen Kumar (eds.), pp. 81–105.

where ϵ is the turbulent energy dissipation rate

$$\epsilon = \frac{\nu}{2} \left(\frac{\partial u_i}{\partial x_j} + \frac{\partial u_j}{\partial x_i} \right)^2 \qquad (2)$$

and u_i are the velocity components ($i = 1, 2, 3$), n is the order of the structure function, ν is the kinematic viscosity, K_n is a universal constant independent of the flow but dependent on n, r is the separation distance that is much smaller than the energy containing length scale or integral length scale (L) but much larger than the Kolmogorov microscale $\eta (= [\nu^3/\langle\epsilon\rangle]^{1/4})$, and $\langle\cdot\rangle$ is the ensemble averaging operator. The scaling laws of Equation (1) have been confirmed by many experiments for $n = 2$ (e.g. the existence of the $2/3$ law for the structure function or $-5/3$ law for the power spectrum) as originally proposed by Kolmogorov [25] and discussed in Monin and Yaglom ([42], pp. 453–527). However, the scaling laws in Equation (1) are not accurate for $n > 2$, as evidenced by many other experiments (see e.g. [1]). Deviations from these scaling laws have classically been attributed to the intermittency in ϵ. This intermittency results in an $\epsilon(x)$ behavior that resembles an on-off process. That is, the dissipation of turbulent kinetic energy occurs only in a small fraction of the fluid volume. Hence, the breakdown of Equation (1) is attributed to the inequality between $\langle\epsilon^n\rangle$ and $\langle\epsilon\rangle^n$, as noted by Landau (see footnote in [29]; p.126). As a result, many phenomenological models for intermittency corrections to K41 have been proposed. Example phenomenological models include the lognormal model [26], the β-model [17], and other multifractal models ([40], [39] and [2]).

Many atmospheric surface layer (ASL) flows exhibit an inertial subrange that extends over many decades so that intermittency effects on inertial subrange scaling becomes important (see [22]; [42], Ch.8). Refined intermittency studies in the natural environment encounter difficulties due to: 1) the limited sampling period over which steady state mean meteorological conditions exist, 2) the need for instrumentation that is free of atmospheric contamination and possible calibration drifts due to temperature and humidity changes, 3) the need for instrumentation that is field robust and capable of providing all three velocity components (since changes in wind direction are inevitable), and 4) the need for turbulence conditions that allow the application of Taylor's frozen hypothesis with minimum wavenumber distortion.

The first difficulty severely limits the number of data points that can be used to evaluate the ensemble average in Equation (1). Typically, the ergodic hypothesis is used to evaluate the ensemble average in Equation (1) from the measured time averages. The convergence of the time average and the ensemble average requires a very large number of measurements that may not be available in many field studies due to unsteadiness in the mean meteorological conditions. The second and third difficulties limit the use of many laboratory fast response sensors such as hot wire probes that are capable of resolving scales as small as the dissipation scales but difficult to operate for long periods in the natural environment. We note here that some success in using such instruments (e.g. triaxial hot wire probes) for extended periods were reported [43]. However, hot wire probes are not suited for ASL measurements in arid and semi-arid environments since the air temperature fluctuation can be very large (up to 6°C in seconds).

The development of analyzing tools that allow the study of intermittency effects in the ASL from limited number of field measurements is necessary. The purpose of this paper is to investigate the usefulness of orthonormal wavelet transforms to quantify intermittency effects on inertial subrange scaling from ASL velocity measurements. For this purpose, we develop a conditional sampling scheme that is capable of identifying dissipation events in the space-scale wavelet domain. This conditional sampling scheme can identify the location of large dissipation events that contribute to inertial subrange intermittency.

The wavelet transform is applied to 56 Hz triaxial sonic anemometer velocity measurements in the ASL that exhibit an inertial subrange for three decades. Since intermittency investigations typically utilize Fourier power spectra and structure functions, we first establish a relation between the wavelet coefficients and these statistical measures. Then, we introduce conditional wavelet statistics that are developed to isolate events responsible for deviations from K41. However, before we discuss these approaches, we offer a brief review of wavelet transforms with emphasis on applications to turbulence measurements.

§2. Analysis of Turbulence using Wavelet Transforms

Wavelet transforms are recent mathematical tools that can unfold turbulence signals into space and scale [15]. Continuous wavelet transforms have been applied to many turbulence measurements and proved to be successful in identifying local scaling exponents ([4], [14] and [3]), intermittency visualization [30], and coherent motion in ASL flows ([10], [19], [18] and [33]). Orthonormal wavelets are a discrete form of the continuous wavelets; however, they have the added feature of forming a complete basis with the analyzing wavelet functions orthogonal to their translates ([41]; [12]; [9], Chapter 1). The application of orthonormal wavelets has yielded important new techniques in the study of turbulence ([53], [54], [55], [37], [38] and [23]). For completeness, a brief review of continuous and orthonormal wavelet transforms is given.

Analogous to Fourier transforms, wavelet transforms can be classified as either continuous or discrete. The continuous wavelet transform is first introduced followed by a motivation for using discrete wavelet transform.

2.1. Continuous wavelet transforms

As shown by Grossmann *et al.* [20], the continuous wavelet transform $W(b,a)$ of a real square integrable signal $f(x)$ (i.e. the integral of $[f(x)]^2$ from $-\infty$ to $+\infty$ is finite) with respect to a real integrable analyzing wavelet $\psi(x)$ (i.e. the integral of $\psi(x)$ from $-\infty$ to $+\infty$ is finite) can be defined as

$$W(b,a) = C_g^{-\frac{1}{2}} \frac{1}{\sqrt{a}} \int_{-\infty}^{+\infty} \psi(\frac{x-b}{a}) f(x) \, dx \tag{3}$$

where a is a scale dilation, b is a position translation, and C_g is defined by

$$C_g = \int_{-\infty}^{+\infty} |K|^{-1} |\psi^*(K)|^2 \, dK < \infty \tag{4}$$

where K is the wavenumber and ψ^* is the Fourier transform of $\psi(x)$ given by

$$\psi^*(K) = \int_{-\infty}^{+\infty} \psi(x) e^{-ikx} \, dx. \tag{5}$$

The condition in Equation (4) ensures the locality of C_g in the Fourier domain. The continuous wavelet transform is commonly viewed as a numerical microscope whose optics, magnification, and position are given by $\psi(x)$, a, and b, respectively ([30] and [10]). The function $\psi(x)$ has to satisfy the following conditions:

1. The admissibility condition, which requires that

$$\int_{-\infty}^{+\infty} \psi(y) \, dy = 0. \tag{6}$$

   Simply stated, Equation (6) requires that the average of $\psi(x)$ be zero.

2. The invertability condition, which requires at least one reconstruction formula for recovering the signal exactly from its wavelet coefficients (see [15]).

The function $f(x)$ can be retrieved from the wavelet coefficients by

$$f(x) = C_g^{-\frac{1}{2}} \int_0^{+\infty} \int_{-\infty}^{+\infty} a^{-\frac{1}{2}} \psi(\frac{x-b}{a}) W(a,b) \frac{db \, da}{a^2}. \tag{7}$$

Further details regarding the continuous wavelet transform theory can be found in many references (e.g. [13], [11], [9] and [15]).

2.2. Orthonormal wavelet expansions

For actual turbulence measurements, discrete wavelet transforms are preferred since $f(x)$ is known at only discrete points x_j (whose spacing de-

pends on the resolution of the sensor and the sampling frequency). Therefore, for discrete measurements, it is necessary to discretize the scale (a)
and space (b) domains of Equation (3). If $f(x_j)$ is defined by N points, one
may consider simply discretizing the space domain of (3) by N nodes and
the scale domain of (3) by N nodes (i.e. discretized by a series of dirac-
delta functions). In this manner, the wavelet transform of $f(x_j)$ requires
N^2 wavelet coefficients. Hence, the discretization discussed above forms an
over-complete description of $f(x_j)$ in the wavelet domain. Because of the
over-complete description of $f(x_j)$, some redundant information is injected
into the wavelet transform of $f(x_j)$. This redundant information may or
may not be advantageous depending on the application. If statistical analysis is to be performed on the wavelet coefficients, then the redundant
information can produce artificial correlations that are a property of the
wavelet transform and not of turbulence.

As shown by Yamada and Ohkitani ([53], [54] and [55]), for space-
scale statistical relations, it is recommended that the discretization of the
space and scale domains form a complete basis so that N measurements are
characterized by N wavelet coefficients. Orthonormal wavelet transforms
are suited for this purpose since the basis function are orthogonal and the
mutual independence of the wavelet coefficients is guaranteed.

Daubechies ([11]; [12], pp. 10), Mallat ([34] and [35]) and Meyer [41]
demonstrated that using a logarithmic uniform spacing for the scale discretization with increasingly coarser spatial resolution at larger scale, a
complete orthogonal wavelet basis can be constructed. We choose the Haar
wavelet basis for its differencing characteristics, since we are interested in
developing explicit relations between the nth order structure function and
the wavelet coefficients. Other important features of the Haar wavelet basis
are discussed in [33].

The Haar basis $\psi(x) = (a^{-1/2})\psi((x-b)/a)$, where $a = 2^m$ and $b = 2^m i$
for $i, m \in \mathbf{Z}$, is given by

$$\psi(x) = \begin{cases} 1 & \text{for } 0 < x < 1/2 \\ -1 & \text{for } 1/2 \leq x < 1 \\ 0 & \text{elsewhere.} \end{cases} \tag{8}$$

For this basis function, the wavelet coefficients $WT^{(m+1)}(k)$ and the coarse
grained signal $S^{(m+1)}(k)$ at scale $m+1$ can be determined from the signal
$S^{(m)}$ at scale m by using

$$WT^{(m+1)}(i) = \frac{1}{\sqrt{2}}[S^{(m)}(2i-1) - S^{(m)}(2i)] \tag{9}$$

$$S^{(m+1)}(i) = \frac{1}{\sqrt{2}}[S^{(m)}(2i-1) + S^{(m)}(2i)] \tag{10}$$

for $m = 0$ to $M - 1$, $i = 0$ to $2^{M-m-1} - 1$, and $M = \log_2(N)$, N is the

number of samples (integer power of 2) (see [5] and [6]). For the Haar
wavelet, the coarse grained signal defined by Equation (10) is a low-pass
filtered function obtained by a simple block moving average, while the
wavelet coefficients are computed from the high-pass filter in Equation (9).
The wavelet coefficients and coarse grained signal may be calculated by the
following pyramidal algorithm:

1. Beginning with $m = 0$, use Equations (9) and (10) to calculate the
   signal $S^{(1)}$ and the coefficients $WT^{(1)}$ at the first scale by looping
   over i from 0 to $2^{M-1} - 1$. This will result in S and WT vectors of
   length $N/2$. The turbulence measurements are stored in $S^{(0)}$.

2. Repeat step 1 with $m = 1$ to calculate the next coarser scale's pair
   of vectors $S^{(2)}$ and $WT^{(2)}$ (each of length $N/4$).

3. Repeat for larger scale m up to $M - 1$ to produce a series of S
   and WT vectors of progressively decreasing length. Note that at
   $m = M - 1$ the coarse grained signal converges to a point.

This algorithm yields $N-1$ wavelet coefficients that define the orthonormal
Haar wavelet transform of the measured turbulence signal. The above pyramidal procedure, which is the basis for Fast Wavelet Transforms (FWT),
requires about N computations vis-a-vis the $N \log_2 N$ computations for
Fast Fourier Transforms (FFT). The set $N - 1$ discrete Haar wavelet coefficients satisfies the conservation of energy condition

$$\sum_{j=0}^{N-1} f(j)^2 = \sum_{m=1}^{M} \sum_{i=0}^{2^{M-m}-1} WT^{(m)}(i). \tag{11}$$

Equation (11) states that the sum of the square of the wavelet coefficients
conserves the norm of the signal and is similar to Parseval's identity in
Fourier expansions ([9], pp. 12).

§3. Experiment

The data presented here were collected during an experiment carried
out on June 27, 1993 over a uniform dry lakebed (Owens lake) in Owens
valley, California. The lakebed is contained in a large basin bounded by
the Sierra Nevada range to the east and the White and Inyo Mountains
to the west. The instrumentation site is located on the northeast end
of the lakebed (elevation=1,100 m). The site's surface is a heaved sand
soil extending uniformly 11 km in the North-South direction and 4 km in
the East-West direction. The three velocity components were measured
at $z = 2.5$ m above the surface using a triaxial ultrasonic anemometer
(Gill Instruments/1012R2) to an accuracy of $\pm 1\%$. Sonic anemometers

Table I

Summary of energy, meteorological, turbulence, and surface roughness conditions during the experiment. The net radiation was measured by a Q6 Fritshen type net radiometer, the soil heat flux was measured by two REBS soil heat flux plates, the sensible heat flux was measured by a Campbell Scientific eddy correlation system (sampling at 10 Hz), and the friction velocity was measured by a triaxial sonic anemometer. The momentum roughness length was determined from 6 near-neutral runs of $\langle U \rangle$ and u_* using the logarithmic velocity profile.

Energy Conditions	
Net radiation (R_n)	173 W m^{-2}
Soil Heat Flux (G)	78 W m^{-2}
Sensible Heat Flux (H)	90 W m^{-2}
Meteorological Conditions	
Mean Horizontal Wind Speed ($\langle U \rangle$)	2.68 m s^{-1}
Mean Air Temperature (T_a)	31.6°C
Turbulence Conditions	
Friction Velocity (u_*)	0.165 m s^{-1}
Root-Mean Square Velocity (σ_u)	0.516 m s^{-1}
Atmospheric Stability Conditions	
Height above ground surface (z)	2.5 m
Obukhov Length (L)	−3.98 m
Surface Roughness	
Momentum Roughness Length (z_o)	0.13 mm

achieve their frequency response by sensing the effect of wind on the transit times of sound pulses traveling in opposite directions across a known path length d_{sl}(= 0.149 m in this study). The sonic anemometer is well suited for these experiments since it is free of calibration nonlinearities, atmospheric contamination drifts, temperature effects, and time lag. The main disadvantage of sonic anemometers is the wavenumber distortion due to averaging over d_{sl}. This distortion is restricted to wavenumbers in excess of $2\pi/d_{sl}$(= 42.2 m^{-1}) as discussed in [51] and [16].

The sampling frequency f_s was 56 Hz and the sampling period T_p was 9.75 minutes. The short sampling period was necessary to achieve steady state conditions in the mean meteorological conditions. Recall also that a long sampling period may not be necessary in this case since our intent is to resolve interial subrange scales that are much smaller than the integral length scale. The sampling at $f_s = 56$ Hz for $T_p = 9.75$ minutes resulted in $N = 32,768$ points which are displayed in Figure 1. Notice in Figure 1 that the velocity fluctuated by as much as 3 m s^{-1} in a few seconds. A summary of the mean meteorological and turbulence conditions is presented in Table 1.

From Table 1, the ratio of the root mean square velocity $\sigma_u (= \langle (U -$

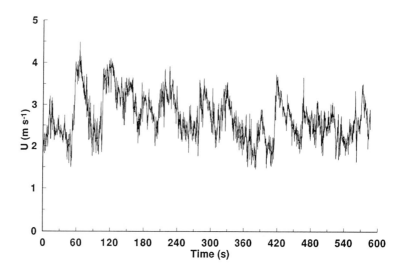

Figure 1. Time variation of the measured longitudinal velocity (U) at $z = 2.5$ m above the ground surface. The sampling frequency and sampling period are 56 Hz and 9.75 minutes, respectively.

$\langle U \rangle)^2 \rangle^{1/2})$ to the mean horizontal wind speed $\langle U \rangle$ is $0.52/2.68 = 0.194$ which is smaller than 0.5. Hence, Taylor's hypothesis [48] can be used to convert time increments to space increments without significant distortion, at least for the inertial subrange scales (e.g. [31], [44], [50] and [45]). The resolvable wavenumber $K_{NY}(= 2\pi/(\langle U \rangle (f_s/2)^{-1}))$ by the sonic anemometer corresponding to the Nyquist frequency= $(f_s/2)$ is 65.2 m^{-1}. We note that this wavenumber is larger than 42.2 m^{-1}, and therefore, we restrict our statistical analysis to wavenumbers smaller than 42.2 m^{-1} but display the spectra for all measured wavenumbers.

§4. Wavelet Statistics

In this section, we first develop relations between the Haar wavelet coefficients and the Fourier power spectrum and the structure function, then we discuss how wavelet transforms can be used to investigate intermittency effects on the inertial subrange.

4.1. Relation between wavelet coefficients and Fourier power spectrum

In Fourier analysis, the fundamental tool used to characterize turbulence is the power spectral density function $E(K)$. The function $E(K)$ represents the energy density contained in each wavenumber band dK, and thus provides information regarding the importance of each scale of motion. However, important spatial information regarding location of events become implicit in the phase angle of Fourier transform. In this section, we relate the Haar wavelet coefficients to the Fourier power spectrum and show how spatial information can be expressed in an explicit manner.

The variance of the turbulence measurement, in terms of the wavelet coefficients, can be deduced from the conservation of energy

$$\sigma^2 = N^{-1} \sum_{m=1}^{M} \sum_{i=0}^{2^{M-m}-1} (WT^{(m)}[i])^2. \tag{12}$$

The total energy T_E contained in scale $R_m (= 2^m \, dy)$ can be computed from the sum of the squared wavelet coefficients at scale index (m) so that

$$T_E = N^{-1} \sum_{i=0}^{2^{M-m}-1} (WT^{(m)}[i])^2 \tag{13}$$

where $dy (= f_s^{-1} \langle U \rangle)$ is the measurement spacing in physical space. In order to compare the wavelet power spectrum to the Fourier spectrum, we define a wavenumber K_m, corresponding to scale R_m, as

$$K_m = \frac{2\pi}{R_m}. \tag{14}$$

Hence, the power spectral density function $E(K_m)$ is computed by dividing T_E by the change in wavenumber $\Delta K_m (= 2\pi 2^{-m} \, dy^{-1} \ln[2])$ so that

$$E(K_m) = \frac{\langle (WT^{(m)}[i])^2 \rangle \, dy}{2\pi \ln(2)} \tag{15}$$

where $\langle \cdot \rangle$ is averaging in space over all values of (i) for scale index (m) (see [37] and [53]). The adequacy of Equations (14) and (15) are discussed in Katul ([24], pp. 149, 186–187). In Equation (15), the wavelet power spectrum at wavenumber K_m is directly proportional to the average of the square of the wavelet coefficients at that scale. Because the power at wavenumber K_m is determined from averaging many squared wavelet coefficients, we expect the wavelet power spectrum to be smoother than its Fourier counterpart. This is apparent in Figure 2 which displays good agreement between Fourier and wavelet power spectrum for all three decades of inertial subrange wavenumbers.

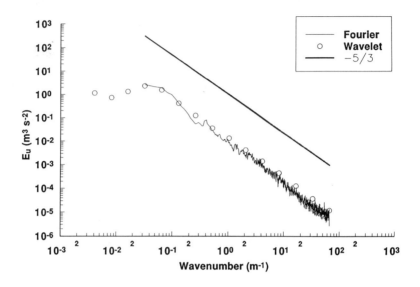

Figure 2. A comparison of the Fourier and Haar wavelet power spectra. Taylor's hypothesis is used to convert time to wavenumber domain. The $-5/3$ power law is also shown.

The Fourier power spectrum was computed by square windowing 8192 points, cosine tapering 5% on each window edge, and averaging the resultant 4 power spectra ($N = 32,768$). The wavelet power spectrum was computed by: 1) Using the pyramidal algorithm in Equations (9) and (10) to obtain the wavelet coefficients over position index (i) and scale index (m), and 2) Using Equation (15) in conjunction with computed wavelet coefficients from step 1 to obtain the wavelet power spectrum. Windowing is unnecessary for the wavelet spectrum. We note that similar comparisons were reported by Hudgins *et al.* [21] and Barcy *et al.* [4] using the continuous wavelet transform.

Since the wavelet power spectrum is directly proportional to the average of the squared wavelet coefficients, we can also calculate the spatial standard deviation using

$$SD_E(K_m) = \frac{dy}{2\pi \ln(2)}[\langle WT^{(m)}[i]^4 \rangle - \langle (WT^{(m)}[i]^2)\rangle^2]^{\frac{1}{2}}. \quad (16)$$

In Meneveau ([37] and [38]), it was pointed out that a plot of $E(K_m)$ and $E(K_m) + SD_E$ gives a compact representation of the energy and its spatial

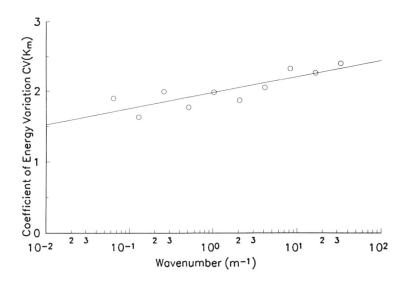

Figure 3. The evolution of the coefficient of energy variation $CV(K)$ as a function of wavenumber (K). The solid line is the best fit regression line through the data points.

variability at each scale which is referred to as the "dual spectrum". However, as shown by Katul and Parlange [23], a better dimensionless indicator for the spatial energy variance is given by the coefficient of variation CV_E defined as

$$CV_E(K_m) = \frac{SD_E(K_m)}{E(K_m)}. \tag{17}$$

An example of the variation of CV_E is shown in Figure 3.

Notice in Figure 3 that CV_E increases as the wavenumber increases indicating increased turbulent activity at smaller scales. The increase in CV_E can be related to the increased spottiness or intermittency in the dissipation rate and is discussed next. In Tennekes and Lumley ([46] pp. 66), the dissipation rate for locally isotropic turbulence is given by

$$\langle \epsilon \rangle = 15\nu \left\langle \left[\frac{\partial u}{\partial x} \right]^2 \right\rangle \tag{18}$$

where u is the velocity fluctuation in the longitudinal direction, and x is identical to x_1. Recall that deviations from K41 are due to the spottiness or intermittency in ϵ. Tennekes [47] suggested that intermittency results in a large spatial variance in the dissipation rate $(\sigma_\epsilon)^2$ around its mean value $\langle \epsilon \rangle$. Hence, we arrive at the desired physical meaning of CV_E: This dimensionless ratio can be interpreted as $(\sigma_\epsilon / \langle \epsilon \rangle)$ if the following arguments are adopted:

1. For locally isotropic turbulence, the standard deviation σ_ϵ around $\langle \epsilon \rangle$ is given by

$$\sigma_\epsilon = 15\nu \left[\left\langle \left[\frac{\partial u}{\partial x} \right]^4 \right\rangle - \left\langle \left[\frac{\partial u}{\partial x} \right]^2 \right\rangle^2 \right]^{0.5} \tag{19}$$

   Notice here the similarity between Equations (16) and (19).

2. The derivatives in Equations (18) and (19) (e.g. $\partial u / \partial x$) can be approximated by differencing operations (e.g. $\Delta u / \Delta x$).

3. Using step (2), the Δx in the ratio $(\sigma_\epsilon / \langle \epsilon \rangle)$ cancels out from the numerator and denominator and we are left with the differencing operations.

4. The differencing (Δu) at scale (m) can be directly related to the wavelet coefficients using Equation (9). This point is the subject of section 4.2.

Hence, the increase in CV_E, as shown in Figure 3, is a direct indication that ϵ is becoming more and more intermittent at smaller scales. What is also important to note here is that this result cannot be inferred from the usual Fourier power spectrum. From a turbulent energy point of view, a key difference between wavelet and Fourier transforms is that Fourier transforms are nonlocal and therefore distribute the energy uniformly in space (i.e. $SD_E = 0$).

4.2. Relation between wavelet coefficients and structure function

Using Equation (9) the wavelet coefficients can be related to the nth order structure function, for any flow variable ϕ, using

$$\langle |\phi(x+r) - \phi(x)|^n \rangle \sim \frac{\langle |WT(i)^{(m)}|^n \rangle}{\left(2^{\frac{m}{2}} \, dy \right)^n}. \tag{20}$$

In the above equation, we made use of the following: 1) the separation distance $r = 2^m \, dy$, 2) the wavelet coefficients are proportional to $\phi(x + r) - \phi(x)$ at position $x = (2^m i) \, dy$, 3) the amplitudes of the Haar wavelet coefficients are proportional to $(2^m)^{1/2}$, and 4) $\langle \cdot \rangle$ is the averaging operator of the wavelet coefficients over all values of the position index (i) at scale index (m). To study intermittency effects on Equation (1), we modify the above relation and propose a conditional structure function to be discussed next.

4.3. Conditional sampling and intermittency effects on K41

In general, intermittency of turbulent fluids is symbolized by an on-off process in the dissipation rate so that at a certain time, the turbulent energy is only active in a small fraction of the fluid volume [36]. In a one dimensional cut (namely along the x direction), the signature of intermittency is large isolated dissipation events within an overall passive surrounding fluid ([36], pp. 104–109). For that purpose, we classify the wavelet coefficients as either "dissipative" or "passive". The dissipative wavelet coefficients are the coefficients directly influenced by the large localized dissipation events discussed above, while the passive coefficients are not. The distinction between dissipative and passive must be based on some minimum dissipation threshold criterion. Such a criterion is difficult to construct without some relation between the velocity and the dissipation. Before we establish the conditioning criteria, let us first define the indicator function $I^{(m)}$ at scale index (m) by

$$I^{(m)} = \begin{cases} 0 & \text{if } [WT^{(m)}(i)]^2 > F \langle [WT^{(m)}(i)]^2 \rangle \\ 1 & \text{otherwise} \end{cases} \qquad (21)$$

where F is a conditioning criteria that allows discrimination between the dissipative and passive wavelet coefficients. Based on Equation (21), F may be interpreted as the ratio of the dissipation at scale index (m) and position index (i) to the mean dissipation, if the following arguments are adopted:

1. The dissipation at position index (i) and scale index (m) is directly proportional to the square of the velocity gradient at that position. That is ϵ at scale index (m) is proportional to $(\partial u / \partial x)^2$ at scale index (m). This statement may be inferred from Equation (18) without the averaging operation. Possible sources of deviation from this statement are presented in [47].

2. The mean dissipation at scale index (m) is directly proportional to $\langle (\partial u / \partial x)^2 \rangle$ at scale index (m). This statement is a direct consequence of Equation (18).

3. Arguments resulting in Equation (20) are all valid.

Arguments (1) and (2) may not be valid for turbulence that is locally anisotropic. However, based on the numerous studies of local isotropy in ASL flows (e.g. [22]; [42], Ch.8), we adopt the working hypothesis that turbulence is locally isotropic in the inertial subrange. These three arguments result in a sequence of equalities given by:

$$\frac{\epsilon^{(m)}}{\langle \epsilon \rangle^{(m)}} = \frac{\left(\frac{\partial u^{(m)}}{\partial x}\right)^2}{\left\langle \left(\frac{\partial u^{(m)}}{\partial x}\right)^2 \right\rangle} = \frac{(u^{(m)}(x+\Delta x) - u^{(m)}(x))^2}{\langle [u^{(m)}(x+\Delta x) - u^{(m)}(x)]^2 \rangle} = \frac{(WT^{(m)}(i))^2}{\langle (WT^{(m)})^2 \rangle}$$

(22)

where $\epsilon^{(m)}$ is the dissipation at position index (i) and scale (m), $\langle \epsilon \rangle^{(m)}$ is the mean dissipation at scale (m), and $[u^{(m)}(x+\Delta x) - u^{(m)}(x)]/\Delta x$ is the finite difference approximation of the velocity gradient at scale (m) and position x. The first equality follows from arguments (1) and (2). The second equality is due to the fact that Δx required to convert gradients to differences (in a finite difference approximation of the derivative) cancels out in the numerator and denominator, respectively. The third equality is a direct consequence of Equation (20). For example, if $F = 5$, then all squared wavelet coefficients that are in excess of 5 times the average squared wavelet coefficient at scale index (m) are set to zero. This conditioning criterion allows us to consider a conditional power spectrum E^c given by

$$E^c(K_m) = \frac{\langle (I^{(m)} WT^{(m)}[i])^2 \rangle \, dy}{2\pi \ln(2)}$$

(23)

where $\langle \cdot \rangle$ is now averaging in space over all non-zero values of $[I^{(m)} WT^{(m)}(i)]^2$. Hence, E^c represents the power spectrum of the passive fluid fraction. Also, we can define the conditional n^{th} order structure function by

$$\langle |\phi(x+r) - \phi(x)|^n \rangle^{(c)} \sim \frac{\langle \langle |I^{(m)} WT(i)^{(m)}|^n \rangle \rangle}{\left(2^{\frac{m}{2}} dy\right)^n}$$

(24)

where $\langle \langle \cdot \rangle \rangle$ is averaging over all non-zero values of $[I^{(m)} WT^{(m)}(i)]$. These conditional statistics can be computed by:

1) Using the pyramidal algorithm to calculate the Haar wavelet coefficients at each scale index (m) and position index (i); 2) squaring these coefficients to obtain the energy content at each scale index (m) and position index (i); 3) averaging the squared wavelet coefficients for each scale index (m); 4) dividing the squared wavelet coefficient (at space index i) by the value computed in step (3); 5) If this ratio is larger than some preset value for F, then set this coefficient to zero, otherwise leave as is; 6) Use Equation (23) or (24) to determine the power spectrum or the nth order structure function with averaging performed over all non-zero values at scale index (m). Repeat the above steps for all values of (m) within the inertial subrange. The adequacy of this conditional sampling criteria for recovering K41 and characterizing intermittency is considered next.

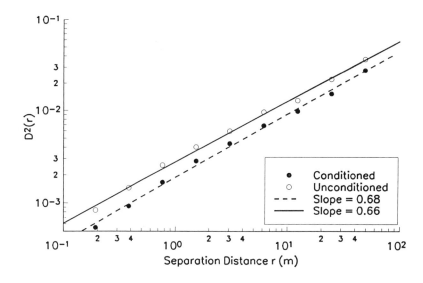

Figure 4a. Comparison between the conditioned ($F = 5$) and unconditioned wavelet structure functions $[D^2(r)]$ for $n = 2$. The solid lines are regression fits through the data points.

§5. Results and Discussion

This section discusses the effects of intermittency on K41 using the conditional wavelet analysis for three cases: 1) $n = 2$, 2) $n = 3$, and 3) $n = 6$. In each case, we check whether K41 is recovered when intermittency is suppressed, and then we investigate the statistical structure of the events responsible for deviations from K41 scaling. We do not present theoretical details regarding intermittency models, but we focus more on the contrast between the conditioned (intermittency suppressed) and unconditioned (intermittency active) statistics.

Case 1: $n = 2$

It is known that intermittency effects are generally small and may not be detectable for the structure function with $n = 2$ ([1] and [52]). We test this hypothesis by comparing the unconditioned and conditioned ($F = 5$) structure functions of Equation (24). The results are presented in Figure 4aa.

Both conditioned and unconditioned second order structure functions exhibit scaling laws that are in agreement with K41 (slope= $2/3$) indicating that intermittency effects may not be significant for $n = 2$ (see [7] and [8] for a possible physical explanation). We also present a summary of the regression statistics for the regression model $\log[D^2(r)] = A \log[r] + B$ in Table 2. Notice in Table 2 that the coefficient of determination (R^2) for the regression model is in excess of 0.99; hence, the determination of scaling laws from wavelet structure functions appears to be very reliable.

Table II

Summary of the regression statistics for the model $\log[\langle|\Delta U|^n\rangle] = A \log[r] + B$. The coefficient of determination (R^2) and the standard error of estimate (SEE) are also shown. The conditioned statistics are for the conditioning criterion $F = 5$.

n	Slope (A)	Intercept (B)	R^2	SEE	Conditioned(C)/ Unconditioned(U)
2	0.680	−2.73	0.996	0.038	C
	0.660	−2.55	0.996	0.036	U
3	1.010	−3.88	0.995	0.056	C
	0.950	−3.53	0.995	0.055	U
6	2.008	−7.10	0.993	0.150	C
	1.690	−5.76	0.987	0.170	U

One must note that certain scale aliasing occurs due to the use of orthogonal and complete basis since the number of Haar modes characterizing the frequency domain is relatively small ($m = 15$). This aliasing may influence the indicator function $I^{(m)}$. Hence, intermittency characterization by the indicator function can be overestimated or underestimated based on the value of F. For that purpose, we performed the same analysis for $F = 4, 5, 7,$ and 10. The slope variation (for $n = 2$) did not differ by more than 0.004. This analysis clearly reveals the robustness of the proposed conditional sampling scheme. Similar results were also obtained by Yamada and Ohkitani [54].

Case 2: $n = 3$

As shown by Landau and Lifshitz ([29], pp. 123–128), a relation between the third order structure function and (r) is given by

$$|(\Delta U)^3| = \frac{4}{5}\langle\epsilon\rangle r. \tag{25}$$

The above relation was explicitly derived using the Navier-Stokes equations for locally isotropic turbulence, and thus, is independent of any assumptions implicit in K41 or any intermittency corrections to K41. If intermittency does not alter the third order structure function behavior, our conditioned

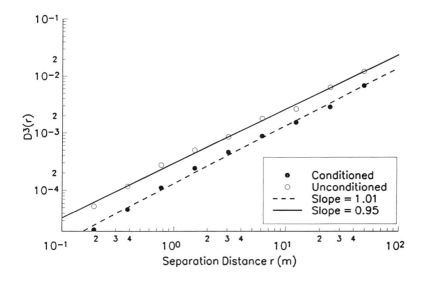

Figure 4b. Same as Figure 4a but for $n = 3$.

and unconditioned statistics should both reproduce the r^1 dependence of Equation (25). Using Equations (20) and (24) with $F = 5$ and $n = 3$, we compare the unconditioned and conditioned third order structure functions in Figure 4b.

Both unconditioned and conditioned slopes (see Table 2) are in good agreement with Landau and Lifshitz [29] predictions (which are also consistent with K41). This analysis supports the insensitivity of the third order structure function to intermittency corrections as noted by Anselmet *et al.* [1]. Yamada and Ohkitani [54] obtained similar results using an analogous conditional analysis procedure, but a different wavelet.

Case 3: $n = 6$

The sixth order structure function can be related to the dissipation correlation function from

$$\frac{\langle (\Delta U)^6 \rangle}{r^2} \sim \langle \epsilon(x)\epsilon(x+r) \rangle \tag{26}$$

where the dissipation correlation function is given by

$$\langle \epsilon(x)\epsilon(x+r) \rangle \sim \left(\frac{L}{r} \right)^{\mu} \tag{27}$$

Table III
Some values of the intermittency parameter μ from various sources.

Source	Measurements	μ
Mahrt [32]	Atmospheric mixed layer: Convective and Nocturnal	0.3–0.4
Anselmet et al. [1]	Turbulent and Duct flow	0.2 ± 0.05
Kuznetsov et al. [27]	Wind-tunnel boundary layer (Depends on external intermittency).	0.15–0.25
Monin and Yaglom [42]	See Ch.8 for an extensive review	0.2–0.5

and μ is the intermittency parameter ([26], [17] and [1]). The value of μ has been the subject of extensive research and its value appears to vary between 0.15 and 0.5 (see Table 3). Thus, from Equations (26) and (27), we see that the sixth order structure function is related to μ by

$$\langle (\Delta U)^6 \rangle \sim r^{2-\mu}. \tag{28}$$

We now evaluate the performance of the conditional wavelet analysis for reproducing K41 scaling and suppressing intermittency (i.e. $\mu = 0$). Using Equations (20) and (24) with $F = 5$ and $n = 6$, we compare the unconditioned and conditioned sixth order wavelet structure function in Figure 4c.

The slope of conditioned sixth order structure function is $2.0(= 2 - \mu)$ indicating that intermittency is well suppressed ($\mu = 0$) for the higher-order statistics (and K41 is recovered). Recall that the conditioning criteria is based on second-order statistics (wavelet power spectrum), yet intermittency was suppressed even in the sixth-order statistics.

The value of μ from the unconditioned sixth-order structure is $0.31(= 2 - 1.69)$, which agrees with many published values from laboratory and atmospheric turbulence studies (see Table 3 for some examples). The consistency in our value of μ with other laboratory experiments, in spite of limited data and sampling resolution, demonstrates the effectiveness and robustness of orthonormal wavelet expansions for intermittency studies.

K41 and Non-Gaussian Statistics

Kraichnan [28] suggested that K41 is consistent with the concept of an inertial cascade if the velocity statistics within the inertial subrange do not differ significantly from Gaussian. We therefore consider the velocity statistics and their relation to K41 scaling. This relation is achieved by noting that conditional wavelet analysis recovers K41 and eliminates any intermittency effects from the inertial subrange. Here Gaussian behavior is tested by computing the conditioned and unconditioned wavelet flatness factor for inertial subrange scales. The wavelet flatness factor (FF) at scale

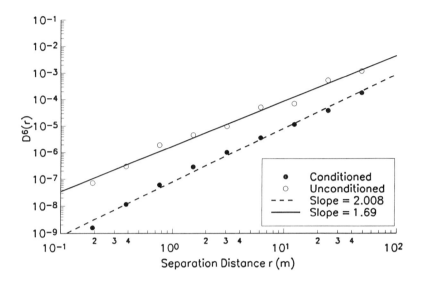

Figure 4c. Same as Figure 4a but for $n = 6$

index (m) is defined as

$$FF(R_m) = \frac{\langle (WT^{(m)}[i])^4 \rangle}{\langle (WT^{(m)}[i])^2 \rangle^2}. \tag{29}$$

Using Equation (29) and the conditional wavelet analysis for $F = 5$, we compute FF for all (m) corresponding to inertial subrange scales. The results are summarized in Figure 4d.

Notice that the conditioned $FF(R_m)$ are nearly Gaussian (i.e. $FF \sim 3$) for all inertial subrange scales while the unconditioned case is clearly non-Gaussian with wavelet flatness factors up to 7. Also, note that the unconditioned flatness factor increases with decreasing separation distance. This analysis clearly indicates that K41 is associated with near-Gaussian statistics and intermittency effects result in non-Gaussian statistics.

Finally, the wavelet flatness factor proposed in Equation (29) can also be used to infer the statistical properties of the horizontal velocity gradients if the following arguments are adopted:

1. The differencing nature of the Haar wavelet transform, as can be noted from Equation (9), results in direct proportionality between wavelet coefficients and horizontal velocity differences.

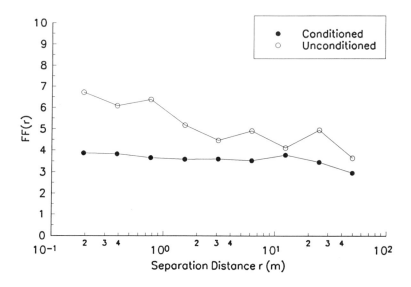

Figure 4d. The wavelet flatness factor (FF) as a function of separation distance (r) for inertial subrange scales. Both conditioned and unconditioned cases are shown.

2. The dimensionless ratio in Equation (29) is the same for differences and gradients since the division by the wavelet width or separation distance that is required to convert differences to gradients in numerator and denominator cancels out. These two arguments are due to Mahrt [33]. Using these two arguments, $FF(R_m)$ in Equation (29) can be interpreted as a gradient flatness factor at scale R_m. Notice in Figure 4d that the unconditioned wavelet flatness factor increases as the separation distance decreases. The flatness factor is commonly used to measure the importance of the tails of the probability density function. Hence, the increase of FF in Figure 4d indicates that the tails of the horizontal velocity gradient probability density function become more important as the scale decreases. This analysis also indicates that the increase in the gradient probability density tails, at smaller scales, is due to intermittency within the inertial subrange (see Figure 4d). In contrast, no such increase is noted for the passive wavelet coefficients.

§6. Summary and Conclusions

Triaxial sonic anemometer velocity measurements at 2.5 m above a uniform dry lakebed (11 km by 4 km) were used to investigate intermittency in the inertial subrange. The power spectrum of the horizontal velocity measurements exhibited a −5/3 power law for three decades allowing detailed investigation of scaling laws in the inertial subrange. In order to describe space-scale relations in the inertial subrange, we utilized the orthonormal wavelet representation. The orthonormal wavelet representation was well suited for this investigation since the basis function are orthogonal and mutual independence of the expansion coefficients is guaranteed. In addition, it was shown that the expansion coefficients can be related directly to quantities commonly used in conventional turbulence analysis. Relations between the orthonormal wavelet coefficients and the Fourier power spectrum, as well as relations with the nth order structure function were established. A comparison between Fourier and wavelet power spectra was also carried out. Good agreement between the two spectra was noted eventhough the Haar wavelet has poor localization in the frequency domain.

Since intermittency build up in the inertial subrange was due to localized dissipative events, a conditional wavelet scheme was developed. The conditional wavelet scheme relied on an indicator function that identified the wavelet coefficients directly influenced by large and localized dissipation events. The conditional wavelet scheme efficiently suppressed intermittency within the inertial subrange by removing these wavelet coefficients. K41 statistics, up to sixth order, were recovered when intermittency in the dissipation was suppressed from the wavelet coefficients. It was also found that intermittency did not significantly influence second and third order statistics in agreement with many other studies. The robustness of the conditional wavelet scheme was also verified.

Finally, we demonstrated that intermittency was directly responsible for non-Gaussian statistics in the inertial subrange, while K41 was associated with near Gaussian statistics. The wavelet transform produced intermittency factors comparable to values obtained from laboratory and other field experiments.

References

1. Anselmet, F., Y. Gagne, E. J. Hopfinger, and R. A. Antonia, High-order velocity structure functions in turbulent shear flows, *J. Fluid Mech.*, 140, 63–89, 1984.
2. Aurell, E., U. Frisch, J. Lutsko, and M. Vergassola, On the multifractal properties of the energy dissipation derived from turbulence data, *J. Fluid Mech.*, 238, 467–486, 1992.
3. Argoul, F., A. Arneodo, G. Grasseau, Y. Gagne, E. J. Hopfinger, U.

Frisch, Wavelet analysis of turbulence reveals the multifractal nature of the Richardson cascade, *Nature*, 338, 51–53, 1989.

4. Barcy, E., A. Arneodo, U. Frisch, Y. Gagne, and E. Hopfinger, Wavelet analysis of fully developed turbulence data and measurement of scaling exponents, in *Turbulence and Coherent Structures*, ed. Metais, O., and M. Lesieur, Kluwer Academic Publishers, 450 pp., 1991.

5. Beylkin, G., R. Coifman, and V. Rokhlin, Fast wavelet transforms and numerical algorithms I., *Comm. Pure and Appl. Math.*, Vol.XLIV, 141–183, 1991.

6. Beylkin, G., R. Coifman, and V. Rokhlin, *Wavelets in numerical analysis, Wavelets and their Applications*, ed. Ruskai, M. B., Beylkin, G., Coifman, R., Daubechies, I., Mallat, S., Meyer, Y., Raphael, L., Jones and Bartlett Publishers, Boston, 474 pp., 1992.

7. Chorin, A. J., Spectrum, dimension, and polymer analogies in fluid turbulence, *Phys. Rev. Lett.*, 60, 1947–1949, 1988.

8. Chorin, A. J., Constrained random walks and vortex filaments in turbulence theory, *Commun. Math. Phys.*, 132, 519–536, 1990.

9. Chui, C. K., *An Introduction to Wavelets*, Academic Press, Inc., 264 pp., 1992.

10. Collineau, S., and Y. Brunet, Detection of turbulent coherent motion in a forested canopy, part I: Wavelet analysis, *Bound. Layer Meteor.*, 65, 357–379, 1993.

11. Daubechies, I., Orthonormal bases of Compactly supported wavelets, *Comm. Pure and Appl. Math.*, Vol.XLI, 909–996, 1988.

12. Daubechies, I., *Ten Lectures on Wavelets*, CBMS-NSF Regional conference series in applied mathematics, S.I.A.M., 61, 357 pp., 1992.

13. David, Wavelets and Singular Integrals on Curves and Surfaces, Springer-Verlag, 109 pp., 1992.

14. Everson, R., L. Sirovich, K. R. Sreenivasan, Wavelet analysis of the turbulent jet, *Physical Letters A*, 145, 314–322, 1990.

15. Farge, M., Wavelet transforms and their applications to turbulence, *Annu. Rev. Fluid Mech.*, 24, 395–457, 1992.

16. Friehe, C. A., Fine-scale measurements of velocity, temperature, and humidity in the atmospheric surface layer, *Probing the Atmospheric Boundary Layer*, ed. Lenschow, D., American Meteorological Society, 269 pp., 1986.

17. Frisch, U., P. L Sulem, and M. Nelkin, A simple dynamical model of intermittent fully developed turbulence, *J. Fluid Mech.*, 87, 719–736, 1978.

18. Gamage, N., and C. Hagelberg, Detection and analysis of microfronts and associated coherent events using localized transforms, *J. Atmos. Sci.*, 50, 750–756, 1993.

19. Gao, W., and B. L. Li, Wavelet analysis of coherent structures at the

atmosphere-forest interface, *J. Appl. Meteor.*, 32, 1717–1725, 1993.

20. Grossmann, A., R. Kronland-Martinet, and J. Morlet,Reading and understanding continuous wavelet transforms, *Wavelets: Time-Frequency Methods and Phase Space*, ed. Combes, J. M., Grossmann, A., Tchamitchian, Ph., Springer-Verlag, 315 pp., 1989.

21. Hudgins, L. H., M. E. Mayer, and C. A. Friehe, Fourier and wavelet analysis of atmospheric turbulence, Paper presented at the *Wavelets and Applications Conference*, Toulouse, France, June 8–13, 1992.

22. Kaimal, J. C., J. C. Wyngaard, Y. Izumi, O. R. Cote, Spectral characteristics of surface layer turbulence, *Quart. J .R . Met. Soc.*, 98, 563–589, 1972.

23. Katul, G. G., and M. B. Parlange, On the active role of temperature in surface layer turbulence, *J. Atmos. Sci.*, 1993,In press.

24. Katul, G. G., *Coupled Processes Near the Land-Atmosphere Interface*, Ph.D Dissertation, University of California at Davis, 193 pp., 1993.

25. Kolmogorov, A. N., The local structure of turbulence in incompressible viscous fluid for very large Reynolds numbers, *Dokl. Akad. Nauk SSSR*, 4, 299–303, 1941.

26. Kolmogorov, A. N., A refinement of previous hypotheses concerning the local structure of turbulence in a viscous incompressible fluid at high Reynolds number, *J. Fluid Mech.*, 13, 82–85, 1962.

27. Kuznetsov, V. R., A. A. Praskovsky, and V. A. Sabelnikov, Fine-scale turbulence structure of intermittent shear flows, *J. Fluid Mech.*, 243, 595–622. 1992.

28. Kraichnan, R., Turbulent cascade and intermittency growth, *Turbulence and Stochastic Processes: Kolmogorov's Ideas 50 Years On*, ed. Hunt, J., Phillips, M., and Williams, D., Roy. Soc., 240 pp., 1991.

29. Landau, L. D., and E. M. Lifshitz, *Fluid Mechanics*, Pergamon Press, 536 pp., 1986.

30. Liandrat, J., and F. Moret-Baily, The Wavelet Transform: some applications to fluid dynamics and turbulence, *European Journal of Mechanics, B/ Fluids*, 9, 1–19, 1990.

31. Lumley, J., Interpretation of time spectra measured in high intensity shear flows, *Phys. Fluids*, 6, 1056–1062, 1965.

32. Mahrt, L., Intermittency of atmospheric turbulence, *J. Atmos. Sci.*, 46, 79–95, 1989.

33. Mahrt, L., Eddy asymmetry in the shear-heated boundary layer, *J. Atmos. Sci.*, 48, 472–492, 1991.

34. Mallat, S, A theory for multiresolution signal decomposition: The wavelet representation, *IEEE Trans. Pattern Analysis and Machine Intelligence*, 11, 674–693, 1989a.

35. Mallat, S., Multiresolution approximations and wavelet orthonormal bases of $L^2(R)$, *Trans. Amer. Math. Soc.*, 315, 69–87, 1989b.

36. McComb, W. D., *The Physics of Fluid Turbulence*, Oxford Science Publications, 572 pp., 1992.
37. Meneveau, C., Analysis of turbulence in the orthonormal wavelet representation, *J. Fluid Mech.*, 232, 469–520, 1991a.
38. Meneveau, C., Dual spectra and mixed energy cascade of turbulence in the wavelet representation, *Physical Review Letters*, 11, 1450–1453, 1991b.
39. Meneveau, C., and K. R. Sreenivasan, The multifractal nature of turbulent energy dissipation, *J. Fluid Mech.*, 224, 429–484, 1991.
40. Meneveau, C., and K. R. Sreenivasan, Simple multifractal cascade model for fully developed turbulence, *Phys. Rev. Lett.*, 59, 1427–1427, 1987.
41. Meyer, Y., Orthonormal Wavelets, *Wavelets: Time-Frequency Methods and Phase Space*, ed. Combes, J.M., Grossmann, A., Tchamitchian, Ph., Springer-Verlag, 315 pp., 1989.
42. Monin, A. S., and A. M. Yaglom, *Statistical Fluid Mechanics Vol.II*, ed. Lumley, J., MIT Press, 874 pp., 1975.
43. Oncley, S. P., TKE dissipation measurements during the FLATS experiment, *Proc. 10th AMS Symp. on Turbulence and Diffusion*, Portland, Oregon, pp.165–166, 1992.
44. Powell, D., and C. E. Elderkin, An investigation of the application of Taylor's hypothesis to atmospheric boundary layer turbulence, *J. Atmos. Sci.*, 31, 990–1002, 1974.
45. Stull, R., *An Introduction to Boundary Layer Meteorology*, Kluwer Academic Press, 666 pp., 1988.
46. Tennekes, H., and J. L. Lumley, *A First Course in Turbulence*, MIT Press, 300 pp., 1990.
47. Tennekes, H., Intermittency of the small-scale structure of atmospheric turbulence, *Bound. Layer Meteorol.*, 4, 241–250, 1973.
48. Taylor, G. I., The spectrum of turbulence, *Proc. Roy. Soc., A*, Vol. CLXIV, 476–490, 1938.
49. Willis, G. E., and J. Deordorff, On the use of Taylor's translation hypothesis for diffusion in the mixed layer, *Quart. J. Roy. Meteor. Soc.*, 102, 817–822, 1976.
50. Wyngaard, J. C., and S. F. Clifford, Taylor's hypothesis and high-frequency turbulence spectra, *J. Atmos. Sci.*, 34, 922–929, 1977.
51. Wyngaard, J. C., Cup, propeller, vane, and sonic anemometer in turbulence research, *Ann. Rev. Fluid Mech.*, 13, 399–423, 1981.
52. Yakhot, V., Z. S. She, and S. A Orzag, Deviations from the classical Kolmogorov theory of the inertial range of homogeneous turbulence, *Phys. Fluids A*, 289–293, 1989.
53. Yamada, M., and K. Ohkitani, Orthonormal expansion and its application to turbulence, *Prog. Theor. Phys.: Progress Letters*, 86, 819–823,

1990.

54. Yamada, M., and K. Ohkitani, Orthonormal wavelet analysis of turbulence, *Fluid Dynamics Research*, 8, 101–115, 1991a.

55. Yamada, M., and K. Ohkitani, An identification of energy cascade in turbulence by orthonormal wavelet analysis, *Prog. Theor. Phys.*, 86, 799–815, 1991b.

The authors would like to thank Scott Tyler for his assistance and support at Owen's lake, and Teresa Ortenburger and Mike Mata for their help in the data collection. We are grateful for the funding support from the National Science Foundation (NSF) grant (EAR-93-04331), United States Geological Survey (USGS), Water Resources Center (WRC) grant (W-812), Kearney Foundation, and UCDAVIS superfund grant (5 P42ES04699-07).

Gabriel G. Katul
School of the Environment
Duke University
Durham, NC 27708-0328
USA
e-mail: *gaby@acpub.duke.edu*

John D. Albertson
Hydrologic Science
University of California at Davis
Davis, CA 95616
USA

Chia R. Chu
Department of Civil Engineering
National Central University
Chungli, Taiwan

Marc B. Parlange
Hydrologic Science &
Department of Agricultural and Biological Engineering
University of California at Davis
Davis, CA 95616
USA

An Adaptive Decomposition:
Application to Turbulence

J. F. Howell and L. Mahrt

Abstract. Nine hours of 45 meter tower anemometer measurements are analyzed to demonstrate an adaptive method for decomposing a time series into
orthogonal modes of variation. In conventional partitioning (or filtering) the cutoff scales are specified *a priori* to be constant throughout the record. Applying
a constant cutoff scale is less effective if two different physical modes vary on
overlapping scales, since the statistical partitioning is then physically ambiguous.
For the turbulence data analyzed in this study, motions leading to a majority of
the momentum flux intermittently occur on small scales which otherwise lead to
little flux. To better separate the transporting motions from the more random
motions, the cutoff scale separating these two modes is allowed to vary with record
position.

The entire record of data is partitioned into four modes of variation using
a piece-wise constant (Haar) decomposition. The two larger scale modes are
characterized as the mesoscale and large eddy modes. Similar to conventional
partitioning, these two modes are determined by spatially constant cutoff scales.
The two smaller scale modes on the other hand, are separated by a scale which
depends on the local transport characteristics of the flow. Local extremes in
the spatial distribution of momentum flux determine the partitioning between
the two small scale modes. This leads to an adaptive cutoff scale, which better
isolates the transport mode responsible for a majority of the momentum flux.
The spatial variation of this cutoff scale allows the time series to be decomposed
into modes which are physically more pure as is verified in terms of traditional
statistics for each mode. For example, the gradient skewness in the longitudinal
wind component for the main transporting eddies is -0.23 using a constant cutoff
scale and -0.85 using the adaptive cutoff scale indicating that the shear driven
transporting eddies are more completely isolated using the variable cutoff scale.
The remaining small scale deviations define the fine scale mode, which consist of
non-transporting nearly isotropic motions.

Wavelets in Geophysics
Efi Foufoula-Georgiou and Praveen Kumar (eds.), pp. 107–128.
Copyright 1994 by Academic Press, Inc.
All rights of reproduction in any form reserved.
ISBN 0-12-262850-0

§1. Introduction

A common issue in time series analysis is sorting out the different modes of variation. If such modes overlap in Fourier space or are primarily local or event-like, then traditional Fourier and eigenvector decompositions are generally less effective in separating the different modes. As an alternative, nine hours of turbulence data are orthogonally decomposed into piece-wise constants. Based on the decomposition statistics, four modes of variation are defined. One of these modes includes a majority of the momentum flux and is described by a subrange of scales which depends on the record position.

The Haar wavelet is the underlying basis in the decomposition applied in this study. Decomposing the turbulence in terms of higher order wavelets which are more compact in Fourier space yield similar results. Higher order wavelets however, are slightly less efficient in capturing the sharp gradients associated with the transport physics which is the primary goal of this study. On the other hand, higher order wavelets do appear better suited for representing the larger scale, smoother variations. Additional reasons for specifically using the Haar wavelet are provided at the beginning of Section 2.1, and results of applying alternative orthogonal wavelets are discussed at the end of Section 2.3.

In general, wavelets decompose global variance or energy in terms of scale and position within the record. Wavelets are used extensively in seismic analysis and are beginning to find their way into other geophysical disciplines [3, 11, 12, 13, 18, 21]. Related to this study, Farge [9] and Meneveau [23] discuss a variety of wavelet applications to turbulence. Ways of constructing, describing, and implementing wavelet tools are many. Daubechies [8] constructs wavelets which are compact in physical space, and yet still provide a complete basis for decomposing the total sampled variance. Viewing wavelet bases from a linear algebra perspective [24] can be useful in solving numerical equations [1]. Spline wavelets are generally effective for interpolating and filtering data [2, 7, 25]. Another common view is that a wavelet basis set describes a multiresolution analysis [10, 19, 22].

A multiresolution analysis is a convenient setting for locally describing the different scales of variation in the data. Multiresolution techniques are used in image analysis, for example, to store and transmit images compactly [5, 6]. The current development could be posed in terms of a multiresolution analysis [16] or alternatively in terms of the wavelet analysis referred to above. However, the methods in this study (Section 2) are kept simplified such that it is unnecessary to explicitly appeal to these closely related topics.

The strategy in this study is to first decompose the turbulence time series into distinct modes. Because small scale variations in the turbulence record correspond to two types of motions, an adaptive technique is used to separate them (Section 2.2). For a simple demonstration of the adaptive technique, only the scale separating these two small scale modes is allowed to spatially vary. Generally, the adaptive technique allows for the separation of physically distinct modes with overlapping scales. This separation is not possible in conventional filtering with a specified response function or distribution of weights. After decomposing the turbulence, the spatial distribution of momentum flux will be reconstructed for the different modes. Additional statistics are then computed for the different modes (Section 2.3). In Section 3 a brief physical interpretation of the results is provided.

§2. Partitioning the Time Series

The data used in this study consist of 9.1 hours of the three velocity components measured 45 meters above flat terrain in near neutral conditions [17]. The wind speed fluctuates about a mean value of $\bar{u} = 12.8$ m/s throughout the 9.1 hours, and most of the energy is concentrated in the 1 minute or 1 km eddy motions (Section 2.3). For a turbulence depth of roughly 500 m the Reynolds number is more than 10^8. Additional statistics describing this data set can be found in [17] and [21].

The value of the longitudinal wind component at the ith record position is denoted as

$$u_i \; ; \; i = 1, 2, ..., 2^M \tag{1}$$

where $\delta = \frac{1}{16}$ s is the width of a sampling interval and in the current analysis $M = 19$, corresponding to 524,288 data points. There are equivalent time series of the cross stream component v and the vertical wind component w.

2.1. Decomposition method

Haar [14] first presented this decomposition method over 80 years ago, yet the utility of such a basic decomposition is only beginning to be realized, primarily in the context of wavelets. Indeed, Daubechies [8] appealed to the Haar basis set in introducing a fundamental set of equations describing a unique sequence of orthogonal wavelet bases.

Two coefficients (-1 and $+1$) essentially describe a Haar basis element. The Haar decomposition involves multiplying differences of arithmetic sums of the data by dyadic numbers ($= 2^p$ where p is an integer). Orthogonal wavelet transforms usually involve more coefficients and irrational numbers. The Haar wavelet is (odd) symmetric while all other compactly supported orthogonal wavelets are apparently asymmetric [8]. The Haar basis set is sometimes discounted because a Haar transform value corresponds to a Fourier spectral window which decays away from a central wavelength at a slower rate than higher order wavelets, an important consideration in many applications. For the turbulence data analyzed in this study, the main transporting eddies often consist of sharp boundaries or microfronts. Microfronts are efficiently decomposed with the Haar basis set because they both contain sharp gradients. Compactness in Fourier space seems less important in this case. The Haar decomposition is now reviewed.

The longitudinal wind averaged over the entire record is defined as

$$\overline{u} = \frac{1}{2^M} \sum_{i=1}^{2^M} u_i. \tag{2}$$

Variations in the longitudinal wind are described by difference terms defined as

$$\mathbf{\Delta}u(a_m; n) = \frac{1}{2^m} \sum_{j=1}^{2^{m-1}} \left(u_{2^m(n-\frac{1}{2})+j} - u_{2^m(n-1)+j} \right). \tag{3}$$

These difference terms correspond to Haar transform values and characterize the average change in the u signal on the scale $a_m = 2^m \delta$ across the interval $[a_m(n-1), a_m n]$. Specifically, (3) is half the difference between the half interval averages, so for example, $\mathbf{\Delta}u(a_1; n) = \frac{1}{2} \times (u_{2n} - u_{2n-1})$. Intervals of length a_m are enumerated starting with $n = 1$ (the first translate) and ending with $n = 2^{M-m}$ (the last translate). These non-overlapping translated intervals completely cover the time series of length a_M. The differences (3) are then computed for each translate and dyadic scale such that the sampled variations are completely decomposed.

The signal deviation from the record average at the ith record position is reconstructed as

$$u_i - \overline{u} = \sum_{m=1}^{M} (-1)^{\ell} \mathbf{\Delta}u(a_m; n) \tag{4}$$

$$n = 1 + int(\frac{i-1}{2^m}) \, ; \, \ell = 1 + int(\frac{i-1}{2^{m-1}})$$

for $i = 1, 2, ..., 2^M$. The second translation number ℓ represents the enumerated half intervals of length $a_m/2$. The translation numbers n and ℓ correspond to the difference terms (3) which are necessary for reconstructing the data value at the ith record position. Specifically, n corresponds to the interval $[a_m(n-1), a_m n]$ which includes the ith record position. The integer ℓ is even (odd) depending on whether the datum at the ith record position was added (subtracted) in the associated difference term. The operator $int(*)$ yields the integer part of the number $*$.

The total sampled variance is obtained by squaring the left hand side of (4) and averaging over the 2^M different record positions. This operation corresponds to a sum over dyadic scales of mean square Haar transform values. The total sampled variance is written as

$$\frac{1}{2^M} \sum_{i=1}^{2^M} (u_i - \overline{u})^2 = \sum_{m=1}^{M} \frac{1}{2^{M-m}} \sum_{n=1}^{2^{M-m}} [\Delta u(a_m; n)]^2. \tag{5}$$

The total covariance between the time series of the longitudinal and vertical wind components is obtained by multiplying the u reconstruction (4) by an equivalent expression for w, and averaging over all record positions. The resulting covariance quantity is written as

$$\frac{1}{2^M} \sum_{i=1}^{2^M} (u_i - \overline{u})(w_i - \overline{w}) = \tag{6}$$

$$\sum_{m=1}^{M} \frac{1}{2^{M-m}} \sum_{n=1}^{2^{M-m}} \Delta u(a_m; n)\Delta w(a_m; n).$$

The identities (4)–(6) can be verified by substituting in the expression for $\overline{u}$ from (2), the expression for $\Delta u(a_m; n)$ from (3), and equivalent expressions for $\overline{w}$ and $\Delta w(a_m; n)$. Verifying (4)-(6) is facilitated by the two identities:

$$\sum_{m=1}^{M} 2^m = 2^{M+1} - 2 \tag{7}$$

$$\sum_{m=1}^{m} \frac{1}{2^m} = 1 - \frac{1}{2^M}. \tag{8}$$

2.2. Modes of variation

Variations contributing to the local signal deviation from the record average (4) are now partitioned into different modes. The sum of the different modes will be equal to (4), and the individual modes will locally correspond to a continuous subrange of scales. Accordingly, the longitudinal wind data associated with a given mode c $(= 1, ..., C)$ at the ith record position is described by a range of scales between $a_{m_c^-}$ and $a_{m_c^+}$ so that

$$u_i - \overline{u} = \sum_{m=m_c^-}^{m_c^+} (-1)^\ell \Delta u(a_m; n) \tag{9}$$

$$n = 1 + int(\frac{i-1}{2^m}) \; ; \; \ell = 1 + int(\frac{i-1}{2^{m-1}})$$

where $m_1^+ = M$, $m_C^- = 1$, and $m_c^+ = m_{c-1}^- - 1$. A given mode c at the ith record position includes all the difference terms (3) associated with the range of dilation scales $a_{m_c^-}$ through $a_{m_c^+}$. The small scale cutoff for the smallest scale mode is $a_1 = 2\delta$, so $m_C^- = 1$ is fixed. This leaves $C - 1$ scales unspecified. In this study $C = 4$, so the values of m_1^-, m_2^- and m_3^- are to be determined.

If the value of m_c^- is set to a constant, then the scale separating the two corresponding modes will be independent of record position. For the atmospheric wind data, the cutoff scales for the two largest scale modes, namely the mesoscale ($c = 1$) and large eddy ($c = 2$) modes, are defined in this manner. These two modes are defined to span the scales from 800 m – 420 km. Here we have converted time to distance (Taylor's hypothesis) using the mean wind speed $\bar{u} = 12.8$ m/s.

Because the horizontal variance drops off rapidly with scale at about 5 km (Figure 5, Section 2.3), 5 km is chosen as the small scale cutoff for the mesoscale mode $c = 1$. The smallest dyadic scale greater than 5 km is $a_{13} = 6.5$ km, so accordingly, $m_1^- = 13$. At the 500 m scale the u-w covariance (momentum flux) increases rapidly with decreasing scale. Therefore, 500 m is chosen as the small scale cutoff for the large eddy mode, $c = 2$, in order that most of the flux remains to be captured. The smallest dyadic scale greater than 500 m is $a_{10} = 800$ m, so accordingly $m_2^- = 10$. The large eddy mode consist of motions which may generally span the entire depth of the boundary layer whereas the smaller scale ($\leq a_9 = 400$ m) motions are likely to be not as deep.

Since there is no obvious constant scale separating the two smaller scale modes, the scale (or $m_{3,i}^-$) is allowed to depend on the record position according to the local physics of the flow. This allows random small scale fluctuations to be effectively separated from small scale variations associated with a transporting eddy. In other words, the transport mode is allowed to locally capture smaller scales if such scales account for significant flux. For example, significant momentum flux may be associated with wind gusts or so called microfronts. Including small scales in the reconstruction of the transport mode at the position of a microfront more effectively captures the flux. Between the gusts, the small scale motions do not significantly transport momentum and therefore are not included in the reconstruction of the transport mode.

Local transport occurs at locations where variations in the longitudinal wind component u are in phase with variations in the vertical component w. The analyzed eddies are centered according to the largest magnitudes of the product $\Delta u(a_m; n) \times \Delta w(a_m; n)$ within each of the 4,096 intervals of width $a_7 \approx 100$ m. Specifically, flux maxima are determined within the non-overlapping intervals $[a_7(n-1), a_7 n]$ for $n = 1, 2, 3, ..., 2^{12}$. These local maxima statistically represent local maxima in the spatial distribution of momentum flux, and thus determine the position and scale of a transporting eddy. In this study we do not use a minimum threshold value for conditioning the local flux maxima. The strongest momentum flux event within each non-overlapping 100 m interval is included in the transport mode, regardless of the sign or strength of the event.

The values of $m_{3,i}^-$ are determined by first requiring $1 \leq m_{3,i}^- \leq 7$ for all record positions. The upper bound on $m_{3,i}^-$ ensures that the transport mode ($c = 3$) includes, as a minimum, all scales between $a_7 = 100$ m and $a_9 = 400$ m. This choice is based on the observed global dependence of covariance on scale (Section 2.3).

The next step in determining $m_{3,i}^-$ is to identify the largest value of $\Delta u(a_m; n) \times \Delta w(a_m; n)$ within in each $a_7 \approx 100$ m interval. Say that within a particular interval of length a_7, this largest product corresponds to the index values m^* and n^* associated with the transform interval $[a_{m^*}(n^* - 1), a_{m^*} n^*]$. The transport mode is then reconstructed *down* to the scale a_{m^*} for all of the record positions contained in the interval $[a_{m^*}(n^* - 1), a_{m^*} n^*]$. This reconstruction corresponds to including all of the difference terms (3) associated with the telescoping intervals $[a_{m'}(n' - 1), a_{m'} n']$ where $a_{m^*} \leq a_{m'} \leq a_7$ and the translation number is $n' = 1 + int(\frac{n^*-1}{2^{m'-m^*}})$.

The difference terms corresponding to the telescoping intervals are used in reconstructing the transport mode $c = 3$. These terms determine $m_{3,i}^-$. Considering the ith record position contained within the given interval of length a_7, this position will necessarily be contained in the interval $[a_{m'}(n' - 1), a_{m'} n']$ for one or more values of m'. For example, this position (by definition) is always contained in the interval corresponding to $m' = 7$. The value of $m_{3,i}^-$ at this given i^{th} record position is simply the smallest of the m'. This value corresponds to the smallest of the telescoping intervals in which this given i^{th} record position is contained. A more complete description (with figures) of a similar adaptive technique can be found in [16].

As an example, a 4 km segment of the m_c^- distributions are plotted in Figure 1. The straight horizontal lines indicate the spatially constant scales separating modes 1 & 2 and 2 & 3. The scale separating modes 3 & 4 ($m_{3,i}^-$) varies from several meters to 100 m depending on the record position. The local minima in $m_{3,i}^-$ occur at locations where the smaller scale variations are identified as belonging to the transport mode. At these locations, some of the flux occurs on small scales usually associated with microfronts.

2.2.1. Definitions

The particular modal definitions, detailed below, are based on an interpretation of the variance and covariance spectra (Section 2.3). The modal reconstructions (Section 2.2.2) substantiate the partitioning. The largest scale (mesoscale) mode leads to very little flux and consist of predominantly horizontal motions. The large eddy mode leads to some flux, and for this

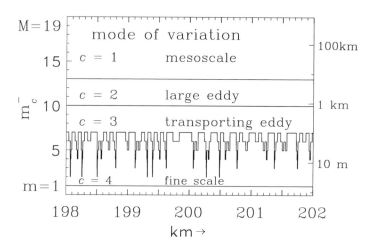

Figure 1. The small scale cutoffs for the four different physical modes
of variation

mode the variance in the longitudinal wind component is significantly larger
than the variance in the cross stream and vertical wind components. The
next smallest scale, the transport mode contains a majority of the flux,
and the motions are more 3-dimensional than the large eddy mode. The
smallest resolved (fine) scale mode is responsible for very little net flux and
consist of nearly isotropic motions.

(i) The mesoscale mode, $c = 1$, at the ith record position is defined as

$$u_{1,i} = \sum_{m=13}^{19} (-1)^{\ell} \boldsymbol{\Delta} u(a_m; n) \tag{10}$$

$$n = 1 + int(\frac{i-1}{2^m}) \; ; \; \ell = 1 + int(\frac{i-1}{2^{m-1}})$$

where $\boldsymbol{\Delta} u(a_m; n)$ is defined in (3). The dilation scale $a_m = 2^m \delta = 2^{m-4}$ s
$\approx 2^{m+3} \times 0.1$ m, so this mode includes scales between $a_{13} \approx 6.5$ km and
$a_{19} \approx 420$ km.

(ii) The large eddy mode, $c = 2$, at the ith record position is defined as

$$u_{2,i} = \sum_{m=10}^{12} (-1)^{\ell} \boldsymbol{\Delta} u(a_m; n) \tag{11}$$

where n and ℓ are defined in (10). This mode includes scales between $a_{10} \approx 800$ m and $a_{12} \approx 3.3$ km.

(iii) The transport mode, $c = 3$, at the ith record position is defined as

$$u_{3,i} = \sum_{m=m_{3,i}^-}^{9} (-1)^{\ell} \Delta u(a_m; n) \qquad (12)$$

where $1 \leq m_{3,i}^- \leq 7$ is determined for each interval of length $a_7 \approx 100$ m according to the above discussion. This mode includes scales between $a_1 \approx 1.6$ m and $a_9 \approx 400$ m.

(iv) The fine scale mode, $c = 4$, at the ith record position is defined as

$$u_{4,i} = \sum_{m=1}^{m_{4,i}^+} (-1)^{\ell} \Delta u(a_m; n) \qquad (13)$$

where $m_{4,i}^+ = m_{3,i}^- - 1$. At record positions where $m_{3,i}^- = 1$ the fine scale mode is zero. This mode includes scales between $a_1 \approx 1.6$ m and $a_6 \approx 50$ m.

2.2.2. Reconstructions

The sum of the different modes is equal to (4), so it follows that

$$u_i - \overline{u} = \sum_{c=1}^{C} u_{c,i} \qquad (14)$$

where $C = 4$ is the total number of modes in this case. Figure 2 shows the different orthogonal modes of the longitudinal wind. The sum of the different modes including the record average 12.8 m/s is equivalent to the original data (Figure 3). The large differences in the scales of variation for the different modes is made apparent by plotting the modes on the same horizontal scale. The transport mode $c = 3$ consist of variable width blocks because the scale separating the two small scale modes depends on the record position. The local minima in the separation scale occur at locations where the block widths are more narrow as can be seen by comparing Figures 1 and 2.

Segment lengths along the horizontal axes are now adjusted according to the different scales of variations in order for the flux $u_c \times w_c$ to be plotted for different c (Figure 4). For example, the mesoscale mode, $c = 1$, can be viewed for the entire (≈ 420 km) record while the fine scale mode, $c = 4$, is suitably viewed on a segment only 400 m in length.

Figure 4 shows that the transport mode captures most of the downward momentum flux while the fine scale mode leads to small flux with nearly equal probability of upward or downward flux. In this sense, the

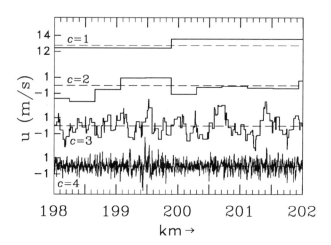

Figure 2. The four orthogonal modes of the longitudinal wind component u plotted on the same segment of the record shown in Figure 1.

partitioning is successful. In order to quantify the differences between the four modes, globally averaged statistics are now computed.

§3. Variance and Covariance Spectra

A global estimate of the variance on the scale a_m is the square of the Haar transform (3) on the scale a_m averaged over all the translation positions, written as

$$var(u; a_m) = \frac{1}{2^{M-m}} \sum_{n=1}^{2^{M-m}} [\Delta u(a_m; n)]^2 \tag{15}$$

where $\Delta u(a_m; n)$ is defined in (3). From (5) the total variance summed over all scales can then be expressed as

$$Var(u) = \sum_{m=1}^{M} var(u; a_m) \tag{16}$$

where $Var(u)$ corresponds to the left hand side of (5).

In order to obtain a more continuous estimate of the variance as a function of scale, the dilation a is allowed to take on values between orthogonal

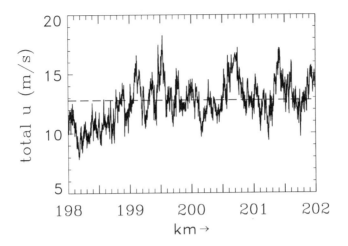

Figure 3. The sum of the four different modes which equals the original data.

scales, and the difference (transform) intervals are allowed to overlap. This leads to a difference term defined as

$$\Delta u(a;b) = \frac{1}{2m}\sum_{j=1}^{m}(u_{i+j} - u_{i+j-m}) \qquad (17)$$

where the dilation scale is $a = 2m\delta$, and $b = i\delta$ is the central position of the transform interval. Provided there are N equally spaced data points, (17) is defined for $i = m, ..., N - m$ so there are a total of $N - 2m + 1$ transform values on the scale $a = 2m\delta$. This generalization leads to a non-orthogonal estimate of the variance on the scale a defined as

$$var^{*}(u;a) = \frac{1}{N - 2m + 1}\sum_{i=m}^{N-m}[\Delta u(a;b)]^{2}. \qquad (18)$$

Depending on the sampling, the non-orthogonal and orthogonal global estimates of a local variance may be nearly equal at a fixed dyadic scale, that is $var^{*}(u;a) \approx var(u;a_m)$ for $a = a_m$. The non-orthogonal estimate based on the spatially overlapping transform intervals leads to smoother spectral estimates with respect to scale.

An orthogonal estimate for the covariance between the longitudinal

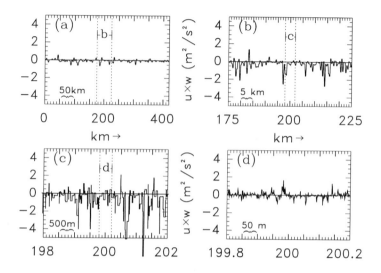

Figure 4. The product $u_c \times w_c$ of the four orthogonal modes. The four modes are (a) the mesoscale mode $c = 1$, (b) the large eddy mode $c = 2$, (c) the transport mode $c = 3$, and (d) the fine scale mode $c = 4$.

and vertical wind components on the scale a_m is written as

$$cov(u, w; a_m) = \frac{1}{2^{M-m}} \sum_{n=1}^{2^{M-m}} \Delta u(a_m; n) \Delta w(a_m; n), \qquad (19)$$

so from (6) the total covariance can be alternatively expressed as

$$Cov(u, w) = \sum_{m=1}^{M} cov(u, w; a_m) \qquad (20)$$

where $Cov(u, w)$ corresponds to the left hand side of (6).

A non-orthogonal estimate of the covariance on the scale a is defined as

$$cov^*(u, w; a) = \frac{1}{N - 2m + 1} \sum_{i=m}^{N-m} \Delta u(a; b) \Delta w(a; b) \qquad (21)$$

which provides better scale resolution for estimating the scales of motion dominating the vertical flux.

The orthogonal and non-orthogonal covariance spectra are plotted for the time series of the longitudinal and vertical wind components (Figures

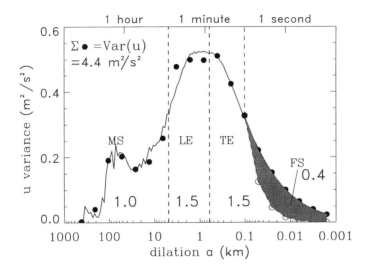

Figure 5. Variance in the longitudinal wind component u. The circles are orthogonal values while the curve represents non-orthogonal estimates. The regions between the vertical dashed lines are denoted MS (mesoscale), LE (large eddy), TE (transporting eddy), and FS (fine scale) and correspond to four orthogonal modes of variation. The numbers in each region are the contributions of the corresponding mode to the total variance. The fine scale mode corresponds to the shaded region. The transport and fine scale modes both include variance on dilation scales less than 100 m.

5-7). The numbers for each mode in the lower portion of the figures are the relative contributions to the total covariances. These numerical values, in addition to other statistics, are summarized in Table I.

Included in these statistics are gradients in the longitudinal wind components. Gradients in a piece-wise constant signal are readily defined as the difference between adjacent constant values divided by the distance between the center positions of the associated segments. Consequently, for the transport mode the gradients are computed over variable distances, since the piece-wise constants are of variable width (Figure 2, $c = 3$).

The skewness values of the gradients in the longitudinal wind components for the different modes are listed in Table I. A positive (negative) value of the skewness for the u gradients indicates the wind speed increases (decreases) more rapidly in the downstream direction. Thus, the results

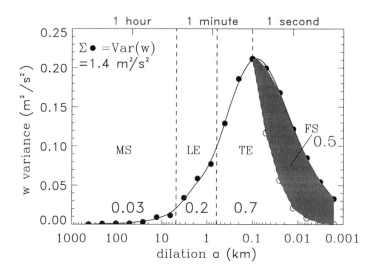

Figure 6. As in Figure 5 except for the variance in the vertical wind component w.

Table I

Haar wavelet decomposition statistics. Variances (var and cov) are in m^2/s^2. The values in the parenthesis are the results if the adaptive step is not taken to include additional small scale motions in the transport mode.

Mode	Mesoscale	Large Eddy	Transport	Fine Scale	Total
u var	1.043	1.474	1.456 (1.263)	0.411 (0.604)	4.3841
v var	0.779	0.407	1.019 (0.852)	0.524 (0.691)	2.7294
w var	0.027	0.170	0.731 (0.526)	0.457 (0.662)	1.3845
u-w cov	−0.049	−0.300	−0.438 (−0.368)	−0.068 (−0.138)	−0.8552
u-w corr	−0.29	−0.60	−0.42 (−0.45)	−0.16 (−0.22)	−0.35
isotropy	0.03	0.18	0.59 (0.50)	0.98 (1.02)	0.39
skewness of u gradient	0.048	−0.227	−0.846 (−0.230)	−0.006 (−0.019)	−0.220

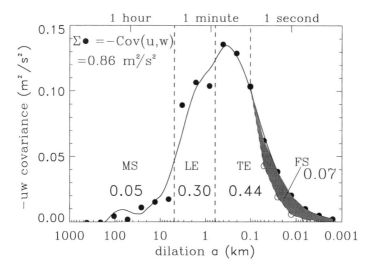

Figure 7. As in Figure 5 except for the covariance between the time series of the longitudinal wind component u and the vertical wind component w.

show that the transport mode is associated with rapid decreases in the wind speed in the downstream direction, a characteristic of shear driven transport. The variable cutoff scale leads to a gradient skewness of -0.846 where as a constant cutoff scale results in a gradient skewness of only -0.230. Thus, the spatially adaptive cutoff scale defines a more complete physical partitioning.

Also listed in Table I are the correlation coefficients, defined in the usual way as

$$\frac{Cov(u, w)}{\sqrt{Var(u)Var(w)}} \qquad (22)$$

and an isotropy coefficient which is calculated as

$$\frac{2 \times Var(w)}{Var(u) + Var(v)} \qquad (23)$$

If all three velocity components contain the same amount of variance, this isotropy coefficient is unity.

The Haar is but one possible wavelet to use in decomposing the turbulence. As mentioned at the beginning of Section 2.1, the Haar wavelet is an element in a sequence of wavelets introduced by Daubechies [8]. Another

Table II

As in Table I except the numbers are D4 wavelet decomposition statistics.

Mode	Mesoscale	Large Eddy	Transport	Fine Scale	Total
u var	1.052	1.556	1.364 (1.207)	0.372 (0.529)	4.3441
v var	0.783	0.432	1.023 (0.891)	0.490 (0.622)	2.7282
w var	0.029	0.176	0.737 (0.563)	0.442 (0.616)	1.3838
u-w cov	−0.069	−0.311	−0.421 (−0.363)	−0.059 (−0.117)	−0.8605
u-w corr	−0.40	−0.59	−0.42 (−0.44)	−0.15 (−0.20)	−0.35
isotropy	0.03	0.18	0.62 (0.54)	1.02 (1.07)	0.39

simple element of that sequence is referred to as a D4 wavelet since it is constructed from four coefficients whereas the Haar requires only two. A distinction between different wavelets in the Daubechies sequence is locality in physical space versus locality in Fourier space. For example, the D4 wavelet is more local in Fourier space than the Haar whereas as the Haar wavelet is more local in physical space. In order to study the sensitivity of the present results to a specific wavelet, the D4 wavelet basis is applied to the turbulence time series in the same manner as the Haar basis. The lack of symmetry in the D4 wavelet was not a significant factor in this case.

The results of applying the Haar and D4 wavelet bases (Tables I and II) are similar, though in the D4 wavelet decomposition (Table I) there is a little more variance and covariance captured in the larger scale modes and there is a little less variance and covariance in the two smaller scale modes. Adapting the D4 wavelet decomposition to the spatial distribution of the covariance has about the same effect as adapting the Haar decomposition. The effect of adapting the decompositions is quantifiable in terms of the differences between the values in parenthesis and the values immediately above those numbers in Tables I and II. For example, when applying the D4 wavelet (Table II), adapting the decomposition increases the u-w covariance (flux) in the transport mode by about 16%. Adapting the Haar decomposition (Table I) leads to a 19% increase in the flux associated with the transport mode. Applying wavelets which are even more compact in Fourier space lead to similar results. Though with higher order wavelets, the ratio of the variance and covariance in the larger scales to that of the smaller scales tends to be greater.

Higher order wavelets appear to be better suited for representing the larger scale modes, which normally have a smoother spatial structure. For example, the variance and covariance spectra generated from decomposing the turbulence data with higher order wavelets contain peaks at the larger (> 1 km) scales, which are better defined than in the Haar spectra. The smaller (< 1 km) scale turbulence and associated sharp gradients, however, are efficiently captured by the Haar decomposition.

Another result is that the total covariances captured by the D4 wavelet (Table II) are slightly different than the totals listed in Table I obtained from the Haar decomposition. Total sampled covariances directly computed according to the left hand sides of (5) and (6) agree with the Haar decomposition totals to more digits than shown in Table I.

§4. Physical Interpretation

The turbulence data analyzed in this study has been decomposed into deviations of small scale averages from larger scale averages. The turbulence measurements include 9.1 hours of wind tower data which translates to about 420 km using Taylor's hypothesis and a mean wind speed $\bar{u} = 12.8$ m/s. The flow is partitioned primarily according to the scale dependence of the u-w covariance (Figure 7). Deviations of 3.2 km averages from the entire record average (variations on scales ≥ 6.4 km) define the mesoscale mode. Deviations from the 3.2 km averages represent the turbulence, which in turn are partitioned into three different modes of variation (Section 2).

Based on the u-w covariance (see shoulder in Figure 7 around $a = 2$ km), the large eddy mode is defined by deviations of 400 m averages from the 3.2 km averages, that is variations on the scales between 800 m and 3.2 km. The exact scales defining the large eddy mode are somewhat arbitrary, since the physics changes between the small scale and large scale ends of this regime. At the small scale end of the large eddies, the vertical motions become more significant. The large eddy mode corresponds to motions which are more horizontal at the 45 m observation level compared to the smaller scale motions as indicated by the small value of the isotropy coefficient (Table I).

The large eddies may be the lower part of boundary layer scale motions (on the order of 1 km deep) where the lower part of the eddies observed at the tower level are forced by the ground to be more horizontal. In this case, the transport by the large eddies would increase with height. Roll vortices are an example of such eddies [4]. When observed from time series measured from towers, such larger scale motions are sometimes referred to as inactive eddies because they contribute significantly to the horizontal variance but contribute little to the momentum flux at levels closer to the ground. In this case the motions are not considered to be traditional turbulence and must be removed from the signal before traditional similarity arguments can be applied [15]. This concept may be most descriptive of the larger scale part

of the large eddies in the present partitioning. The weak vertical motions
that occur on the large eddy scales, however, are well correlated with the
variations of the longitudinal wind component leading to significant (35%
of total) momentum flux. While the mean shear seems to exert a greater
influence on the large eddy mode compared to the mesoscale mode (greater
gradient skewness, Table I), the shear effect on the large eddies is still small
compared to that of the transporting eddies.

Deviations of the raw time series from the 400 m averages define the
two smallest scale turbulence modes. In order to distinguish the two types
of small scale motions, the scale separating the two modes varies depending
on the local behavior of the transport as discussed in Section 2. Specifically,
averages are computed over a sufficiently small scale ($<$ 400 m) in order
to resolve the local vertical transport of momentum. Deviations of these
smaller scale averages from the 400 m averages are included in the transport
mode to capture a majority of the momentum flux, including local extremes
(Figure 4, panel c).

As a result, the fine scale structure is also determined by the variable
cutoff scale. This is in contrast to conventional high pass filtering where
the cutoff scale is constant throughout the record. A constant cutoff scale
is not used in this case because transporting eddies intermittently occur
on scales which are traditionally assigned to the fine scale structure. From
another point of view, the fine scale mode includes relatively larger scale
motions only at locations where small scale transport is absent. This means
that motions occurring on a range of scales between 1.6 m and 50 m are of
the transporting eddy type at some positions while at other positions these
motions make up the fine scale structure.

The transport mode is characterized by strong gradient skewness (Table I) reflecting the strong influence of the mean shear on the transporting
eddies. According to conventional expectations, the mean shear generates
eddy motions which transport higher momentum toward the surface. In
terms of energetics, this momentum transport corresponds to conversion of
mean kinetic energy to turbulence kinetic energy. **The data decomposition in this study provides a definition of the transporting eddies
which allows quantitative verification of classical concepts.** Also
as expected, the value of the isotropy coefficient for the transporting eddies
is between that of the large eddies and the fine scale structure. The main
transporting eddies are characterized by significant vertical motions which,
however, remain smaller than the horizontal velocity components. This
is because the mean shear directly generates variance in the u-component
which subsequently induces vertical velocity fluctuations through pressure
fluctuations.

The remaining small scale deviations make up the fine scale structure.
Very little transport is associated with the fine scale mode, and the energy

in the horizontal and vertical wind components are nearly equal. More specifically, the isotropy coefficient is close to unity. The gradient skewness is very small, since the fine scale structure does not receive energy directly from the mean shear. Instead the fine scale structure receives energy from the transporting eddies through the so called energy cascade to smaller scales.

§5. Conclusion

Geophysical time series generally consist of physically distinct modes of variation, each occurring on a subrange of scales which depend on space and time. Conventional decomposition or filtering techniques divide the time series according to scales which are constant in space and time. In this study, distinct modes were isolated using a piece-wise constant (Haar wavelet) decomposition which allows the scales defining a particular mode to vary with record position. With this approach sampled covariances are completely and orthogonally decomposed. Partitioning the flow in this manner allows assessment of the relative contributions of the different modes to traditional statistics.

The turbulence data analyzed in this study has been partitioned into four modes of variation. Each mode is defined locally in terms of an upper and lower cutoff scale. The cutoff scales for the two larger scale modes are specified to be constant with respect to record position. The scale separating the two smaller scale modes varies with position according to the local maxima in the spatial distribution of momentum flux. A local momentum flux is quantified in terms of the product of the difference terms $\Delta u(a_m; n) \times \Delta w(a_m; n)$, which is equivalent to a product of wavelet coefficients at a fixed scale and position. Using a wavelet decomposition to examine the spatial or temporal distribution of the scale dependent flux is a promising approach for distinguishing distinct physical modes of variation.

Adapting the decomposition to the spatial distribution of momentum flux leads to an improvement in the small scale partitioning as interpreted in terms of globally averaged statistics. The spatial dependence of the cutoff scale allows the computed transport mode to capture more of the momentum flux; some of the flux occurring on scales traditionally included as fine scale structure is now more correctly included in the transport mode. For the transport mode, the gradients in the longitudinal wind component are negatively skewed in the downstream direction, which verifies that this mode is primarily shear driven. If a constant cutoff scale were used, the skewness of the gradients in the longitudinal wind for the main transporting motions would be only -0.230. After the additional small scale variations are included by varying the cutoff scale, the gradient skewness is -0.846. This change is a result of further resolving the microfronts associated with momentum transport. Moreover, the adaptive step leading to the spatially varying cutoff scale reduces the u–w correlation for the computed fine scale structure leading to a physically more pure decomposition.

Application of the adaptive technique to other geophysical time series requires that the variable cutoff scale be posed in terms of the physical process of interest. The physics of the decomposition might also be posed in terms of sudden changes of a quantity associated with larger scale variations. With traditional filtering, for example, sudden climatic changes, sharp frontal boundaries, or any near discontinuities in the time series which lead to large scale changes will be partially partitioned into the small scale part; the corresponding low pass filtered signal will include only a smoothed version of the sharp changes. The adaptive decomposition applied here can be constructed to include sharp changes in the larger scale part of the signal, avoiding undue smoothing [16]. The basic goal of the adaptive technique is to partition the flow according to the physics when the subrange of scales describing a given physical mode varies spatially or temporally. Since the present approach partitions the original time series into separate time series for each mode, the coherent structures associated with a particular physical mode can in turn be analyzed.

References

1. Alpert, B. K., Wavelets and other bases for fast numerical linear algebra, *Wavelets: A Tutorial in Theory and Applications* (C. K. Chui, ed.), Academic Press, 181–216, 1992.

2. Battle, G., Cardinal Spline Interpolation and the Block Spin Construction of Wavelets, *Wavelets: A Tutorial in Theory and Applications* (C. K. Chui, ed.), Academic Press, 73–90, 1992.

3. Bradshaw, G. A., and T. A. Spies, Characterizing canopy gap structure in forests using wavelet analysis, *J. Ecology*, **80**, 105–215, 1992.

4. Brown, R. A., Longitudinal instabilities and secondary flows in the planetary boundary layer: A review, *Rev. Geoph. Space Phys.*, **18**, 683–697, 1980.

5. Burt, P. J., The Pyramid as a Structure for Efficient Computation, *Multiresolution Image Processing and Analysis* (A. Rosenfeld, ed.), Springer Verlag, 6–35, 1984.

6. Burt, P. J., The Laplacian pyramid as a compact image code, *IEEE Trans. Commun.*, **31**, 532–540, 1983.

7. Chui, C. K., *An Introduction to Wavelets*, Academic Press, 1992.

8. Daubechies, I., Orthonormal bases of compactly supported wavelets, *Commun. Pure Appl. Math*, **41**, 909–996, 1988.

9. Farge, M., Wavelet transforms and their applications to turbulence, *Annu. Rev. Fluid Mech.* **24**, 395–457, 1992.

10. Feauveau, J.-C., Nonorthogonal multiresolution analysis using wavelets, *Wavelets: A Tutorial in Theory and Applications* (C. K. Chui, ed.), Academic Press, 153–178, 1992.

11. Gamage, N. and C. Hagelberg, Detection and analysis of microfronts and associated coherent events using localized transforms, *J. Atmos. Sci.*, **50**, 750–756, 1993.

12. Gamage, N. and W. Blumen, Comparative analysis of low level cold fronts: wavelet, Fourier, and empirical orthogonal function decompositions, *Mon. Wea. Rev.* **121**, 2867–2878, 1993.

13. Gambis, D., Wavelet transform analysis of the length of the day and the El-Niño/Southern Oscillation variations at intraseasonal and interannual time scales, *Ann. Geophysicae*, **10**, 429–437, 1992.

14. Haar, A., Zur Theorie der orthogonalen Funktionensysteme, *Mathematische Annalen* **69**, 331–371, 1910.

15. Högström, U., Analysis of turbulence structure in the surface layer with a modified similarity formulation for near neutral conditions, *J. Atmos. Sci.*, **47**, 1949–1972, 1990.

16. Howell, J. F. and L. Mahrt, An adaptive multiresolution data filter: application to turbulence and climatic time series, *J. Atmos. Sci.*, **51**, 1994.

17. Kristensen, L., D. H. Lenschow, P. Kirkegaard, and M. Courtney, The spectral velocity tensor for homogeneous turbulence, *Bound.-Lay. Meteor.*, **47**, 149–193, 1989.

18. Kumar, P. and E. Foufoula-Georgiou, A new look at rainfall fluctuations and scaling properties of spatial rainfall using orthogonal wavelets, *J. Appl. Meteor.*, **32**, 209–222, 1993.

19. Madych, W. R., Some elementary properties of multiresolution analyses of $L^2(R^n)$, *Wavelets: A Tutorial in Theory and Applications* (C. K. Chui, ed.), Academic Press, 259–294, 1992.

20. Mahrt, L., Eddy Asymmetry in the Sheared Heated Boundary Layer, *J. Atmos. Sci.* **48**, 472–492, 1991.

21. Mahrt, L. and J. F. Howell, The influence of coherent structures and microfronts on scaling laws using global and local transforms, *J. Fluid Mech.*, **260**, 143–168, 1994.

22. Mallat, S., The theory of multiresolution signal decomposition: the wavelet representation, *IEEE Trans. Pattern Anal. Machine Intell.*, **7**, 674–693, 1989.

23. Meneveau, C., Analysis of turbulence in the orthonormal wavelet representation, *J. Fluid Mech.* **232**, 469–520, 1991.

24. Strang, G., Wavelets and dilation equations: a brief introduction, *SIAM Review*, **31**, 614–627, 1989.

25. Unser, M. and A. Aldroubi, Polynomial splines and wavelets—a signal processing perspective, *Wavelets: A Tutorial in Theory and Applica-*

tions (C. K. Chui, ed.), Academic Press, 91–122, 1992.

This material is based upon work supported by Grant DAA H04-93-G-0019 from the Army Research Office and Grant ATM-9310576 from the Physical Meteorology Program of the National Science Foundation.

J. F. Howell
Oceanic and Atmospheric Sciences
Oregon State University
Corvallis, Oregon
USA
e-mail: *howell@ats.orst.edu*

L. Mahrt
Oceanic and Atmospheric Sciences
Oregon State University
Corvallis, Oregon
USA
e-mail: *mahrt@pbl.ats.orst.edu*

Wavelet Analysis of Diurnal and Nocturnal Turbulence Above a Maize Crop

Yves Brunet and Serge Collineau

Abstract. A set of daytime and nighttime turbulence data acquired above a maize crop is analyzed, using various properties of the wavelet transform. Wavelet variances and covariances provide characteristic duration scales for the turbulent motions contributing most to the signal energy. Subsequent detection of their signatures in the raw time series allows a characteristic gust frequency to be determined. Once normalized with friction velocity and canopy height, this frequency appears to be similar to that observed above a pine forest, which supports the postulate that transfer processes over plant canopies are dominated by populations of canopy-scale eddies, with universal characteristics. Conditional averaging of the corresponding patterns provides a clear picture of the ejection-sweep processes. Around the inversion time of the mean vertical temperature gradient, two scales of motions are inferred from the observation of wavelet variances and covariances. A technique based on the inverse wavelet transform is then developed for partitioning the original signals into small- and large-scale components, using a cut-off scale deduced from wavelet variances and covariances. Despite their wavelike aspect, the extracted large-scale components are not gravity waves. Their nature is unclear but the proposed methodology has potential applications for studying wave-turbulence interactions.

§1. Introduction

Daytime turbulence in the vicinity of vegetation canopies has been extensively investigated during the past ten years. It is now widely recognized that, within and just above vegetation canopies, turbulent transport processes are to a large extent dominated by intermittent, energetic coherent structures, with length scales of the order of the canopy depth ([1], [18], and [20]). A dynamic model for the formation and development of these 'canopy eddies' has been proposed recently ([11] and [19]).

Wavelets in Geophysics
Efi Foufoula-Georgiou and Praveen Kumar (eds.), pp. 129–150.

These structures may be seen in time series of wind velocity components as periodic patterns revealing the existence of ejection-sweep processes (a slow upward movement of air, followed by a strong downward motion associated with an acceleration of horizontal velocity). In the presence of a mean vertical temperature gradient, periodic ramp patterns are then apparent in time series of temperature (gradual rises followed by relatively sharp drops). Collineau and Brunet [4] developed a methodology based on the wavelet transform for studying such time series. From measurements acquired above and within a pine forest in slightly unstable conditions, Collineau and Brunet [5] determined a mean duration scale for the active part of the coherent motions, using the wavelet variance, and estimated the frequency distribution of time intervals between contiguous structures, using the wavelet transform as a jump-detector. The latter approach enabled them to perform conditional sampling on the time series. This provided a clear picture of the ejection-sweep process, as well as an evaluation of the relative contributions of coherent motions to momentum and heat fluxes.

Much less attention has been paid so far to nighttime turbulence. Paw U *et al.* [18] presented some evidence that similar processes occur, revealed by inverse temperature ramps. Also, gravity waves have been observed in a variety of plant canopies, just after sunset [17]. Both phenomena are intermittent. Over plant canopies, turbulent coherent motions appear quasi-periodically at a frequency depending, to a first approximation, on friction velocity u_* and canopy height h ([18] and [19]). Gravity waves can be seen in time series of temperature and vertical velocity as trains of oscillations characterized by the Brunt-Vaisala frequency. Discriminating between propagatory and turbulent events is a matter of importance. Their respective behaviour towards pollutant diffusion, for example, has consequences on diffusion modelling: turbulent processes induce transport of air, while linear waves do not. However, both contribute to the variance of vertical velocity, which is a critical input parameter in diffusion models. Not discriminating between waves and turbulence can thus lead to an overestimation of the actual rates of diffusion [7].

Given these considerations, the aim of this paper is fourfold:

1. show that the methodology proposed by Collineau and Brunet ([4] and [5]) is also applicable over shorter plant canopies, where the 'signal' (i.e., organized turbulence) to 'noise' (i.e., random small-scale motions) ratio is much smaller;

2. show that organized turbulence in slightly unstable conditions exhibits the same structure above a forest and a maize crop, a scale factor apart;

3. extend the analysis to stable conditions;

4. propose a methodology for performing scale separation in measured time series, using the filtering capabilities of the wavelet transform.

For this, we use a micrometeorological data set acquired above a maize canopy. After a presentation of the site and the experimental procedures, we briefly describe the characteristics of mean flow and turbulence. We then recall the concept of wavelet variance (see [4] and [12]) and introduce the wavelet covariance. Both are used to determine characteristic duration scales of the energy-containing events in the time series. This enables an objective detection scheme to be designed, using the good localization in time of the wavelet transform. Conditional sampling is then performed on the time series, leading to a clear picture of the ejection-sweep process. Finally, we present a few preliminary results of wavelet decomposition of turbulent data into large and small scale components.

§2. Experimental Procedures and Flow Characteristics

The experiment was carried out over a maize crop at Grignon, France ($48°51'$N, $1°58'$E), from July to September 1990. The data presented here were obtained on August 13. At this date, the canopy height (evaluated as the mean height of a set of 100 individual plants) was $h = 1.55$ m, with a zero-plane displacement height $d = 1.15$ m (d is the apparent level of momentum absorption by the plants, used as the origin of heights in crop studies). The leaf area index (LAI, or total leaf area per unit soil surface) was 4.1. A meteorological mast was equipped with slow-response temperature (shielded, aspirated thermocouples) and windspeed (cup anemometers) sensors at 7 heights z ranging from $z/h = 0.3$ to $z/h = 4.6$. The distance from the leading edge of the field was larger than 200 m in the direction of prevailing winds. Two three-dimensional sonic anemometers-thermometers (model PAC-100, Dobbie Instruments), were also set up on the mast at heights $z/h = 0.6$ and 1.6. All three velocity components (u, streamwise; v, lateral; w, vertical) and air temperature T were sampled at 20 Hz, a rate high enough for exploring a significant part of the inertial subrange, and stored on a micro-computer.

Only the data acquired above the canopy are used here. Linear trends are first removed from all original time series, splitted into contiguous 30 min runs. In what follows, prime notations u', w' and T' stand for the fluctuations of u, w and T around their mean values $\overline{u}$, $\overline{w}$ and $\overline{T}$, calculated for each run.

Two sets of samples are used in this study, one ('diurnal') acquired from 13:30 to 15:30 UT and the other ('nocturnal') from 18:00 to 20:00 UT, just before and after sunset (which occurred at 19:10 UT). An illustration of the micrometeorological conditions encountered during these periods is given in Table 1. For daytime samples mean horizontal windspeed is between 1.5 and 2 m s^{-1}, and later drops to about 1 m s^{-1}. In the first case the friction velocity $u_* = (-\overline{u'w'})^{1/2}$ is of the order of 0.4 m s^{-1}, whereas in the second case it drops from 0.35 m s^{-1} at 18:00 UT to about 0.1 m s^{-1} after 19:00. As a result of radiative cooling of the underlying surfaces, the sensible heat flux $H = \rho c_p \overline{w'T'}$ (where ρ is the air density and c_p the specific heat at constant pressure) is directed towards the canopy throughout the second period, with values between about -10 and -40 W m^{-2}. For diurnal samples H is positive, between 100 and 170 W m^{-2}. The values taken by $(z - d)/L$ (where L is the Monin-Obukhov length) indicate that slightly unstable (or near-neutral) conditions are prevailing in the afternoon, whereas light to moderate stability occurs from 18:00 to 20:00 UT.

Table I

Summary of experimental conditions: statistical moments of u, w and T (means, standard deviations, covariances) and stability parameter $(z - d)/L$.

Time	$\overline{u}$ (m/s)	$\overline{T}$ (°C)	σ_u (m/s)	σ_w (m/s)	σ_T (K)	$\overline{u'w'}$ (m²/s²)	$\overline{w'T'}$ (K m/s)	$\frac{z-d}{L}$
13:30	1.482	28.68	0.984	0.497	0.765	-0.111	0.134	-0.064
14:00	1.706	28.44	0.868	0.517	0.616	-0.139	0.113	-0.038
14:30	1.904	28.40	0.994	0.598	0.597	-0.193	0.116	-0.024
15:00	1.776	28.19	0.815	0.529	0.507	-0.170	0.080	-0.020
18:00	1.660	25.90	0.786	0.444	0.203	-0.113	-0.027	0.013
18:30	0.912	24.20	0.399	0.218	0.422	-0.030	-0.030	0.101
19:00	0.851	21.76	0.179	0.106	0.346	-0.006	-0.010	0.404
19:30	0.992	20.84	0.220	0.137	0.353	-0.010	-0.014	0.230

Figure 1 shows typical 200 s samples of instantaneous fluctuations, extracted from the two data sets. All time series exhibit some degree of intermittency, with sporadic active periods particularly visible in the traces of instantaneous cross-products. Closer inspection of Figure 1b (night) reveals typical recurrent features characterized by sharp increases in temperature, associated with positive (respectively negative) fluctuations in u (respectively w). These temperature patterns are commonly called 'inversed ramps' by analogy with their daytime counterpart (slow increases followed by sudden drops), readily visible in Figure 1a. These temperature discontinuities can be unambiguously identified as the signatures of turbulent coherent motions sweeping down into the canopy air space. In stable conditions, this results in warmer air being imported from aloft by energetic gusts, yielding a sudden increase in time series of temperature.

From this example, it therefore seems that in nightime and daytime

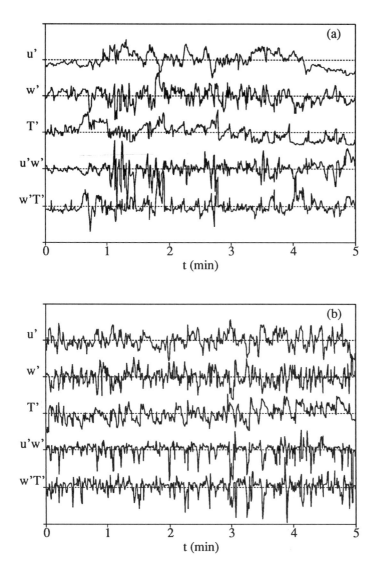

Figure 1. Typical examples of time series of velocity components, air temperature and cross-products $u'w'$ and $w'T'$ above the maize crop ($z/h = 1.25$) on August 13. (a): diurnal sample, (b): nocturnal sample. For the clarity of these graphs the original signals (recorded at 20 Hz) have been low-pass filtered at 2 Hz.

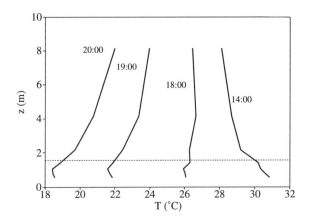

Figure 2. Mean air temperature profiles at various times, within and above the canopy (the horizontal dashed line represents the canopy top).

conditions momentum and sensible heat fluxes are transferred by the same dynamic process, the only difference (sharp rises or drops in temperature) being due to the inversion of the mean temperature profiles (Figure 2). No regular wavelike oscillations are visually detectable in the nocturnal time series. However, this may not always be the case and further analysis is desirable, in order to investigate the possible existence of gravity waves. For this, we need advanced data processing tools, such as those provided by the wavelet transform.

§3. Wavelet Variance and Covariance

In this paper, dilation over the non-dimensional scale a and translation at b of the analyzing wavelet ψ are performed with the following normalization factor (preserving the L_1-norm):

$$\psi^{(a,b)}(t) = \frac{1}{a}\psi(\frac{t-b}{a}). \tag{1}$$

We use only the continuous version of the wavelet transform of a function f, noted $\langle f, \psi^{(a,b)} \rangle$ and defined as:

$$\langle f, \psi^{(a,b)} \rangle = \int_{-\infty}^{+\infty} f(t)\psi^{(a,b)}\, dt. \tag{2}$$

Let f be a given experimental time series with energy $E_f = \langle f, f \rangle$. We define the wavelet variance WV_f of f for the analyzing wavelet ψ as:

$$WV_f(a) = \int_{-\infty}^{+\infty} |\langle f, \psi^{(a,b)} \rangle|^2 \, db. \tag{3}$$

Because energy is conserved by wavelet transformation, the area under a wavelet variance is proportional to the signal energy, when appropriate logarithmic axes are used (see [4]). Due to the 'built-in' smoothing process of the wavelet variance (integration over the translation parameter), ensemble averaging is not usually required to obtain stable, readable variance graphs.

Collineau and Brunet [4] showed that when a wavelet variance graph exhibits a single peak at a scale $\hat{a}$, a characteristic *duration scale* $\hat{D}$ can be defined as:

$$\hat{D} = \hat{a} D_\psi, \tag{4}$$

where D_ψ is a *duration constant* which was shown to be an intrinsic property of the wavelet ψ. D_ψ can be calculated as the solution of a simple, first-order differential equation, which was done analytically for a few common wavelets (Haar, Mexican hat, first derivative of a Gaussian function...). It was demonstrated that $\hat{D}$ can be interpreted as the mean duration of the elementary events contributing most to the signal energy ($\hat{D} = \tau/2$ for a sine function of period τ).

The wavelet transform conserves not only the energy $\langle f, f \rangle$, but also the inner product $\langle f, g \rangle$ of two functions [6], since:

$$\langle f, g \rangle = \frac{1}{C_\psi} \int_{-\infty}^{+\infty} \int_0^{+\infty} \frac{1}{a} \langle f, \psi^{(a,b)} \rangle \langle g, \psi^{(a,b)} \rangle \, da \, db, \tag{5}$$

where

$$C_\psi = 2\pi \int_{-\infty}^{+\infty} |\Psi(\omega)|^2 \, \frac{d\omega}{\omega} \tag{6}$$

has a finite value, by definition of a wavelet (Ψ being the Fourier transform of ψ). Just as a wavelet variance can be considered as the wavelet equivalent of a Fourier spectrum, one can define a *wavelet covariance* $WC_{fg}(a)$ from Equations (2), (3) and (5), the wavelet equivalent of a cospectrum:

$$WC_{fg}(a) = \int_{-\infty}^{+\infty} \langle f, \psi^{(a,b)} \rangle \langle g, \psi^{(a,b)} \rangle \, db. \tag{7}$$

To our knowledge, this property has not been used yet on experimental data sets[1].

---

[1] After this work was completed, it was brought to the authors' knowledge that the concept of wavelet covariance has been independently described by L. H. Hudgins (Wavelet Analysis of Atmospheric Turbulence, Thesis, University of California, Irvine, March 1992).

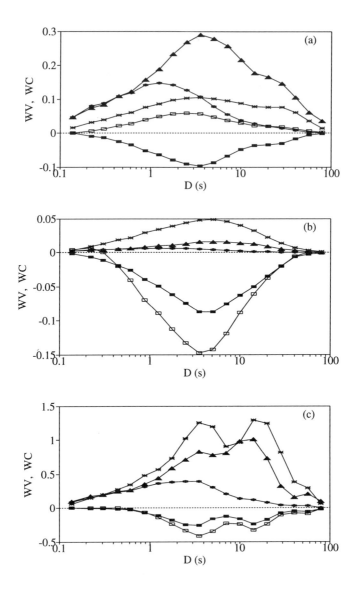

Figure 3. Wavelet variances and covariances: WV_u (triangles), WV_w (plus), WV_T (crosses), WC_{uw} (full squares) and WC_{wT} (open squares). (a): 14:00-15:00 UT, (b): 19:00-20:00 UT, (c): 18:00-18:30 UT.

We have calculated the wavelet variances of u, w and T, and the wavelet covariances of (u, w) and (w, T), for all the available data samples. The Mexican hat wavelet was chosen for this, because of its good localization in frequency. Figure 3a shows the variances and covariances obtained from 14:00 to 15:00 UT, under slightly unstable conditions.

Single, well-defined peaks can be seen on the wavelet variance and covariance curves. The wavelet variance for vertical wind velocity WV_w peaks at $\hat{D}_w \approx 1$–2 s, whereas WV_u and WC_T peak at $\hat{D}_u \approx \hat{D}_T \approx 3$–4 s. From a qualitative point of view, these results are very similar to those obtained by Collineau and Brunet [5] over a pine forest in similar stability conditions. Only the magnitudes of the $\hat{D}$-scales differ: in the latter case typical values of 5–7 s and 10–15 s respectively were obtained for the w-variance peaks, and the peaks for u and T. However, normalizing the duration scales $\hat{D}_u$ or $\hat{D}_T$ by u_* and h as suggested in the introduction give the same mean value $\hat{D}u_*/h \approx 0.85$ for both canopies. It therefore seems that over the maize crop the motions contributing to these peaks are also typical canopy-scale eddies, of the same type as those depicted in [5]. This will be confirmed further.

The covariances WC_{uw} and WC_{wT} also exhibit unambiguous peaks, at the same $\hat{D}$ scales as WV_u and WV_T (Figure 3a). Instead of using WC_{uw} and WC_{wT}, Collineau and Brunet [5] calculated the wavelet variance of series of instantaneous cross-products $u'w'$ and $w'T'$. They were shown to display a peak at the same scale as WV_w. This is very similar to what happens with Fourier spectra and cospectra in the surface layer: the cospectra peak at the same frequency as the u and T spectra (see [13] for a review), whereas the spectra of the time series of cross-products such as $u'w'$ and $w'T'$ peak at the same frequency as the w spectra.

Figure 3b shows the wavelet variances and covariances obtained from 19:00 to 20:00 UT, under conditions of moderate stability. The wT co-variance is now negative, following the inversion in sensible heat flux, but all curves are qualitatively similar to those in Figure 3a, with identical $\hat{D}$ scales. The variances of u and w appear flatter than in Figure 3a but this is just a consequence of a change in the respective magnitude of the wavelet variances and covariances. Full-scale graphs of WV_u and WV_w indeed show well-marked peaks at the same $\hat{D}$-scales as in Figure 3a. This confirms the visual impression given by Figure 1b, that the same kind of structures as those commonly seen in diurnal time traces are also present in this noctur-nal sample. The *normalized* peak scales $\hat{D}u_*/h$ are now about 3–4 times larger than in Figure 3a since u_* has dropped from 0.3–0.4 m s^{-1} to about 0.1 m s^{-1}. Using mean wind speed as a velocity scale would lead to a smaller difference since $\bar{u}$ has only dropped by a factor of 2.

Considering now the period from 18:00 to 18:30 UT (Figure 3c), two peaks are apparent on all curves but WV_w. Consequently, there seems to be two superimposed types of motions in this data sample, which both contribute to the signal energy. The motions responsible for the left-hand side peak are probably of the same nature as those visible in Figures 3a and 3b, since here also $\hat{D} \approx 3.5$ s. The second peak is located at $\hat{D} \approx 13$ s and its nature is yet unknown. We will focus on this in Section 5.

§4. Jump Detection and Conditional Sampling

Having determined a typical event duration, we are now interested in how frequently these events occur. Several methods for detecting sharp edges in digital time series, based on the use of localized transforms, have been proposed recently ([12] and [14]). The combined localization properties of wavelets in both time and frequency can be exploited together usefully for designing efficient jump-detection algorithms. Collineau and Brunet [5] showed that using $\langle T, \psi^{(\hat{a},b)} \rangle$ (or $\langle u, \psi^{(\hat{a},b)} \rangle$) as detection functions of ramps (or large excursions from the mean) in time series of temperature (or streamwise wind velocity), after determination of the scale $\hat{a}$ of the wavelet variance peak, leads to an accurate detection of the structures visible in the time series. The final decision step in the detection algorithm requires an empirical threshold for first derivative like wavelets (e.g., the Haar wavelet), but only a slope sign for the Mexican hat wavelet, which yields zero-crossings in $\langle T, \psi^{(\hat{a},b)} \rangle$ (or $\langle u, \psi^{(\hat{a},b)} \rangle$) whenever large, sharp rises or drops occur in the series. With a comparable reliability, the zero-crossing algorithm is therefore simpler than those based on thresholds, which is the case of most standard jump-detection algorithms used in turbulence, such as the Variable Interval Time-Averaging technique (see for example [21]). However, it has been made clear in [15] that detection through first or second derivative like wavelets are equivalent algorithmic problems when the objective is to classify the 'sharpness' of variations of the signal, which, in the latter case, requires thresholding the slope of the wavelet transform.

Here, detection was performed on all time series of streamwise wind velocity, using the Mexican hat wavelet. Wind velocity was chosen because it has a higher signal-to-noise (as defined in the introduction) ratio than temperature in this particular data set, and also because in this case the slope sign criterion does not depend on whether the data are diurnal or nocturnal. Figure 4 shows histograms of the time intervals Δ between the detected events, for two typical half-hour samples with similar values of u_*, in both diurnal and nocturnal conditions. The two distributions are strikingly similar and provide the same mean value $\overline{\Delta} \approx 9$ s. Normalizing $\overline{\Delta}$ as above, we obtain a mean value $\overline{\Delta} u_*/h \approx 2.1$ over the whole data set, which is again in fairly good agreement with the value of 1.8 obtained over the forest canopy already mentioned [5].

Finnigan ([9] and [10]) found that the peak frequency of the Fourier

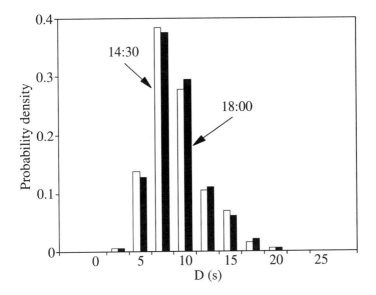

Figure 4. Histograms of the time intervals between the events detected in diurnal (14:30-15:00 UT) and nocturnal (18:00-18:30 UT) samples.

spectrum of u just above a wheat canopy was very close to the directly measured arrival frequency of canopy-scale sweeps. So, if the time scale $\overline{\Delta}$ really is a measure of the mean time interval between successive canopy-scale eddies, the inverse $1/\overline{\Delta}$ should be directly comparable to the peak frequency of the u spectrum. It should also be close to the peak frequency of the T spectra and uw cospectra, which were shown in [13] to have all the same normalized value $\hat{f}h/\overline{u}_h \approx 0.15$ (± 0.05), within and just above a plant canopy ($\overline{u}_h$ being the mean horizontal wind speed at canopy top). Fourier spectra and cospectra for the sample 14:00-15:00 UT are shown in Figures 5a and 5b, respectively. Except for the w spectrum, they all peak at a frequency $\hat{f} \approx 0.1$ Hz, which is in excellent agreement with $1/\overline{\Delta} = 0.11$ Hz. As the mean wind speed at the canopy top is $\overline{u}_h = 1.1$ m s^{-1} for this sample, we also find $\hat{f}h/\overline{u}_h \approx 0.14$, well in the range of expected values.

Having identified the relevant patterns, we can perform conditional sampling and averaging in order to obtain the mean 'signature' of the detected motions. Collineau and Brunet [5] showed that conditional averaging requires the use of a wavelet very well localized in time, which is not the

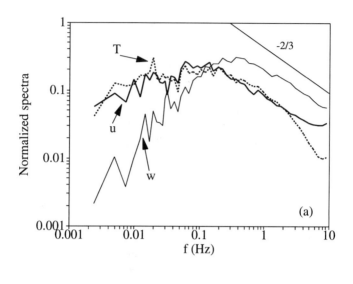

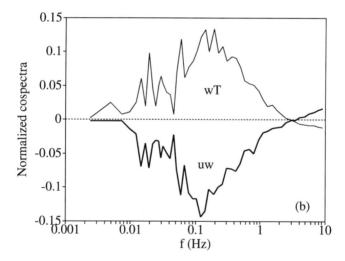

Figure 5. *u*-, *w*- and *T*- Fourier spectra (a), and *uw*- and *wT*-cospectra (b) for the sample 14:00-15:00 UT. Spectral densities are multiplied by the frequency and divided by the signal variance.

case of the Mexican hat. Instead, we use the 'Ramp' wavelet defined as:

$$\psi(x) = \begin{cases} 0 & -\infty < x < -0.5 \\ 2x + 1 & -0.5 \leq x < 0 \\ 2x - 1 & 0 \leq x < 0.5 \\ 0 & 0.5 \leq x < +\infty \end{cases}$$

which was shown by these authors to lead to slightly better results than the Haar wavelet, because its shape matches more closely the signature of the turbulent structures (see Figure 6). Since this type of wavelet exhibits a peak at jump times of the input signal, we had to calibrate the required detection threshold. This was done so that the number of detected events matches the number of events seen by the Mexican hat in the same time series. For this exercise the detection was performed on the temperature signal as in [5], but quite similar results would be obtained from the velocity signals.

For each turbulent variable f, we first define N time-windows of length $\overline{\Delta}$, centered at the N points detected in the temperature series. Then, for each variable (u, w, T), the raw signal in each window is normalized by the local mean $\overline{f_i}$ and the local standard deviation σ_{f_i}. Finally, the conditional (normalized) average $\{f(t)\}$ is calculated over the N windows as:

$$\{f(t)\} = \frac{1}{N} \sum_{i=1}^{N} \frac{f(t + \tau_i) - \overline{f_i}}{\sigma_{f_i}} \tag{8}$$

with $\tau_i - \overline{\Delta}/2 \leq t \leq \tau_i + \overline{\Delta}/2$, τ_i being the ith detection time.

The averaged patterns are presented in Figures 6a (diurnal samples) and 6b (nocturnal samples). Both figures show a characteristic slow ($\{u\} < 0$) upward ($\{w\} > 0$) movement of air, rapidly switching to a strong downward ($\{w\} < 0$) motion associated with an acceleration of horizontal velocity ($\{u\} > 0$). In the diurnal case a sharp temperature drop follows a slow rise, whereas the opposite occurs in the nocturnal case. Such patterns have the characteristics of typical 'ejection-sweep' processes. It has to be emphasized that detection was performed on time series of temperature *only*, which suggests a strong coupling between temperature, vertical and streamwise velocities.

These results invite several comments. Firstly, they show that above our maize crop the motions responsible for the peaks in wavelet variances and covariances, and objectively detected by the wavelet transform, do have the 'signatures' of coherent eddies, such as those observed over a variety of plant canopies or rough surfaces ([18], [19], and [20]). Secondly, the normalized patterns are very similar to those obtained by the same technique above a forest canopy ([5]), with $\{T\}$ peaking between ± 0.6 and ± 0.8, and $\{u\}$ and $\{w\}$ having equally smaller amplitudes, both comprised between

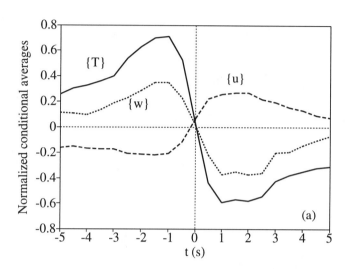

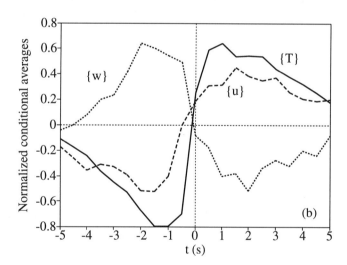

Figure 6. Normalized conditional averages $\{u\}$, $\{w\}$ and $\{T\}$, as defined in Equation (8), for the diurnal (a) and nocturnal (b) series.

about ± 0.2 and ± 0.5. Thirdly, apart from the difference in the temperature signature due to the opposite sign of the mean vertical temperature gradient, turbulence seems to be dominated by the same processes in our daytime and nighttime samples. However, this should not be generalized since a relatively small stability range has been explored ($(z-d)/L$ between -0.064 and 0.404, as shown in Table 1).

We noticed in Figure 3c that just after the inversion in the mean temperature gradient, another peak was visible in the wavelet variances and covariances, corresponding to longer duration scales. As gravity waves have already been observed in similar conditions [17], one may reasonably wonder whether those peaks could be due to such phenomena. This is the objective of the next section.

§5. Wavelet Decomposition

The occurrence of linear gravity waves can be predicted theoretically by a linear stability analysis of the dynamical flow equations, as the result of Kelvin-Helmoltz instabilities (see [7] and [8]). This analysis provides a characteristic wave frequency depending on the mean vertical temperature gradient. This is the Brunt-Vaisala frequency $\mathbf{N}$, defined as:

$$\mathbf{N}^2 = \frac{g}{\overline{T}} \frac{\partial \overline{T}}{\partial z}, \qquad (9)$$

where g is the acceleration due to gravity (see for example [22]). As the mean temperature gradient at the height of the sonic anemometer is of the order of 0.3 K m^{-1} in the period 18:00-18:30, the Brunt-Vaisala frequency is $\mathbf{N} \approx 0.1$ Hz (which corresponds to a characteristic time scale of about 10 s). On the other hand, the $\hat{D}$-scale corresponding to the second peak in Figure 3c is 13 s; as the wavelet variance of a sine function peaks at a frequency corresponding to half the period ([5]), the associated time scale would be of 26 s, a value more than twice as large as the former.

The linear stability analysis also predicts a phase shift of $\pm\pi/2$ between w and T [22]. In fact, most experimental approaches for studying gravity waves have relied on Fourier analysis, performed on time series of streamwise and vertical velocity, pressure and temperature. Cospectra, variance, phase, quadrature and coherence spectra have been shown to exhibit specific features in the presence of gravity waves ([2], [7], and [16]). However this is not always the case, for several reasons: (i) there is not always a clear separation in the power spectra between the regions of waves and turbulence, especially for w near the ground [16]; (ii) factors such as vertical wind shear, advection or proximity to the surface may complicate the situation and, for instance, perturbate the phase angles and decrease the levels of coherence [3]; (iii) non-stationary waves cannot be easily detected with Fourier-based methods if they only last for a few oscillations.

In our case, no specific feature could be visually detected in the various spectra and cospectra. Nevertheless, it is still desirable to investigate the nature of the second peak in the variances and covariances. For this, the combined time- and frequency-localization properties of the wavelet transform provide powerful tools.

From the definition of the wavelet transform, one can define an *inverse* wavelet transform as:

$$f(t) = \frac{1}{C_\psi} \int_{-\infty}^{+\infty} \int_0^{+\infty} \frac{1}{a} \langle f, \psi^{(a,b)} \rangle \psi^{(a,b)}(t) \, da \, db, \tag{10}$$

by which the original signal can be retrieved from the whole set of wavelet coefficients $\langle f, \psi^{(a,b)} \rangle$. This 'reconstruction' formula enables one to filter the original function, by splitting the integration domain into specific ranges of a scales. For instance, the wavelet transform can be used as a low-pass, high-pass or band-pass filter, provided that adequate cut-off scales are imposed in Equation (10).

In the present case, decomposition of the signal into two components can be simply performed by splitting the reconstruction integral over a into large and small a scales. If the wavelet variance of the signal exhibits two peaks separated by a gap (the wavelet equivalent of a spectral gap), an obvious choice for the cut-off scale a_c is the gap scale. A textbook example, using the sum of two sine functions with different periods, was given in [4]: the reconstructed small and large components of the signal allowed the authors to retrieve accurately the respective original sine functions.

We applied this method to the series acquired from 18:00 to 18:30 UT, after they were band-averaged at 0.5 s in order to facilitate the computation. The Mexican hat wavelet was chosen for its good frequency localization. A cut-off duration scale $D_c = 7$ s was inferred from Figure 3c, corresponding, for this particular wavelet, to a cut-off dilation scale $a_c = D_c/D_\psi = D_c\sqrt{2}/\pi \approx 3.15$. Figure 7 presents a small sample of the extracted small-scale components u_s, w_s and T_s of the respective input signals. Roughly speaking, what we see here is a representation of the original series after low-pass filtering at $f_l = 1/0.5 = 2$ Hz and high-pass filtering at $f_h = 1/D_c = 0.14$ Hz. All traces exhibit pseudo-periodic oscillations, sometimes combined into two characteristic types of events: one set consists in slow, upward movements of cold air and the other in fast, downward motions of warm air, each individual event lasting for about 3 s. These patterns correspond respectively to the ejection and sweep processes described above. As they occur repeatedly all over the half-hour sample with the appropriate time scales, it is likely that most of them are signatures of the above-mentioned coherent structures, even if each individual pattern is not a replicate of the *averaged* patterns showed in Figure 6b. The large-scale components u_l, w_l and T_l shown in Figure 8 exhibit similar

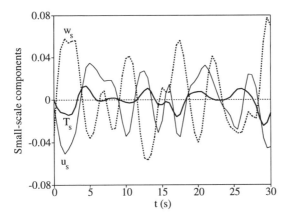

Figure 7. A sample of small-scale components (u_s, w_s and T_s) extracted by the inverse wavelet transform (from the run 18:00-18:30 UT).

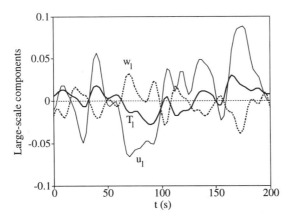

Figure 8. A sample of large-scale components (u_l, w_l and T_l) extracted by the inverse wavelet transform (from the run 18:00-18:30 UT).

features, at larger time scales. It is now clear that the underlying motions are not gravity waves. In particular, w_l and T_l are not out of phase, but in phase opposition (just as w_s and T_s). The scale difference between the two types of components is illustrated in Figure 9a and 9b for vertical velocity and temperature, respectively. The two large-scale components are strongly linked. Both exhibit wavelike motions, with a periodicity of order 20–25 s, e.g. twice the duration scale associated with the second peak.

This scale separation is only based upon scale criteria. However, as all large-scale components show repeatable features, they are not likely to be due to unpredictable low-frequency trends, but rather result from a deterministic, dynamic process, perhaps linked with the establishment of the thermal inversion. At the present stage, the reasons for these features are unclear. More data collected during such transition regimes must be analyzed, as well as data acquired in much more stable conditions.

Nevertheless, this analysis emphasizes the filtering capabilities of the wavelet transform, which appears to have great potential for extracting specific components and investigating phase aspects.

§6. Summary and Conclusions

The results of this experimental study confirm the importance of coherent structures in turbulent transfer processes between plant canopies and the atmosphere. The features observed in time series of velocity components and air temperature recorded above a maize crop are very similar to those observed above other plant canopies, and quite consistent with the picture of canopy turbulence acquired over the past few years. Wavelet analysis has enabled us to detect characteristic signatures of coherent motions in all series, revealing the occurrence of ejection-sweep processes. The scale of these motions is canopy-dependent and a comparison between the pine forest and the maize crop suggests that the occurrence frequency of these motions scales with friction velocity and canopy height. In both cases a value of $\overline{\Delta}u_*/h \approx 2$ was obtained for the mean time interval between successive structures. Also, a unique value $\hat{D}u_*/h \approx 0.85$ was obtained for the duration scale corresponding to the peak in the wavelet variances of streamwise velocity and temperature. The wavelet covariances of uw and wT were also observed to peak at the same scale. These results support the postulate that 'universal' mechanisms are responsible for the structure of turbulence in the vicinity of plant canopies, in near-neutral conditions ([1], [11], and [19]).

Analysis of nighttime samples in light to moderate stability also proved the existence of similar processes. In moderate stability $((z - d)/L \approx 0.3)$ the time scale $\overline{\Delta}u_*/h$ was found to be somewhat larger, but this needs further confirmation since only one hour long sample was available for this

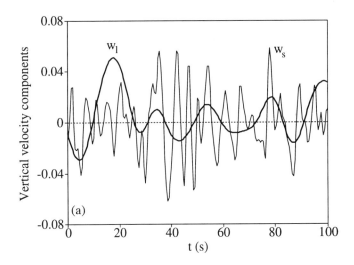

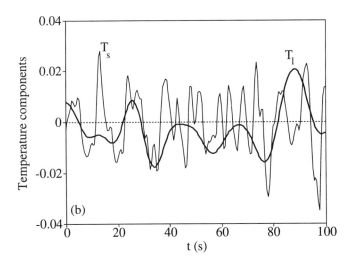

Figure 9. A sample of small- and large-scale components of vertical velocity (a) and temperature (b) (from the run 18:00-18:30 UT).

study. In lighter stability, around the inversion time of the mean vertical temperature gradient, another scale of motions was observed, for which no satisfactory explanation can be given at the present time.

These results were obtained by performing a wavelet analysis of turbulent time series, consisting in several steps. Firstly, computation of wavelet variances and covariances provide characteristic duration scales of the events contributing most to the signal energy. Secondly, the wavelet transform is used for detecting these events, by 'looking' at the series at these particular scales. As far as the detection itself is concerned, the use of a wavelet such as the Mexican hat, which yields zero-crossings whenever large discontinuities occur in the series, provides a simple detection scheme which does not require adjustment of an empirical threshold. Conditional averaging of the detected patterns can then be performed, enabling one to extract from the time series a clean picture of the signatures of these motions. Compared to more traditional analysis techniques, this procedure has the major advantage of being based upon a unique, self-consistent line of mathematical treatments, which rely on the combined time- and frequency-localization properties of the wavelet transform.

These properties have also been used for partitioning the original series into large- and small-scale components, using a cut-off scale defined as the gap scale between two peaks in the wavelet variances and covariances. This has promising applications, for instance in the study of wave-turbulence interactions.

References

1. Brunet, Y., J. J. Finnigan and M. R. Raupach, A wind tunnel study of air flow in waving wheat: single-point velocity statistics, *Boundary-Layer Meteorol.*, 1994, In press.
2. Caughey, S. J., Boundary-layer turbulence spectra in stable conditions, *Boundary-Layer Meteorol.*, 11, 3–14, 1977.
3. Caughey, S. J. and C. J. Readings, An observation of waves and turbulence in the earth's boundary layer, *Boundary-Layer Meteorol.*, 9, 279–296, 1975.
4. Collineau, S. and Y. Brunet, Detection of turbulent coherent motions in a forest canopy. Part I: Wavelet analysis, *Boundary-Layer Meteorol.*, 65, 357–379, 1993.
5. Collineau, S. and Y. Brunet, Detection of turbulent coherent motions in a forest canopy. Part II: Time-scales and conditional averages, *Boundary-Layer Meteorol.*, 66, 49–73, 1993.
6. Daubechies, I., The wavelet transform, time-frequency localization and signal analysis, *IEEE Trans. Information Theory*, 36, 961–1005, 1990.
7. De Baas, A. F. and A. G. M. Driedonks, Internal gravity waves in a

stably stratified boundary layer, *Boundary-Layer Meteorol.*, 31, 303–323, 1985.

8. Einaudi, F. and J. J. Finnigan, The interaction between an internal gravity wave and the planetary boundary layer. Part I: The linear analysis, *Quart. J. Roy. Meteorol. Soc.*, 107, 793–806, 1981.

9. Finnigan, J. J., Turbulence in waving wheat. Part I: Mean statistics and honami, *Boundary-Layer Meteorol.*, 16, 181–211, 1979.

10. Finnigan, J. J., Turbulence in waving wheat. Part II: Structure of momentum transfer, *Boundary-Layer Meteorol.*, 16, 181–211, 1979.

11. Finnigan, J. J. and Y. Brunet, Turbulent airflow in forests on flat and hilly terrain, in *Wind and wind-related damage to trees*, M. P. Coutts, J. Grace (eds.), Cambridge University Press, Cambridge, 1994, In press.

12. Gamage, N. K. K. and C. Hagelberg, Detection and analysis of microfronts and associated coherent events using localized transforms, *J. Atmos. Sci.*, 50, 750–756, 1993.

13. Kaimal, J. C. and J. J. Finnigan, *Atmospheric boundary layer flows. Their structure and measurement*, Oxford University Press, New York, 1993.

14. Mahrt, L., Eigen structure of eddies; continuous weighting versus conditional sampling, in *Non-linear variability in geophysics: scaling and fractals*, D. Schertzer, S. Lovejoy (eds.), Kluwer Academic Publishers, Dordrecht, 145–151, 1991.

15. Mallat, S. and S. Zhong, Characterization of signals from multiscale edge, *IEEE Trans. on Pattern Anal. and Mach. Intel.*, 14, 710–732, 1992.

16. Nai-Ping, Lu, W. D. Neff and J. C. Kaimal, Wave and turbulence structure in a disturbed nocturnal inversion, *Boundary-Layer Meteorol.*, 26, 141–155, 1983.

17. Paw U, K. T., R. H. Shaw and T. Maitani, Gravity waves, coherent structures and plant canopies, in *Reprints of the 9th Symposium on turbulence and Diffusion, American Meteorological Society*, Boston, 244–246, 1990.

18. Paw U, K. T., Y. Brunet, S. Collineau, R. H. Shaw, T. Maitani, Q. Jie and L. Hipps, On coherent structures and turbulence above and within agricultural plant canopies, *Agric. For. Meteorol.*, 61, 55–68, 1992. *Corrigendum, Agric. For. Meteorol.*, 63, 127, 1993.

19. Raupach, M. R., J. J. Finnigan and Y. Brunet, Coherent eddies in vegetation canopies, in *Proc. Fourth Australasian Conf. on Heat and Mass Transfer*, Christchurch, 75–90, 1989.

20. Raupach M. R., R. A. Antonia and S. Rajagopalan, Rough-wall turbulent boundary layers, *Appl. Mech. Rev.*, 44, 1–25, 1991.

21. Schols, J. L. J., The detection and measurement of turbulent structures in the atmospheric surface layer, *Boundary-Layer Meteorol.*, 29, 39–58,

1984.
22. Stull, R. B., *An introduction to boundary layer meteorology*, Kluwer
    Academic Publishers, Dordrecht, 1988.

Yves Brunet
Laboratoire de Bioclimatologie
INRA
BP 81
33883 Villenave d'Ornon Cédex
France
email: *brunet@bordeaux.inra.fr*

Serge Collineau
Laboratoire de Bioclimatologie
INRA
BP 81
33883 Villenave d'Ornon Cédex
France

Wavelet Spectrum Analysis
and Ocean Wind Waves

Paul C. Liu

Abstract. Wavelet spectrum analysis is applied to a set of measured ocean wind waves data collected during the 1990 SWADE (Surface Wave Dynamics Experiment) program. The results reveal significantly new and previously unexplored insights on wave grouping parameterizations, phase relations during wind wave growth, and detecting wave breaking characteristics. These insights are due to the nature of the wavelet transform that would not be immediately evident using a traditional Fourier transform approach.

§1. Introduction

Ever since Willard J. Pierson [18] adopted the works of John W. Tukey [22] and introduced the power spectrum analysis to ocean wave studies, Fourier spectrum analysis has been successfully and persistently used in data analysis of wind-generated ocean waves. Over the past four decades, with the increased availability of new instruments for measuring wind and waves, spectrum analysis has continued to be the fundamental standard procedure used for analyzing wind and wave data.

Fourier spectrum analysis generally provides frequency information about the energy content of measured, and presumed stationary, time-series data. Characteristic properties of waves such as total energy and dominant or average frequency can be readily derived from the estimated spectrum. This information, however, pertains only to the time span of the measured data. Changes and variations within a time series cannot be easily unraveled. As stationarity in the data simply represents a mathematical idealization, its validity is usually regarded as an approximation of the real wave field. The effectiveness of applying Fourier spectrum analysis to a rapidly changing wave field, such as during wave growth or decay, is

Wavelets in Geophysics
Efi Foufoula-Georgiou and Praveen Kumar (eds.), pp. 151–166.
ISBN 0-12-262850-0

uncertain. The emergence of wavelet transform analysis which can yield localized time-frequency information without requiring that the time-series be stationary has presented a rewarding and complementary approach to the traditional Fourier spectrum analysis and has advanced significant new perspectives for improved wave data analysis.

While wavelet analysis has been widely recognized as a revolutionary approach applicable to many fields of studies, the application of wavelet transform to wind wave data analysis is still in its infancy. This article mainly presents the author's own attempt at understanding wind-generated waves using the wavelet decomposition.

§2. Wavelet Spectrum

Following a standard formulation [3], we briefly summarize the wavelet transform. We start with a family of functions, the so-called analyzing wavelets, $\psi_{ab}(t)$, that are generated by dilations a and translations b from a mother wavelet, $\psi(t)$, as

$$\psi_{ab}(t) = \frac{1}{\sqrt{|a|}} \psi(\frac{t-b}{a}) \tag{1}$$

where $a > 0$, $-\infty < b < +\infty$, and $\int_{-\infty}^{+\infty} \psi(t)dt = 0$. The continuous wavelet transform of a time-series, $X(t)$, is then defined as the inner product of ψ_{ab} and X as

$$\tilde{X}(a,b) = <\psi_{ab}, X> = \frac{1}{\sqrt{|a|}} \int_{-\infty}^{+\infty} X(t)\psi^*(\frac{t-b}{a})dt \tag{2}$$

or equivalently in terms of their corresponding Fourier transforms

$$\tilde{X}(a,b) = \sqrt{|a|} \int_{-\infty}^{+\infty} \hat{X}(\omega)\hat{\psi}^*(a\omega)e^{ib\omega}d\omega \tag{3}$$

where an asterisk superscript indicates the complex conjugate. In essence the wavelet transform takes a one-dimensional function of time into a two-dimensional function of time and scale (or equivalently, frequency).

In practical applications, the wavelets can be conveniently discretized by setting $a = 2^s$ and $b = \tau 2^s$ in octaves [4] to obtain

$$\psi_{s\tau}(t) = 2^{-s/2}\psi(2^{-s}t - \tau), \tag{4}$$

where s and τ are integers. Then the continuous wavelet transforms (2) and (3) for time series data $X(t)$ become

$$\tilde{X}(s,\tau) = \frac{1}{\sqrt{2^s}} \int_{-\infty}^{+\infty} X(t)\psi^*(\frac{t}{2^s} - \tau)dt \tag{5}$$

and

$$\tilde{X}(s,\tau) = 2^{s/2} \int_{-\infty}^{+\infty} \hat{X}(2^s\omega)\hat{\psi}^*(\omega)e^{i\tau 2^s\omega}d\omega. \tag{6}$$

In general, the studies of wavelet transforms and wavelet analysis are centered on two basic questions [4]: (1) Do the wavelet coefficients completely characterize the time-series data? (2) Can the original time series be reconstructed from the wavelet coefficients? The answers to both of these questions are clearly yes as evidenced by the voluminous literature in recent years. In this paper we rely on the affirmative answer to the first question and concentrate on exploring the wavelet transform of measured wind waves. It is an exciting and fruitful area for practical application of the wavelet transform . As data analysis on wind wave studies comprises mainly of applications of statistics and Fourier transforms, the summary shown in Table 1 indicates that wavelet transform analysis is a logical extension to the currently available analyses.

In analogy with Fourier energy density spectrum, we can readily define a *wavelet spectrum* for a data series $X(t)$ as

$$W_X(s,\tau) = \tilde{X}(s,\tau)\tilde{X}^*(s,\tau) = |\tilde{X}(s,\tau)|^2. \tag{7}$$

There are other designations in the literature for $W_X(s,\tau)$ [21]. Results from the application of short-time Fourier transforms have been called spectrograms, whereas results from the application of wavelet transforms have been called scalograms. Since in practice the scale, s, and translation, τ, can be associated with a corresponding frequency, ω, and time, t, (7) can be considered as a representation of the time-varying, localized energy spectrum for a given time series.

We can similarly define a *cross wavelet spectrum* for the study of two simultaneously measured data sets $X(t)$ and $Y(t)$ as

$$W_{XY}(s,\tau) = \tilde{X}(s,\tau)\tilde{Y}^*(s,\tau) \tag{8}$$

and accordingly,

$$\Gamma(s,\tau) = \frac{W_{XY}(s,\tau)}{\sqrt{W_{X_k}(s,\tau)W_{Y_k}(s,\tau)}} \tag{9}$$

and

$$\Gamma^2(s,\tau) = \frac{[\Re W_{XY}(s,\tau)]^2 + [\Im W_{XY}(s,\tau)]^2}{W_{X_k}(s,\tau)W_{Y_k}(s,\tau)} \tag{10}$$

as the complex-valued wavelet coherency and its square, the real valued wavelet coherence, respectively, between the two data sets. The functions $\Re W_{XY}(s,\tau)$ and $\Im W_{XY}(s,\tau)$ in (10) are respectively the real and imaginary parts of $W_{XY}(s,\tau)$, and hence the co- and quadrature- wavelet spectra of $X(t)$ and $Y(t)$.

Table 1. Analogy of statistics, Fourier transform and wavelet transform analysis.

Statistics	Fourier Transform	Wavelet Transform
<u>Variance</u> $E(X^2)$	<u>Frequency Spectrum</u> $S_X(\omega) = \hat{X}\hat{X}^* = |\hat{X}|^2$	<u>Wavelet Spectrum</u> $W_X(s,\tau) = \tilde{X}\tilde{X}^* = |\tilde{X}|^2$
<u>Covariance</u> $E(XY)$	<u>Cross Spectrum</u> $S_{XY}(\omega) = \hat{X}\hat{Y}^*$	<u>Cross Wavelet Spectrum</u> $W_{XY}(s,\tau) = \tilde{X}\tilde{Y}^*$
<u>Coefficient of Correlation</u> $r = \dfrac{E(XY)}{\sqrt{E(X)E(Y)}}$	<u>Coherency</u> $\gamma = \dfrac{S_{XY}(\omega)}{\sqrt{S_X(\omega)S_Y(\omega)}}$	<u>Wavelet Coherency</u> $\Gamma = \dfrac{W_{XY}(s,\tau)}{\sqrt{W_X(s,\tau)W_Y(s,\tau)}}$
<u>Coefficient of Determination</u> $r^2 = \dfrac{[E(XY)]^2}{E(X)E(Y)}$	<u>Coherence</u> $\gamma^2 = \dfrac{|S_{XY}(\omega)|^2}{S_X(\omega)S_Y(\omega)}$	<u>Wavelet Coherence</u> $\Gamma^2 = \dfrac{|W_{XY}(s,\tau)|^2}{W_X(s,\tau)W_Y(s,\tau)}$

In implementing the applications, there are a number of well-defined continuous wavelet forms available [6]. In this study we choose to use the complex-valued, modulated Gaussian analyzing wavelet known as the Morlet wavelet. This wavelet, originally proposed by Morlet *et al.* [16], ushered in the present wavelet era and is given by

$$\psi(t) = e^{imt}e^{-t^2/2}. \qquad (11)$$

Its Fourier transform is

$$\hat{\psi}(\omega) = \sqrt{2\pi}e^{-(\omega-m)^2/2}. \qquad (12)$$

Here we should point out that this wavelet is not an admissible wavelet since a correction term is needed because $\hat{\psi}(0) \neq 0$. However, in practice choice of a large enough value for the parameter m, (*e.g.* $m > 5$) generally renders the correction term negligible. In this study, we follow Daubechies [5] and use $m = \pi\sqrt{2/ln2}$. While there are admissible wavelets available, the Morlet wavelet has been widely used in signal analysis and sound pattern studies. Aside from its convenient formulation and historical significance, its localized frequency is independent of time, a feature of particular advantage for wind wave studies.

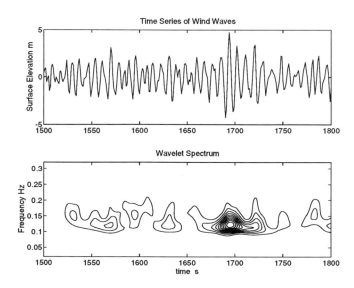

Figure 1. A sample plot of a time series of wind waves and its respective wavelet spectrum.

§3. Applications

In the following three subsections we present three wavelet transform analyses of wind wave data leading to distinct results that would be difficult, if not impossible, to obtain from the usual Fourier transform. The data used in the applications were measured during the recent SWADE (Surface Wave Dynamics Experiment) program [25]. The wind and wave data were recorded from a 3 m discus buoy during the severe storm of October 26, 1990. The buoy was located at latitude $38°22.1'$ N and longitude $73°38.9'$ W, with a water depth of 115 m near the edge of the continental shelf offshore of Virginia in the Atlantic Ocean. Time series of wind and waves were both recorded at 1 Hz from a combined design of a three-axis accelerometer and magnetometer along with the Datawell Hippy system. A total of 100 sets of data, each 1024 s in length, were used in the analyses. The data, predominantly wind-generated waves, covered the entire duration of the storm with wind speeds ranging from calm to 18 m/s and significant wave heights approaching 7 m.

3.1. Wave grouping effects

Wind waves appear in groups; i.e., higher waves occur successively in separated sequences. This phenomenon is well-known to seasoned sailors and can sometimes be seen in wind wave recordings. Apart from being a confirmed natural phenomenon, the existence of wave groups tends to challenge the conventional notion that wave data can be considered as a stationary process.

Wave data analysis, aimed at studying wave group characteristics, has been confined to identifying individual groups by counting the number of wave heights that exceed a prescribed height. A group is simply measured by a group length which is the number of waves counted. While statistics of the group lengths can be assessed, efforts have been generally directed at correlating the mean group length with spectral properties of the data [14].

In a wavelet transform analysis of wave data, an examination of a contour plot of resulting wavelet spectrum of waves shows distinct energy density parcels in the time-frequency domain. Figure 1 presents a simultaneous plot of sea surface elevations and the contours of their corresponding wavelet spectrum. The contour patches shown in the figure clearly indicate wave groupings that are visibly identifiable in the time series. The boundary of a wave group can be readily specified by setting an appropriate threshold energy level in the wavelet spectrum. Essentially there is a localized time-frequency energy spectrum for each group of waves, which is potentially more informative than previous approaches.

Based on the boundary specified for each wave group from the wavelet spectrum, we have at least four relevant group parameters to characterize a wave group:

(i) The group time length, t_g, which is the difference between the maximum and minimum time scales the group boundary covered.

(ii) The total group energy, E_g, which is an integration of the local wavelet spectrum over the time length t_g.

(iii) The dominant group frequency, f_p, which is the frequency of the peak energy over the time length t_g.

(iv) The dominant group wave height, h_p, which can be obtained from the time series as the maximum trough-to-crest wave height over the time length t_g.

The variability of these parameters indicates that wave groups are apparently diverse, irregular, nonperiodic, and independent from each other. The formidable task is to determine the significance and usefulness of these parameters. Here we consider a simplified approach of forming two normalized parameters:

- normalized group time length $= t_g * f_p$, and

- normalized total group energy $= E_g/h_p^2$.

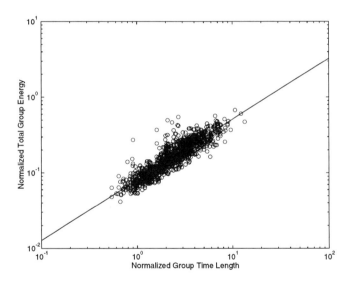

Figure 2. A scatter plot of normalized total group energy versus normalized group time length, the solid line is a linear least square fit for the data.

A scatter plot of these two normalized parameters, shown in Figure 2, indicates a fairly well-defined linear relationship. As the normalized total group energy is a measure of energy content, and the normalized group time length is a measure of the number of waves with possibly the same peak energy frequency, Figure 2 implies, that higher energy content in a wave group tends to generate more waves in the group. This interesting result, while intuitively understandable, is new.

A scatter plot of averages of dominant group wave heights versus significant wave heights is shown in Figure 3. The significant wave height, defined as the average of the highest one-third wave heights in the wave record, is a familiar and widely-used parameter. For practical applications, such as in engineering design, mean dominant group wave height would be more pertinent than the significant wave height. Figure 3 shows that significant wave heights are slightly less than the averages of dominant group wave heights.

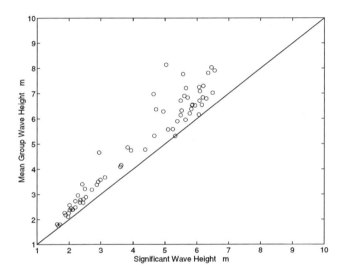

Figure 3.   A scatter plot of mean group wave height versus significant-wave height.

3.2. How do wind waves grow?

The wind and waves measurements in the SWADE program introduced a new data collection practice, namely, that wind and waves were measured at the same resolution simultaneously. Previously, wind data were merely collected as hourly averages. The availability of simultaneous high-resolution wind and wave data has provided an unparalleled opportunity to directly examine detailed wind action on waves, especially during wave growth.

How do wind waves grow? It is a question that several generations of scientists have addressed. In addition to the early work of Jeffreys [11] and Ursell's [23] famous "nothing very satisfying" summary, modern conceptual perceptions of wind waves primarily stem from the theoretical conjectures of Phillips [17], Miles [15], and Hasselmann [8]. The current proliferation of numerical wave models is basically developed from these early theories. Numerous measurements of wave energy spectra with average wind speeds have been conducted for the validation and possible enhancement of the available models. Now with the latest SWADE measurements and the advancement of wavelet transforms, we are able to examine wind wave processes from new perspectives.

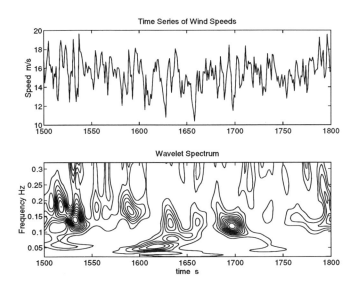

Figure 4. The time series of wind speeds corresponding to the wave data of Figure 1 and its respective wavelet spectrum.

One way of analyzing simultaneously recorded wind and wave measurements is through cross wavelet spectrum analysis. Figure 4 shows a part of the wind speeds and their wavelet spectrum corresponding to the wave data of Figure 1. There is no obvious relationship between the two time series that we can deduce from the top parts of Figures 1 and 4. However, if we consider the wavelet spectrum, a tract of high energy density contours appears in both spectra over the same frequency ranges and during the time when highest wave heights occurred in the wave time series. Qualitatively we might infer that wind and waves interact immediately during wave growth.

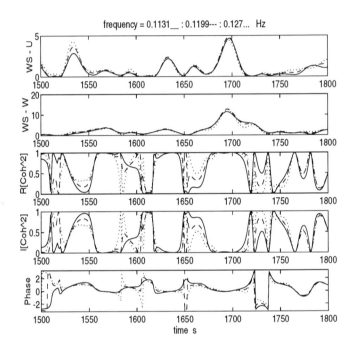

Figure 5. Plots of three peak-energy frequency components versus time. The five subgraphs from top down are, respectively, wavelet spectrum for wind speeds, wavelet spectrum for waves, the real part, the imaginary, and the phase of coherence.

To see if we can verify this inference quantitatively, we calculate the cross wavelet spectrum and their corresponding wavelet coherence for the simultaneous wind and wave data. The results, expressed either in contour or three-dimensional plots, are rather intricate and perplexing. It is not at all clear what we can meanfully deduce. If, however, we plot the results for individual frequencies, we can see some interesting results. Figure 5, corresponding to the same data of Figures 1 and 4, is an example of what these plots can tell us. The five separate graphs in Figure 5 display, respectively from top down, the wavelet spectrum for wind, the wavelet spectrum for waves, the real part, the imaginary part, and the phase of wavelet coherence. All of the plots contain the three frequency components of 0.1131, 0.1199, and 0.127 Hz for which the energy density is highest.

Note that in Figure 1 there are five groups of waves that can be identified from the wavelet spectrum. In the second graph of Figure 5 in which energy densities increase and decrease with respect to time, only three stronger groups (i.e., at time marks 1570, 1630, and 1695) are reflected from the fluctuations of these frequency components. The top graph of Figure 5 shows that the wavelet spectrum components for wind speeds exhibit similar, but more, energy fluctuations with time. Some of the fluctuations correspond closely to those of the waves. By examining the bottom three graphs of Figure 5, it shows quite clearly that for the three wave groups identified with appreciable energy contents, the real part of their coherence is close to[1] 1, their imaginary part close to 0, and their phase is also close to 0. Therefore, during wave growth, the frequency components for peak wave energy between wind and waves are inherently *in phase*. Wave groups constitute the basic elements of wind wave processes, and the wave growth are primarily taking place within the wave group.

As the growth of wind waves is an extremely complicated process, the above results contribute still qualitatively toward an understanding of the nature of how do waves grow. While we are accustomed to correlate wave growth with "average" wind speeds, the results presented here clearly show that waves are in fact responding to wind speeds instantly. Further detailed studies may challenge or counter more familiar notions of wind waves. Using cross wavelet spectrum analysis not only introduces new data analysis techniques, it may also leads to new courses of exploration.

3.3. Detecting breaking waves

Wave breaking is a familiar phenomenon that occurs intermittently and ubiquitously on the ocean surface. It is visible from the appearance of the whitecaps, yet it can not be readily measured with customary instruments. Wave breaking has been recognized as playing a crucial role in accurate estimations of the exchange of gases between the ocean and the atmosphere [24] and in the transfer of momentum from wind to the ocean surface [1]. Most of the practical works on wave breaking [2], both

---

[1] Unlike the Fourier cross spectrum analysis where averaging can be used to avoid coherence being identically one, we have to use the real and imaginary parts of the coherence separately here, since their sum, the coherence, is indeed identically one in this formulation. On the other hand, this approach successfully substantiates the use of the co- and quadrature spectra [19] to signify their "in phase" and "out of phase" properties, respectively.

in the laboratory and in the field, have been done with specialized methods based on radar reflectivity, optical contrast, or acoustic output of the ocean surface. Here we show that with the help of wavelet spectra [12], instead of using specialized measurement devices, a basic wave-breaking criterion can be easily implemented to wind wave time series to distinguish breaking from non-breaking waves. This simple and fairly efficient approach can be readily applied to indirectly estimate wave breaking statistics from any available time series of wind-generated waves.

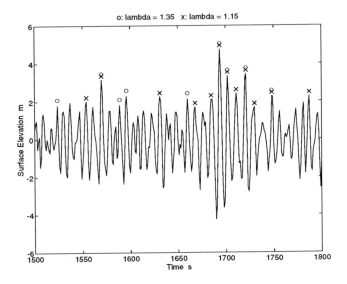

Figure 6. A sample plot of a time series of wind waves, same as Figure 1, with possible breaking waves marked by o's and x's indicating different high frequency ranges $\lambda \omega_p : \omega_n$ defined by the λ values given at the top of the figure.

One of the most frequently used approaches for the study of wave breaking is the use of a limiting value of the wave steepness beyond which the surface cannot be sustained [13]. Alternatively, assuming a linear dispersion relationship, the wave surface will break when its downward acceleration exceeds a limiting fraction, γ, of the gravitational acceleration, g, that is $a\sigma^2 \cong \gamma g$. The quantity $a\sigma^2$ can be calculated for a time series of wave data since the local wave amplitude, a, is available from the measured time series while the local wave frequency, σ, can be obtained from the wavelet spectrum. In classical studies, it has generally been assumed that $\gamma = 0.5$. Recent laboratory studies [10] have shown that γ is closer to 0.4. Some field measurements [9] further indicate that the value of γ should even be lower. In this study we chose to follow the laboratory results and use $\gamma = 0.4$.

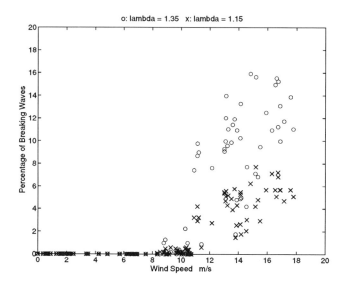

Figure 7. Plot of the percentage of breaking waves with respect to wind speeds. The o's and x's are the same as defined in Figure 6.

Since the wavelet transform provides an equivalent time-frequency spectrum, $W_X(\omega, t)$, for the wind wave time series, then there is a localized frequency spectrum at each data point, $X(t_i)$, given by

$$\Phi_i(\omega) = [W_X(\omega, t)]_{t=t_i}.$$

It is not immediately clear which frequency should be used for σ in calculating $a\sigma^2$. Because breaking events are generally associated with the high frequency part of the spectrum, for each $X(t_i)$ we chose to define a σ_i as the average frequency [20] over the high frequency range, $\lambda\omega_p : \omega_n$, of the localized spectrum at $t = t_i$ as

$$\sigma_i = \left[\frac{\int_{\lambda\omega_p}^{\omega_n} \omega^2 \Phi_i(\omega)d\omega}{\int_{\lambda\omega_p}^{\omega_n} \Phi_i(\omega)d\omega} \right]^{1/2}$$

where ω_p is the localized frequency at the energy peak, ω_n is the cut-off frequency, and λ is a number greater than 1 that denotes the start of the high frequency range beyond ω_p. The exact location of this high frequency range has not been clearly defined. Considering this range as corresponding to the familiar equilibrium range, one frequently used value of λ has been 1.35 [7].

To test this approach, Figure 6 presents an illustration of the analysis where estimated breaking waves are marked on the same time series segment given in figure 1. The x's and o's represent the results with a high frequency range between 1.15 and 1.35 times, respectively, of the local peak energy frequency, ω_p, and cut-off frequency, ω_n. While the λ values of 1.15 or 1.35 has been chosen rather arbitrarily for comparisons, they are clearly not always recognizing the same breaking waves. In general with the same cut-off frequency, the lower end of the frequency range farther away from the local peak frequency, i.e. large λ value, would yield higher local average frequency σ and more breaking waves. Therefore an exploration of breaking waves could potentially serve to resolve the definition of the well-known but still not yet well-defined equilibrium range. Figures 7 present plots of overall percentages of breaking waves from all the data analyzed in this study as a function of wind speed. While the data points are scattered considerably, there is an approximate linear trend indicating an increase in the percentage of breaking waves with an increase in wind speed. The results shown in Figure 7 are in general accord with various available observations [9]. According to these results, breaking waves become prevalent when wind speeds exceed 10 m/s.

At the present, the limiting fraction of downward wave acceleration from the gravitational acceleration, γ, and the parameter locating the local equilibrium range beyond local peak frequency, λ, are both tentative. Therefore, the wavelet transform approach that leads to these results is useful, convenient, and also exploratory. Perhaps a better simultaneous measurement of wind-wave time series and wave breaking would suffice to substantiate the approach. Unfortunately operational and sufficient instrument for this simple purpose is still lacking.

§4. Concluding Remarks

We anticipate two groups of readers who might be benefit from this paper, namely, those interested in wavelet applications and those interested in wind wave studies. As this is a first attempt in applying the wavelet transform to wind waves, the results are inevitably primitive. We hope we have succeeded at least in demonstrating the rich potentials for wavelet analysis. There are many important analysis issues that must yet be addressed, including rigorous basis for the cross wavelet spectrum analysis which we have used here. For the wind wave studies, wavelets certainly provide ample opportunities for data analysis. From the encouraging results we reported here, we are justified in being optimistic.

References

1. Agrawal, Y. C., E. A. Terray, M. A. Donelan, P. A. Hwang, A. J. Williams III, W. M. Drennan, K. K. Kahma, and S. A. Kitaigorodiski, Enhanced dissipation of kinetic energy beneath surface waves, *Nature*, **359**, 219–220, 1992.
2. Banner, M. L., and D. H. Peregrine, Wave breaking in deep water, *Annu. Rev. Fluid Mech.*, **25**, 373–397, 1993.
3. Combes, J. A., A. Grossmann, and Ph. Tchamitchian (Eds.), *Wavelets, Time-Frequency Methods and Phase Space*, 2nd ed. Springer-Verlag, 1989.
4. Daubechies, I., *Ten Lectures on Wavelets*, Society of Industrial and Applied Mathematics, 1992.
5. Daubechies, I., The wavelet transform, time-frequency localization and signal analysis, *IEEE Trans. Inform. Theory*, **36**, 961–1005, 1990.
6. Farge, M., Wavelet transforms and their applications to turbulence, *Annu. Rev. Fluid Mech.*, **24**, 395–457, 1992.
7. Gunther, H., W. Rosenthal, T. J. Weare, B. A. Worthington, K. Hasselmann, and J. A. Ewing, A hybrid parametrical wave prediction model, *J. Geophys. Res.*, **84**, 5727–5738, 1979.
8. Hasselmann, K., On the non-linear energy transfer in a gravity-wave spectrum, Part 1, General theory, *J. Fluid Mech.*, **12**, 481–500, 1962.
9. Holthuijsen, L. H., and T. H. C. Herbers, Statistics of breaking waves observed as whitecaps in the open sea, *J. Phys. Oceanogr.*, **16**, 290–297, 1986.
10. Hwang, P. A., D. Xu, and J. Wu, Breaking of wind-generated waves: measurements and characteristics, *J. Fluid Mech.*, **202**, 177–200, 1989.
11. Jeffreys, H., On the formation of water waves by wind, *Proc. Roy. Soc.*, bf A 107, 189–206, 1925.
12. Liu, P. C., Estimating breaking wave statistics from wind-wave time series data, *Annales Geophysicae*, **4**, 970–972, 1993.
13. Longuet-Higgins, M. S., On wave breaking and the equilibrium spectrum of wind -generating waves, *Proc. Roy. Soc.*, **A 310**, 151–159, 1969.
14. Masson, D. and P. Chandler, Wave groups, a closer look at spectral methods, *Coastal Engineering*, **20**, 249–275, 1993.
15. Miles, J. W., On the generation of surface waves by shear flows, *J. Fluid Mech.*, **3**, 185–204, 1957.
16. Morlet, J., G. Arens, I. Fourgeau, and D. Giard, Wave propagation and sampling theory, *Geophysics*, **47**, 203–236, 1982.

17. Phillips, O. M., On the generation of waves by turbulent wind, *J. Fluid Mech.*, **2**, 417–455, 1957.

18. Pierson, W. J., and W. Marks, The power spectrum analysis of ocean wave record, *Trans. Amer. Geophys. Un.*, **33**, 834–844, 1952.

19. Priestley, M. B., *Spectral Analysis and Time Series*, Academic Press, 1981.

20. Rice, S. O., Mathematical analysis of random noise, in *Noise and Stochastic Processes* (N. Wax, ed.), Dover, New York, 133–294, 1954.

21. Rioul, O., and F. Flandrin, Time-scale energy distribution: a general class extending wavelet transforms, *IEEE Trans. Signal Processing*, **40**, 1746–1757, 1992.

22. Tukey, J. W., The sampling theory of power spectrum estimates, *Symposium on Applications of Autocorrelation Analysis to Physical Problems*, NAVEXOS-0-735, Office of Naval Research, 1949.

23. Ursell, F., Wave generation by wind, in *Surveys in Mechanics*, Cambridge University Press, 216–249, 1956.

24. Wallace, D. W. R., and C. D. Wirick, Large air-sea gas fluxes associated with breaking waves, *Nature*, **356**, 694–696, 1992.

25. Weller, R. A., M. A. Donelan, M. G., Briscoe, and N. E. Huang, Riding the crest: A tale of two wave experiments, *Bulletin Am. Meteorol. Soc.*, **72**, 163–183, 1991.

This is GLERL Contribution no. 873. Partial financial support to initiate this study was provided by the U.S. Office of Naval Research. This study was also benefited from early collaborations with A. K. Liu and B. Chapron.

Paul C. Liu
NOAA Great Lakes Environmental Research Laboratory
Ann Arbor, Michigan
e-mail: *liu@glerl.noaa.gov*

Wavelet Analysis of Seafloor Bathymetry: An Example

Sarah A. Little

Abstract. 1-D wavelet analysis has been shown to be useful in studying bathymetric profiles [7]. 2-D bathymetric maps are less common than 1-D profiles, but offer immensely more information about seafloor generation processes. 2-D wavelet analysis is applied to swath-mapped bathymetric data from the Mid-Atlantic Ridge. Both image enhancement and feature identification are performed with excellent results in the identification of the location and scarp facing direction of ridge-parallel faulting. Wavelet image processing techniques enable computer analysis of distribution and spatial patterns in faults to be performed without the tedious job of transcribing hand picked and ruler-measured fault parameters from printed images to a digital data base.

§1. Introduction

The wavelet decomposition of bathymetric data reveals structures and patterns which are easily overlooked in the raw data. In addition, it can be used to isolate features of interest, such as fault scarps, for use in subsequent quantitative analysis. This paper describes the application of wavelet analysis to seafloor topography. Much of seafloor bathymetry has been collected by ships of opportunity traversing various sections of ocean. These depth measurements are closely spaced in the along-track direction, but the tracks are often many kilometers apart. The resultant data sets are essentially 1-dimensional spatial series. In a few areas of the seafloor, swath bathymetric surveys have been conducted which return detailed, 2-D, maps of limited sections of the ocean. 1-D spatial series are more common, and 1-D wavelet analysis of bathymetric profiles can be used to improve our understanding of the shape of the seafloor. Swath mapped areas have much more information than single topographic profiles, of course, and I present an example of image enhancement and fault scarp identification using 2-D wavelet analysis.

Wavelets in Geophysics
Efi Foufoula-Georgiou and Praveen Kumar (eds.), pp. 167–182.
Copyright 1994 by Academic Press, Inc.
ISBN 0-12-262850-0

§2. 1-D Topographic Profile

Little et al. [7] recently applied a wavelet decomposition to a 1-D, 1600 km long, bathymetric profile from the northeast Pacific near Hawaii. The profile runs parallel to the Murray Fracture Zone near 32°N, 152°W, and crosses directly over a small Pacific bathymetric high [9]. A 200 km anomalous zone was discovered with the wavelet decomposition, which has high-amplitude, long-wavelength topography and lacks the short-wavelength topography which dominates both to the east and west. This "low-frequency" zone is located between two fracture zones and the regional setting for this area includes a bend in the Pioneer Fracture Zone to the north and a break in the Murray Fracture Zone to the south. It is clear that in this topographically anomalous zone either the volcanism, tectonics or crustal structure is significantly different from the surroundings. They interpreted the anomalous crust to be the site of a short-lived, abandoned spreading center which is associated with a regional, thermally induced, topographic high.

The wavelet decomposition was particularly useful in finding the abandoned spreading center because these sites are marked by changes in the roughness of the topography [10] which remain permanently frozen in the crust after the spreading has ceased. The amplitude of the features of this particular site were smaller than other, known abandoned spreading centers, and these features are easily overlooked without the application of the wavelet transform to decompose the signal into events at different scales.

The wavelet transform can be used to decompose a spatial series into energy in a given frequency band and at a given location [4]. This can be contrasted to Fourier analysis which gives the energy in a given frequency band for an entire signal. Although windowed Fourier transforms (or spectrograms) do give an estimate of frequency content as a function of location, the spatial window is fixed, and the spatial resolution generally coarse. The advantage of the wavelet transform is that it matches its spatial localization to the frequency of interest—fine spatial resolution for high frequencies and broad spatial localization for low frequencies. In many cases, especially for transient and variable signals such as faults and volcanic constructions, the wavelet transform is superior to the Fourier transform for space-frequency decomposition and event localization.

For spatial data, such as a bathymetric profile, data are decomposed into energy at a given wavelet scale and at a given distance along track. In the case of the Morlet wavelet used by Little et. al. [7], the wavelet scale has a natural correspondence to spatial frequency.

The Morlet wavelet [5] has the properties of rapid decay in both the space and frequency domains, so as to maximally localize information in both domains, smooth so that it is narrow-band and differentiable, and complex so as to preserve phase information about the signal. The Morlet wavelet is defined as a complex exponential, at some fixed frequency, times a Gaussian with fixed variance:

$$w(t) = e^{(2\pi i k t)} e^{-\frac{t^2}{2\sigma}}$$

where t is space, k is wavenumber, and σ is the Gaussian variance.

A single scale wavelet transform is the convolution of the wavelet at that scale with the data, both of which are functions of the spatial domain. For increased computational speed, this convolution may be computed in the Fourier domain by multiplying the Fourier transform of the data by the Fourier transform of the wavelet, then taking the inverse transform of the product. On a single scale, the wavelet transform can be thought of as a linear filter. For a multiscale analysis, the wavelet shape is rescaled, in powers of 2, to longer and shorter lengths, creating a bank of filters of different sizes. The power in each size wavelet is normalized to one, and then the wavelet is convolved with the data. In this way the series of wavelet filters is used to scan for small- and large-scale events in the data.

When a wavelet of a given size and shape is convolved with a spatial series, the magnitude of the resultant series is a measure of the match between the wavelet and data series—small-scale events will match small wavelets but not large wavelets. Therefore, when a bank of wavelets is applied to a spatial series, one can both identify transient events and distinguish events on different scales. In this way one can scan a data series for interesting features on a broad range of scales, and also identify regions of the data where events of a certain scale are entirely missing. Little et. al. [7] used the Morlet wavelet transform, in this way, to discover the anomalous low-frequency zone in a 1-D bathymetric profile.

§3. 2-D Bathymetry

The following section describes the application of wavelets generated from spline functions to image enhancement and fault scarp identification in 2-D swath-mapped bathymetric data. The bathymetric data cover an approximately 100 km × 70 km section of Mid-Atlantic Ridge near 29°N, 43°W, which is about 1/10 of the data available in this region [13].

Topography near the Mid-Atlantic Ridge (MAR) is characterized by a deep central median valley surrounded by faulted blocks leading out over crestal mountains and down the outer flanks. These faulted blocks tend to be long linear ridges running parallel to the axis of spreading, ranging in width from order of 10 m to 10 km. Swath mapped areas of the MAR reveal scales ranging from order of 100 m up to 100 km, and hence offer a reasonable data set with which to examine this major faulting process at the MAR. Two major questions to ask about the MAR are 1) what does it look like? and 2) where are the faults located?

Two types of spline filters are used, a linear B-spline wavelet for visual image enhancement to improve intuitive understanding, and the derivative of a cubic spline for quantitative fault scarp edge detection.

3.1. Data

The data used in this analysis was collected by a 16-beam Sea Beam swath echo sounder. This instrument is mounted on the hull of a ship and sends out sound in beams whose individual footprints are $2\frac{2}{3} \times 2\frac{2}{3}$ degrees, which translates to about 150 m on a side in a water depth of 3000 m. As the ship steams forward, it continuously collects an approximately 2 km wide swath of depth readings. These data are tied to latitude and longitude via satellite navigation and put together in a regional map of bathymetry. There are places where the swaths do not overlap, and a linear interpolation has been performed in these gaps so that the data can be smoothly represented in a 2-D matrix of depth values. The data are shown in Figure 1 in a contour/gray scale image. There are 755 points of longitude between 43.567°W and 42.552°W, and 495 points of latitude between 28.581°N and 29.917°N. Depths range from 4000 m in the deepest areas (black) to 1800 m in the shallowest areas (white). There are two white areas inside the image which represent no data, a large one centered on longitude point number 200 and latitude point number 50, and a small gap located at longitude point number 240, latitude point number 100. The white areas to the left and right of the image also contain no bathymetric information. The central valley of the MAR runs up the middle of the image at about 20° off vertical. On each side of the central valley are the crestal mountains, and beyond them a small portion of the flanks. This area of the Mid-Atlantic Ridge contains a ridge offset, visible in the upper central part of the image, where the two major valleys are offset from one another. The data used in this paper have previously been analyzed for fault locations using a curvature method [14], [15].

3.2. Filter design

In two-dimensions, the wavelet transform can be used to decompose an image into a series of images, each of which contains information at a specific location of features at a single scale (or spatial frequency). Further, since images are 2-dimensional, each spatial frequency is coupled with an orientation. These two parameters can be varied to produce a filter tuned for features of nearly any width, length and orientation. The filter is a matrix of coefficients which is convolved with a 2-dimensional image. The resolution of location of a given feature in the wavelet transform is matched to the scale of that feature. Precision in location of fault scarps will be comparable to their width.

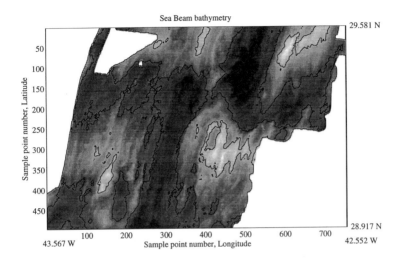

Figure 1. A contour/gray scale image of interpolated Sea Beam bathymetry between 42.552–43.567°W and 28.917–29.581°N, approximately an area of 100 km × 70 km. Depth ranges from approximately 4000 m (darkest areas) to 1200 m (lightest areas) and contours are shown in 560 m intervals. The white areas represent no data, and there are two no-data islands within the image: a large island centered at latitude point 50 and longitude point 200; and a small one centered at latitude point 100, longitude point 250. The nearly-vertical dark central valley runs from south to north and is offset to the east at latitude point number 150. The crestal mountains are the light gray areas to the east and west of the central valley before and after the offset zone.

The first step in the design of a suitable wavelet transform for image analysis is the selection of a proper wavelet which, when convolved with the data, will enhance features of interest. A wavelet needs to be a function which integrates to zero and is square integrable [3]; (see also Kumar and Foufoula-Georgiou, this volume). There are a number of wavelet classes in the literature which have been shown to meet these criteria and have been used in data analysis. These include the Morlet wavelet [11], the Haar wavelet [4], [6], the B-spline wavelets [1], [2], [4], [8], [16], and the Daubechies wavelets [3], [4].

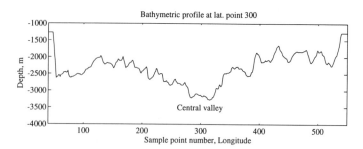

Figure 2. A bathymetric profile running from west to east across the image at latitude point 300 showing the profiles of ridges associated with faulting. The plateaus at the far left and right are zones of no data. From left to right, the profile comes up the flanks of the western crestal mountains, crosses over seven or so major ridges associated with faulting in the mountains, runs down to the central valley, up the eastern crestal mountains and across five or so major ridges.

In the 2-D bathymetry data under analysis, the features of interest are relatively long, linear and narrow ridges with a preferred orientation parallel to the axis of spreading. These fault blocks resemble straight lines with cross sections that look like ridges. An example of a profile across the MAR, perpendicular to the axis of spreading, is shown in Figure 2. The high plateaus at the extreme left and right are artifacts where no bathymetric data exist. Moving from left to right, one travels up the flanks, over the western crestal mountains, across approximately 7 major fault ridges (visible at this Sea Beam resolution), down into the central valley, up the other side and over the eastern crestal mountains, crossing approximately 5 major fault ridges, and finally reaches the edge of available data. The cross-section of these fault ridges suggests that an edge detector would be a good choice for a wavelet. For visual image enhancement it is desirable to "bring out" the ridges, so that they are easily seen by eye. For this, a wavelet which would transform a given size class of ridges into highs and valleys into lows would create an image where the ridges and valleys would visibly stand out. One such detector for this purpose is a linear B-spline wavelet [4], [2], [16]. For quantitative analysis, however, it is desirable to have a wavelet which will transform the location of the fault scarp—that is the transition from a low to a high (or vice versa) into a single high. The filter formed from the derivative of the cubic B-spline achieves this [1] and moreover, it can distinguish between a left facing edge and a right facing edge (in profile), as described below and in Figure 3. This is very useful for the analysis of fault distribution at the MAR because faults facing in towards the central valley have been observed to be structurally different from faults facing outward.

These two types of edge detecting filters will now be discussed.

3.3. Linear B-spline wavelet

Spline wavelets are constructed by a shifting, weighting and summing of the B-spline functions, $N_i(x)$, where x is the spatial variable, and i is the order of the spline. Details of the construction of the linear and other B-spline wavelets can be found in [2].

The linear B-spline function, $N_2(x)$, is defined as:

$$
N_2(x) = \begin{cases} 0 & x < 0 \\ x & 0 \le x < 1 \\ 2 - x & 1 \le x < 2 \\ 0 & x \ge 2. \end{cases}
$$

The linear B-spline wavelet, W_2, is formulated from N_2 as follows:

$$
W_2 = \frac{1}{12}[N_2(2x) - 6N_2(2x-1) + 10N_2(2x-2) - 6N_2(2x-3) + N_2(2x-4)].
$$

Figure 3a shows this wavelet (at 16 points) and the 1-D convolution of this wavelet with three, noise free, 1-D ridges of different sizes. Notice that, in the transform, each step edge is converted to a low and a high. The ridge whose width is most similar to the central portion of the wavelet is transformed into a single high with two lower amplitude lows on either side. This creates contrast which will visually "bring out" ridges.

The 2-dimensional filter is formulated to take advantage of the strongly linear and oriented nature of the faults. Each linear ridge is qualitatively like a long series of 1-D ridge-edges stacked together. Therefore, a 2-D filter can be built by stacking up a number of 1-D ridge-edge detecting wavelets. This is a departure from strict wavelet construction because the ridge-edge parallel direction of this 2-D filter is not a wavelet. The 2-dimensional version of the filter is specifically created in the following way: generate a 1-dimensional linear B-spline wavelet of a desired length; replicate this filter to create a 2-dimensional matrix of identical filters; taper the edge-parallel direction with a Hanning window [12]; and rotate the entire matrix to the desired orientation. A 3-D plot, contour/gray scale plot, and horizontal and vertical profiles are presented in Figure 4 for this filter. Figure 4e shows the filter rotated by -110 degrees from horizontal to match the dominant orientation of the faults in the bathymetry. The length of the 1-dimensional B-spline (BS) determines the width of the faults to be isolated. The number of identical filters (NF) in the 2-dimensional matrix determines the fault length over which to average (or smooth). If NF is large, then very long

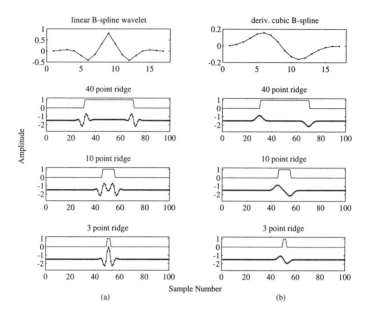

Figure 3. a) The top of the figure shows a 16 point linear B-spline wavelet. Below is shown this wavelet convolved with ridge-edges of three different sizes. The solid lines are the edges, and the solid-dot lines are the resultant convolution. High amplitude is associated with narrow ridges, and the edges of wide ridges. In addition, the high amplitude is preceded by a low amplitude, which enhances visual contrast at the edges. **b)** The top of the figure shows a 16 point derivative of a cubic B-spline function. Below this is shown this filter convolved with the same ridge-edges as shown in a). Notice that left-facing steps transform to highs, while right-facing steps transform to lows. The location of these edges is marked by the local maximum of the transform.

and straight faults will be identified; if NF is small, short and curved faults will be identified as well as long and straight. A small NF, although it picks up both short and long faults, also picks up short curved features which are volcanic rather than faults, and hence there is a trade-off between getting all the curved faults and getting too many short, non-fault features.

The 2-D linear B-spline filter, with BS=16 and NF=7 is convolved, via the 2-D Fourier transform, with the Sea Beam bathymetry and the results shown in gray scale in Figure 5. Light areas correspond to topographic ridges, and dark areas correspond loosely to valleys. The fabric of the

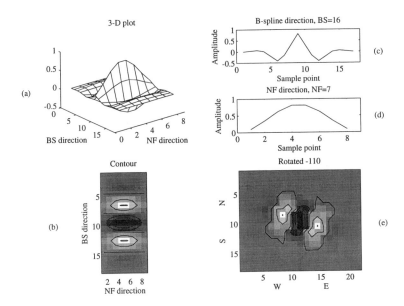

Figure 4. **a)** A 3-D mesh plot of the 2-D linear B-spline filter with BS=16 and NF=7. **b)** Contour/gray scale plot of this filter, dark is high. **c)** Profile of this filter in the B-spline direction. **d)** Profile of this filter in the NF direction. **e)** Contour/gray scale plot of this filter rotated by −110° to match the dominant orientation of MAR-parallel fault ridges.

topography is brought out visually in this image. Note that although this is a small scale wavelet, it is still possible to see large scale topographic trends, e.g. the central valleys and crestal mountains. This is due to the fact that the 2-D filter used was not exactly zero mean. This filter makes the image easier to study because most of the large scale topography has been removed, but just enough allowed through to visually locate small scale features in relation to the large trends such as the median valley and crestal mountains. For instance, approximately 7 large bright linear ridges are clearly on the left of the central valley, and fade out as they reach the offset zone at latitude point 150. This filter transforms the data to an image which helps the viewer gain a qualitative understanding of the topography.

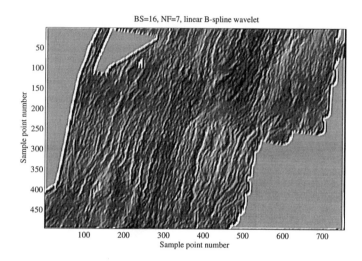

Figure 5. Application of filter shown in Figure 4e to bathymetric data shown in Figure 1. This is a filter for visually enhancing the image to improve intuitive understanding of the topography. The lighter areas correspond to bathymetric ridges. The large scale topography is still slightly visible, the dark central valley and lighter crestal mountains, due to the non-zero mean of the filter. This enhances visual interpretation of the image by enabling correlation between small scale ridges and larger features while removing most of the effects of the large scale topography.

3.4. Derivative of cubic B-spline

Quantitative analysis of fault location, spacing and distribution requires more than the intuitive information gained from image enhancement. The derivative of the cubic B-spline provides a filter which transforms a step up into a positive peak and a step down into a negative peak. This provides a mechanism for locating fault scarps, that is, the transitions from valley to ridge (or vice versa), and classifying their facing direction. The

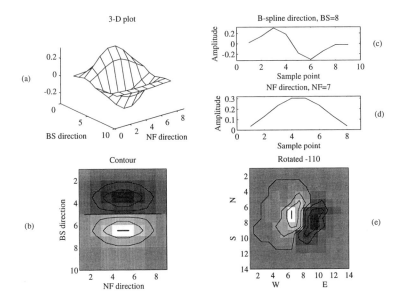

Figure 6. **a)** A 3-D mesh plot of the 2-D derivative of the cubic B-spline function with BS=8 and NF=7. **b)** Contour/gray scale plot of this filter, dark is high. **c)** Profile of this filter in the B-spline direction. **d)** profile of this filter in the NF direction. **e)** Contour/gray scale plot of this filter rotated by $-110°$ to match the dominant orientation of MAR-parallel fault scarps.

cubic B-spline function, $N_4(x)$, is defined as:

$$
N_4(x) = \begin{cases}
0 & x < 0 \\
\frac{1}{6}x^3 & 0 \leq x < 1 \\
-\frac{1}{3} + \frac{1}{2}x + \frac{1}{2}(x-1)^2 - \frac{1}{2}(x-1)^3 & 1 \leq x < 2 \\
\frac{2}{3} - (x-2)^2 + \frac{1}{2}(x-2)^3 & 2 \leq x < 3 \\
\frac{5}{3} - \frac{1}{2}x + \frac{1}{2}(x-3)^2 - \frac{1}{6}(x-3)^3 & 3 \leq x < 4 \\
0 & x \geq 4.
\end{cases}
$$

A sixteen point long, 1-D filter created from the derivative of this function is shown in Figure 3b. This filter is convolved with ridge edges of three different sizes to show how it transforms a rising slope to a numerical high, and a falling slope to a numerical low. In this way it locates fault scarps and identifies the direction they face.

The 2-D filter is created from the derivative of the cubic B-spline the same way as for the linear B-spline wavelet: Generate a 1-D filter of a de-

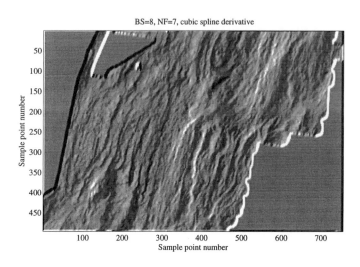

BS=8, NF=7, cubic spline derivative

Figure 7. A gray scale plot of the application of filter shown in Figure 6e to bathymetric data shown in Figure 1. Dark corresponds to right-facing fault scarps and light corresponds to left-facing fault scarps. This image does not correspond to illuminated bathymetry and apparent 3-D texture is misleading. The fault scarps predominantly face in toward the central valley, dark on the left and light on the left. The faults also fade out as they reach the offset zone.

sired length; replicate this filter to create a 2-dimensional matrix of identical filters; taper the edge-parallel direction with a Hanning window; and rotate the entire matrix to the desired orientation. The resulting filter is shown in Figure 6 in a 3-D plot, a contour/gray scale plot, and horizontal and vertical profiles. Figure 6e shows the filter rotated by -110 degrees from horizontal to match the dominant fault scarp orientation.

A BS=8 and NF=7, 2-D derivative of a cubic B-spline filter is now convolved with the bathymetry and the results, shown in Figure 7, show excellent fault scarp identification. Dark gray corresponds to faults which face easterly (to the right) and light gray corresponds to faults which face westerly (to the left). The apparent surface texture in this figure is some-what deceiving because it appears at a glance to be illuminated bathymetry. It is not, as there are no true shadows in which features are hidden and

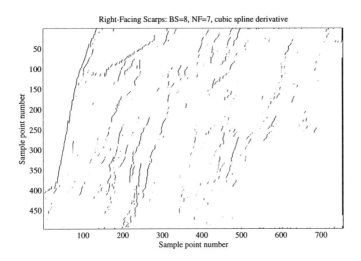

Figure 8. A plot of large-amplitude local minima (in horizontal rows) of the transform shown in Figure 7. The local minima correspond to the centers of right-facing fault scarps. There is a predominance of large right-facing faults on the west side of the central valley, both below and above the offset zone.

the 3-D effect is somewhat misleading. However, this data can be processed quantitatively. A plot of large-amplitude local minima (calculated in west-to-east lines) is shown in Figure 8. This figure shows the location of large faults which face easterly. There are more large east facing faults on the left-hand side of the central valley. Figure 9 shows large-amplitude local maxima, and clearly identifies more large west facing faults on the right-hand side. This implies that there are larger faults facing in toward the central valley than facing away. This characteristic of the MAR has been noted before from visual observations of bathymetric maps. The 2-D derivative filters offers a chance to quantify this qualitative observation. Figure 10 shows both left and right facing fault scarps, which gives a good picture of the distribution of faulting, and also shows that the faults fade out in the offset zone which runs from the upper left, latitude point number 150, to the lower right, latitude point number 200.

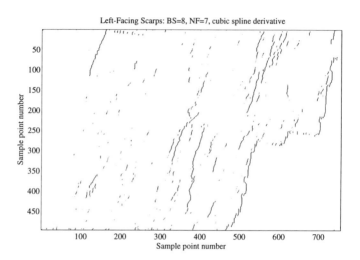

Figure 9. A plot of large-amplitude local maxima (in horizontal rows) of the transform shown in Figure 7. The local maxima correspond to the centers of left-facing fault scarps. There is a predominance of large left-facing faults on the east side of the central valley, both below and above the offset zone.

§4. Conclusions

Bathymetric data from the seafloor contain the superposition of ridges, valleys, and volcanos at many different scales. Wavelet analysis offers a useful method for decomposing the texture of the seafloor to help understand processes which occur at many different scales. Much of the bathymetric data available is in the form of 1-D spatial series. Scale decomposition of these series has been shown to yield an understanding of geophysical processes which originally were overlooked in the raw data. Smaller portions of the seafloor have been swath-mapped, and these 2-D data sets offer an unusually good opportunity to decipher the complexity of seafloor topography. This paper has presented two wavelet techniques for working with swath-mapped bathymetry, one a qualitative image enhancement, and the other a quantitative fault scarp identifier. A linear B-spline wavelet is used to design a 2-D filter which produces an image in which the linear, MAR-parallel fault ridges are easily seen in the context of the broader to-

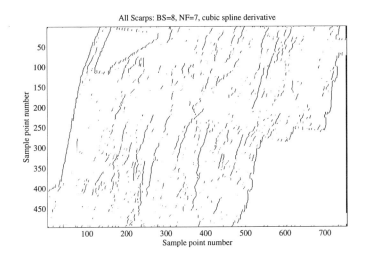

All Scarps: BS=8, NF=7, cubic spline derivative

Figure 10. A plot of both local minima and local maxima, showing the locations of fault scarps, regardless of facing direction. These data can be used for quantitative analysis of fault spacing, orientation and distribution.

pographic undulations. A second filter, designed from the derivative of a cubic B-spline, is used to obtain quantitative information on the locations and facing directions of fault scarps, that is, the zones where the crustal material was actually sheared. This quantitative information can be further processed to obtain fault spacing, preferred orientation, density and distribution, parameters which can be used to understand the underlying geophysical processes which occur at the MAR.

References

1. Canny, J., A computational approach to edge detection, *Trans. on Pattern Anal. and Machine Intelligence*, 8(6), 679-698, 1986.
2. Chui, C. K., *An Introduction to Wavelets*, Academic Press, New York, 1992.
3. Daubechies, I., Orthonormal bases of compactly supported wavelets, *Comm. Pure Appl. Math.*, 41, 909-996, 1988.
4. Daubechies, I., *Ten lectures on wavelets*, CBMS Regional Conference

Series in Applied Mathematics, SIAM, Philadelphia, PA, 1992.

5. Grossman, A. and J. Morlet, Decomposition of a Hardy functions into square integrable wavelets of constant shape, *SIAM J. Math. Anal.*, 15(4), 723-736, 1984.

6. Haar, A., Zur Theorie der orthogonalen Funktionen-Systeme, *Math. Ann.*, 69, 331-371, 1910.

7. Little, S. A., P. H. Carter, and D. K. Smith, Wavelet analysis of a bathymetric profile reveals anomalous crust, *Geophys. Res. Letts.*, 20(18), 1915-1918, 1993.

8. Mallat, S., and Zhong, S., Characterization of signals from multiscale edges, *IEEE Trans. on Pattern Anal. and Machine Intelligence*, 14(7), 710-732, 1992.

9. Mammerickx, J., Depth anomalies in the Pacific: Active, fossil, and precursor, *Earth Plant. Sci. Letts.*, 53, 147-157, 1981.

10. Mammerickx, J., and D. Sandwell, Rifting of old oceanic lithosphere, *J. Geophys. Res.*, 91(B2), 1975-1988, 1986.

11. Morlet, J., Sampling theory and wave propagation, in *Issues in acoustic signal/image processing and recognition*, C.H. Chen (ed.), *NATO ASI Series*, Vol. 1, Springer-Verlag, Berlin, 233-261, 1983.

12. Oppenheim, A. V., and R. W. Schafer, *Digital Signal Processing*, Prentice-Hall, 556, 1975.

13. Purdy, G. M., J.-C. Sempere, H. Schouten, D. DuBois, and R. Goldsmith, Bathymetry of the Mid-Atlantic Ridge, 24°–31°N: A map series, *Mar. Geophys. Res.*, 12, 247-252, 1990.

14. Shaw, P. R., Ridge segmentation, faulting, and crustal thickness in the Atlantic, *Nature*, 358, 490-494, 1992.

15. Shaw, P. R., and J. Lin, Causes and consequences of variations in faulting style at the Mid- Atlantic Ridge, *J. Geophys. Res.*, 98, 21,839–21,851, 1993.

16. Vinskus, M., *The discrete wavelet transform with applications to edge detection*, Master's thesis, Dept. Electrical and Computer Engineering, Worcester Polytechnic Institute, 1993.

This work was supported by grant N00014-93-1-0179 from the Office of Naval Research. WHOI contribution number 8617.

Sarah A. Little
Department of Geology and Geophysics
Woods Hole Oceanographic Institution
Woods Hole
MA 02543, USA
e-mail: *slittle@attractor.whoi.edu*

Analysis of High Resolution Marine Seismic Data Using the Wavelet Transform

Chris J. Pike

Abstract. A method is proposed for the analysis of high resolution acoustic signals using the wavelet transform. For signals with intensities exhibiting frequencies ranging from 100 Hz to 10 kHz, time-frequency displays of the signal can lead to more robust means of estimating attenuation as well as quantification of wavefield scatterers within shallow marine sediments. The time-frequency decompositions of the signals are accomplished using the wavelet transform and a Morlet analyzing wavelet. Zero offset acoustic signals are analyzed and the modulus of the wavelet transform is displayed as a function of depth below seabed versus wavelength. The results are discussed and related to the subseabed soil conditions at an experimental field site.

§1. Introduction

The Centre for Cold Ocean Resources Engineering (C-CORE) is an independently-funded research institute of Memorial University of Newfoundland in St. John's, Newfoundland, Canada. C-CORE is involved in solving engineering problems related to resource development in the ocean environment. The Centre has traditionally focused on a limited number of research areas related to ocean resources and has built up expertise reflected by the four research groups operating at C-CORE. They are the Remote Sensing group, the Geotechnical Engineering group, the Ice Engineering group and the Seabed Geophysics group.

One of the areas of research engaged by the Seabed Geophysics group has been the high resolution acoustic investigation of the sub-seabed. A central goal of the acoustic program is the development of methods for extracting marine soil properties by correlating acoustic properties derived from acoustic signals with the geotechnical properties of the soil. During the course of this research it was evident that the high frequency acoustic

Wavelets in Geophysics
Efi Foufoula-Georgiou and Praveen Kumar (eds.), pp. 183–211.

signals (500 Hz to 10,000 Hz) used were rapidly attenuated, especially the higher frequencies, in those soils that contained many scatterers – boulders and cobbles with dimensions comparable to the wavelengths of the acoustic waves generated by the source. Spectra for the acoustic signals are usually calculated by using the Fourier transform (FT) or the fast Fourier transform (FFT). From the equation for the Fourier transform it is evident that this equation is not well suited to the study of localized (in time) disturbances in a signal nor is it appropriate to apply it to non-stationary time series. The power spectrum of an acoustic signal is often analyzed in an attempt to investigate attenuation of an acoustic signal. Techniques that compare the spectra of the emitted and transmitted or reflected pulse are used to assess attenuation as a function of frequency. The typical frequency range for explortion seismic data is between 10 and 100 Hz. Over this range, and for the length of duration of the seismic record, the basic assumptions of stationarity are not grossly violated. High resolution seismic signals can often exhibit severe loss of high frequencies as they travel down into the earth and return again. A method that reflected the time-frequency structure in high resolution signals would be more diagnostic and more appropriate for the analysis of very high resolution signals.

Experiments designed to measure attenuation within the earth sometimes use down hole geophones to record the direct pulse from a source near the surface or a core sample may be taken to the laboratory where a high frequency acoustic wave is transmitted through the sample. If the outgoing pulse and the transmitted pulse can be isolated then the spectra can be compared. The isolation of the first arrival is usually accomplished by using a short window Fourier transform (SWFT) and often the spectra exhibit properties associated with the window function used as well as the signal. The wavelet transform offers a means of avoiding this problem.

High resolving power of the acoustic wave is desirable for a range of depths but this ability is compromised when multiple scatterers are present. Morlet *et al.* ([21] and [22]) investigated the propagation of plane waves for normal incidence, through periodic multi-layered media for wavelengths ranging from much greater than the spatial period to periods on the order of the spatial period. They found that for large wavelengths (16 times the spatial period of the medium) the composite medium was transparent but phase-delaying; for short wavelengths (2 times the spatial period of the medium) the signal was strongly attenuated and super-reflectivity could occur; for intermediate wavelengths (8 times the spatial period of the medium) velocity dispersion versus frequency appeared.

These observations are quite diagnostic of the medium, offering the geophysical investigators of shallow marine soils a new and potentially powerful means of quantifying subseabed soil conditions through the use of time-frequency methods. Morlet *et al.* ([21] and [22]) conclude that when frequency bandwidth spans four or five octaves, time-frequency methods will be more sensitive to attenuation and velocity dispersion. The wavelet transform ([21], [22], [6], [17], [27], [5] and [7]) is an appropriate tool for investigating the time-frequency nature of high resolution acoustic signals. Displaying the energy distribution in the time-frequency plane describes a signal more uniquely than the FT spectra and could provide a means for evaluating or quantifying the degree of backscatter in different frequency bands and at different times. A complete map of attenuation as a function of depth and frequency or wavelength can be constructed and this type of information may be more readily correlated to selected marine soil properties.

The main objective of this paper is to report on the application of the wavelet transform to high resolution marine acoustic signals and its use in the estimation of attenuation of acoustic wave energy for signals reflected from subseabed reflectors. Usefulness of the wavelet transform as a classifier for backscatter is discussed briefly with reference to wavelet transform plots and depth-wavelength displays.

§2. Acoustic-Geotechnical Correlations: Physical and Historical Context

The motivation for research into correlations between acoustic/seismic responses of a marine soil and the soil's geotechnical properties is to provide the geotechnical engineer with cost-effective and reliable estimates of the geotechnical properties of submerged soils. Reliable estimates of geotechnical properties are needed, for example, in the selection of a site and the design of a structure that is intended to be supported by the seabed. Such structures can include gravity-based oil production platforms or support columns for causeways or bridges. Porosity, density and grain size distribution are some of the soil properties that affect the acoustic response of a signal and are also of interest to the geotechnical engineer [31]. Acoustic wave properties related to these parameters through empirical relationships include acoustic impedance, velocity and attenuation ([1], [9], [20] and [13]). These properties can be determined from digital acoustic data with greater accuracy than was possible in the past using analogue records.

The recent and rapid development of smaller, faster, and more powerful computers has had an impact upon many branches of science but none so dramatic, perhaps, as in the area of high resolution sub-bottom marine seismic reflection profiling. At about the same time as the first papers on wavelet transforms were being written, analogue paper recordings of acoustic returns from the seabed and sub-seabed represented the current practice in data collection and presentation. Practitioners of land-based engineering seismic data acquisition, and the oil and gas exploration industry, have been acquiring and processing digital data for almost as many years as there have been computers. During the past ten years the marine high-resolution (> 500 Hz) or site surveying industry has made the leap from analogue data acquisition systems to digital data acquisition systems. This technological leap can be attributed to the development of high-speed analogue-to-digital computer boards and digital signal-processing chips, which together effectively allow real-time data acquisition and processing of high-resolution acoustic data.

The availability of digital acoustic data has led to the desire for a more quantitative assessment of seismic signals. Empirical relationships between acoustic properties and soil properties have been reported ([1], [32] and [10]). Studies that relate acoustic properties to rock or sediment properties involve the use of frequencies below 100 Hz ([26] and [12]) or above 10 kHz ([10] and [11]) but very few studies provide information in the 100 to 10,000 Hz range which is the typical range for many high-resolution sub-bottom marine surveys. In soils characterized by large scatterers, such as boulders in glacial tills, the 100-1,000 Hz range of frequencies is more efficient in achieving greater depth penetration. Testing carried out in laboratories, usually at very high (> 100 kHz) frequencies, on samples collected in the field or on cone-penetrometer data acquired *in situ*, provide data for correlations that are sometimes biased toward high frequencies. But the means to correlate geotechnical properties of a soil over a more complete range of frequencies is also needed. Some commercial acoustic systems already offer classification schemes based upon how the transmitted pulse is modified by the water-sediment interface upon reflection [18] or internally upon transmission to deeper layers and subsequent return to the receiver [29]. It is the seafloor sediment type that is most easily determined by evaluating the seabed return as in Leblanc *et al.* [18] but characterization derived from deeper returns becomes quite complex due to constructive and destructive interference by scattered waves. Time-frequency analyses would aid in assessing the degree of interference as well as capitalizing upon the tuning or detuning of acoustic waves that can occur within the sub-bottom layers.

There are several factors that contribute to the attenuation of an acoustic wave. Acoustic energy attenuates with increasing distance from the source, a function described by the term geometrical spreading loss. Geometrical spreading losses are not related directly to the properties of the medium in which the energy is propagating. Usually a means of correcting or nullifying this effect is sought before attenuation due to the medium is determined. Attenuation of acoustic energy can be divided into two basic categories, intrinsic attenuation and apparent attenuation. Intrinsic attenuation is the process that degrades the amplitude of the signal such as absorption due to frictional heating or viscous losses [14]. Apparent attenuation is due to processes like scattering, energy partitioning at acoustic impedance contrasts, interbed multiples and mode conversions ([14] and [30]). What is more generally observed, however, is effective attenuation which is a combination of intrinsic attenuation and apparent attenuation.

Estimation of intrinsic attenuation of propagating acoustic pulses can provide information concerning grain size, moisture content, shear modulus and Atterburg limits [13]. Estimates for effective attenuation are useful as they can be related to the degree of scattering, and possibly to quantification of the size of the scatterers within a soil. There are many methods for measuring or estimating attenuation of acoustic energy in sediments. Data collected for the determination of attenuation generally fall into two categories, *in situ* measurements and laboratory measurements, The frequency content of the acoustic energy used to investigate the sediment can be vastly different – from several hertz to millions of hertz. The connection between the range of frequencies and measurements is not always clear.

Laboratory methods generally rely upon free vibration, forced vibration, wave propagation and observation of stress-strain curves [35]. Laboratory studies involving wave propagation methods use frequencies in the 100 kHz to 1 MHz range [34]. *In situ* measurements that rely upon wave propagation can range from a few hertz to hundreds of kilohertz but as noted in the preceding text there is a need for more information in the range of frequencies between 100 Hz to 10 kHz range.

The estimation of attenuation using field or *in situ* measurements has spawned many methods. In some situations it is possible to record the outgoing acoustic pulse as a reference for comparison with the energy from acoustic or seismic waves refracted through the sediments and returned to the surface to be recorded by geophones or hydrophones ([15], [14], [8] and [33]). More reliable measurements can be made by drilling a hole down through the sediment and recording the direct arrival of acoustic energy from a source near the surface to a geophone placed at successively deeper depths within the hole ([12], [24] and [26]). Similarly measurements can be made between two holes drilled through the rock or sediments [25] by placing a source in one hole and a receiver in the other and varying the depth of each.

Attention has been focused on extracting attenuation estimates from seismic reflection data, particularly in the marine setting ([18], [29], [3] and [36]). Schock *et al.* [29] estimate attenuation in the seabed by comparing the correlated chirp signal from a reflector with a synthetically attenuated chirp wavelet. From this measurement the frequency roll-off is determined and plotted as a function of depth. The slope then gives the attenuation coefficient in units of dB/m kHz. LeBlanc *et al.* [18], also using a chirp system, attempt to estimate the surficial soil properties of the seabed by correlating reflection losses at the water sediment interface with soil properties. Tyce [36] took advantage of naturally occurring sedimentary wedges to estimate attenuation. This situation is useful because the decrease in energy is easily observed and calculated as the wedge of sediment thickens.

A common technique for characterizing attenuation of acoustic signals utilizes the spectral ratio between two pulses ([15], [14] and [12]). This technique uses the ratio of the amplitude spectrum of the known unattenuated source function with the amplitude spectrum of the attenuated amplitude spectrum. A slope of this ratio against frequency yields the attenuation, and terms that are assumed to be independent of frequency such as geometrical spreading or reflected energy can be ignored. A more recent technique, introduced by Courtney and Mayer [2], employs a filter correlation method that uses a series of bandpass filters on the attenuated pulse and compares each pulse to the similarly bandpassed source pulse.

§3. The Wavelet Transform for Acoustic Soil Analysis

3.1. The wavelet transform and Morlet wavelet

The form of the wavelet transform used in this application is given by Equation (6) of Goupillaud *et al.* [6]. Adopting the terminology common to this volume the wavelet transform of f with respect to the analyzing wavelet is:

$$(W\,f)(u,\,a) = |a|^{-\frac{1}{2}} \int \Psi(\frac{u-t}{a})f(t)\,dt. \tag{1}$$

Where $f(t)$ is the time series or signal, $\Psi(t)$ is the analyzing wavelet and $\Psi(\frac{u-t}{a})$ represents the dilated (a) and translated (u) wavelets. The $|a|^{-1/2}$ term serves to ensure that the dilated wavelets have the same total energy as the analyzing, or mother wavelet. In the analyses presented in this paper, frequency or the natural logarithm of the frequency, is given by the vertical axis and the time shifts are given by the horizontal axis for the time-frequency displays. The frequency associated with each scale value, a, was determined directly from the equation for the Morlet wavelet [6]. The Morlet wavelet used in these examples is described by the following equation,

$$\Psi_b(t) = e^{ibt}e^{-\frac{t^2}{2}} - \sqrt{2}e^{-\frac{b^2}{4}}e^{ibt}e^{-t^2}. \tag{2}$$

The Fourier transform of Equation (2) is

$$FT(\Psi_b(t)) = \hat{\Psi}_b(\omega) = e^{\frac{-(\omega-b)^2}{2}} - e^{-\frac{b^2}{4}}e^{\frac{-(\omega-b)^2}{4}}. \tag{3}$$

This wavelet is shown in Figure 1. The envelope of the wavelet and its amplitude spectrum are Gaussian-modulated functions. The variable b in Equations (2) and (3) has the same value of 5.336 and is the same as that given by Goupillaud *et al.* [6].

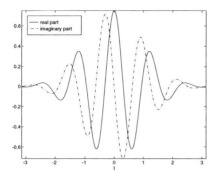

Figure 1. Morlet analyzing wavelet.

3.2. Estimating attenuation: a simple example

The following example demonstrates the mechanics of the attenuation analysis. A synthetic trace with the direct pulse from one of the normal incidence traces collected at an offshore experimental field site in Terrenceville, Newfoundland (described later), was convolved with a series of five spikes placed at the 0.0003, 0.002, 0.004, 0.006 and 0.008 second time marks. A geometrical spreading factor was then applied as well as an attenuation factor, α, as follows:

$$A(t) = G(t)A_0 e^{-\alpha v t}. \tag{4}$$

The geometrical spreading factor, $G(t)$, is given by $1/vt$ where v is a constant velocity of 1500 m/s in this example, but it can also be a function of depth, and t is time in seconds. The definition used for the attenuation coefficient, α, is that given by Johnston and Toksöz [16] and it is assumed to be linear with frequency for the range of frequencies under investigation. For the synthetic example α was assigned a value of 0.25 m^{-1}. In Figure 2a the 0.010 second attenuated trace (without spreading losses) is displayed. Figure 2b depicts the modulus of the wavelet transform, with the horizontal axis representing time and the vertical axis the natural logarithm of the frequency. The values range from -13.55 (black) to 10.35 (white). For all the wavelet analyses presented the axes are labelled in this manner unless otherwise noted.

Figure 2c depicts the spectral ratio for the wavelet transform of the synthetic signal. The spectral ratio is a measure of the ratio between the spectrum for the direct or emitted pulse and the spectrum for the reflected of transmitted pulse. Values range from -22.87 dB (white) to -2.47 dB (black). This display, as well as the method used to calculate the spectral ratios ([34] and [16]), bears further explanation by way of presenting the

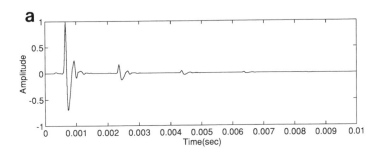

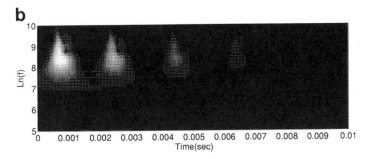

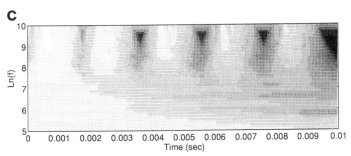

Figure 2. Synthetic example; (a) 0.25 dB/m attenuated synthetic seismic trace (the direct water borne arrival from the sparker was used as the wavelet for the synthetic); (b) modulus of the wavelet transform of synthetic trace in (a); (c) spectral ratio calculated from wavelet transform in (b).

relevant part of the derivation of the spectral ratio equation. After Johnston and Toksöz [16], a plane wave propagating in a homogeneous medium can be described by:

$$A(x,t) = A_0 e^{i(kx-\omega t)} \tag{5}$$

where attenuation may be represented by allowing either k, the wavenumber, or ω, the angular frequency, to be complex. Following [16] and including a geometrical spreading term as in [34], the ratio of the reference amplitude spectrum versus the spectrum at some time, t_1, after taking the natural logarithm of the ratio of the two amplitude spectra can be written as:

$$\ln \frac{A_m}{A_1} = (\gamma_m - \gamma_1)vtf + \ln \frac{G_m}{G_1} \tag{6}$$

where v is velocity, t is one-way travel time, f is frequency, $\alpha(f) = \gamma f$ is the attenuation at f, and the left hand side of Equation (6), the spectral ratio, is represented by the data displayed in Figure 2c. Equation (6) can be interpreted in two ways; first the free variable can be the frequency, f, and then the slope, of a line fitted to the data when the spectral ratio is plotted against frequency, would be:

$$\text{slope} = (\gamma_m - \gamma_1)vt. \tag{7}$$

The geometrical term, $\ln(G_1/G_2)$, can be ignored if one assumes that the geometrical spreading terms are independent of frequency. Alternatively, travel time can be taken as the free variable but in this case the geometrical factor does depend upon t and it will influence the value of the attenuation derived from the slope of a line fitted to the curve of the spectral ratio versus time calculation.

For this study the spectral ratio method was applied by taking the frequency distribution for the time at which the maximum magnitude for the direct pulse is greatest and using this spectrum as the reference spectrum. This occurs at approximately 0.0007 sec. in Figure 2b. Therefore, any determination of the slopes of lines fitted to the data in the direction parallel to the vertical axis satisfies Equation (7) and those that are fitted to the data in the direction parallel to the horizontal axis are subject to the time-dependency condition.

The procedure for estimating the attenuation from normally-incident acoustic waves is demonstrated using the simple synthetic seismic example described in the preceding text. The results of a comparison of the value of the attenuation obtained using the synthetic trace corrected for geometrical spreading losses and an uncorrected synthetic trace are displayed in Figures 3a and 3b respectively. The synthetic trace was corrected for the known geometrical spreading factor and the wavelet transform was then calculated and the modulus displayed (Figure 2b). The mean power at each time and for all frequencies in the transform was calculated and plotted in Figure 3a. There are five broad peaks shown in Figure 3a associated with each of the attenuated pulses in Figure 2a. The largest local maxima for a particular broad peak was chosen interactively and a least squares fit was made to these selected values. The criterion for selecting the local maxima is based upon providing a guideline for consistent picking, and in selecting the large broad peaks it was assumed that boundaries or acoustic reflectors will reflect energy independent of frequency content which would not be the case for random scatterers. The line and the points chosen for the least squares fit are indicated (straight black line and black x's respectively). The value of the slope was found to be 360 nepers/sec which, when converted to meters by dividing out the velocity term, yields a value of 0.24 nepers/meter. The conversion between nepers/unit length and dB (decibels) per unit length is α (dB/unit length) $= 8.686\pi$ (nepers/unit length). The actual value used was 0.25 nepers/meter, a 4% difference between the actual and estimated value. If the geometrical spreading losses are not removed then there should be some effect on the result for the attenuation calculation, and this is observed in Figure 3b. Using the spectral ratio data from Figure 2c, the variance in the direction parallel to the frequency axis was calculated for every time sample. The slope of a least squares fit to the user-selected points from this plot yielded a value of 277 nepers/sec or 0.185 nepers/meter, a 27% difference. When all digital sample values are considered, the least squares estimate for the corrected data is 0.29 nepers/meter, a 16% difference. As a final comparison the primary event was windowed (150-point window) as well as the event at 0.008 sec. The amplitude spectrum was calculated for each event using the fast Fourier transform and then the difference was divided by the separation between the two events yielding a value of 0.22 nepers/meter or a 13% difference between the actual and estimated value.

The results given in Figures 3a and 3b are very encouraging, and in spite of the large difference (16%) for the case where all values for a corrected transform were used an order of magnitude for the attenuation is nevertheless achievable. For real acoustic signals the zones or valleys in Figure 3b, where the spectral ratio falls below 5 nepers with respect to the primary signal, the interference due to scattering of acoustic energy would tend to fill in these valleys possibly resulting in a lower value for the attenuation estimate when including all values in the estimation.

Application of the wavelet transform to the problem of determining attenuation offers the potential for automated order-of-magnitude determinations of attenuation from digitally-recorded acoustic signals. As well, the potential to surgically isolate energy in time and frequency can enhance its application for sub-bottom attenuation measurements using reflected seismic signals.

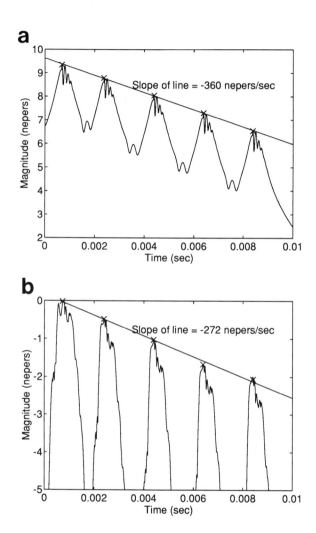

Figure 3. (a) Attenuation curve for synthetic trace with correction for geometrical spreading losses and (b) without correction for spreading losses. (Note: see text for conversion between nepers and decibels)

§4. Experimental Field Program

Three-dimensional multichannel data were collected at a seabed test site at Terrenceville near St. John's, Newfoundland. These data were processed and interpreted conventionally [23]. Within the circular 78 m² test site the marine sediment was between 4 and 5 m thick overlying an undulose and fractured bedrock. The depth of the water at the site was 10 m. Three distinct programs were carried out at the test site, an acoustic program, a physical sampling program and a geotechnical program.

The acoustic program comprised four acoustic lines that intersected at 45 degrees at the centre of the test site (Figure 4a). Figure 4b shows a simple schematic of the source/receiver geometry which was manipulated into the survey grid shown in Figure 4a by scuba divers from the Ocean Science Centre at Memorial University of Newfoundland. The 16 hydrophones were arranged in four banks of four hydrophones per bank for recording purposes and are labelled A through D. The acoustic source was mounted on a tripod and was positioned approximately 40 cm directly above each hydrophone of bank D (stations 5 through 8). These four stations were the shot point locations for each profile of the survey. The hydrophones positioned beneath the source are referred to as normal incidence hydrophones. The horizontal spacing between receiver stations was 1.42 m. The average height above the seabed for hydrophones comprising bank A and C was 1.7 m and for bank D it was 1.3 m. The average source height was the same as that for hydrophone banks A and C.

The sparker source used was developed specifically for the high resolution seismic research program at C-CORE [4] to deliver a repeatable, broadband acoustic pulse. Up to 1080 joules (J) of energy was delivered to this source using a Huntec seismic energy power supply. Brüel and Kjær (B&K) 8105 omnidirectional hydrophones, with a flat frequency response from 0.1 Hz–180 kHz, were used to measure the pressure field. Brüel and Kjær 2635 charge amplifiers, with built-in antialias filters, conditioned the signals prior to digitizing. The data were sampled at 10 microseconds with antialias filters of 0.2 Hz (low cut) and 30 kHz (high cut). The automated acquisition system directed the analogue-to-digital recorder to trigger the seismic energy source and then record the acoustic signals. The emitted acoustic pulse, recorded on the normal incidence hydrophone, was then correlated with a previously recorded reference pulse. If the correlation coefficient exceeded a user-defined threshold, set at 0.95 for this survey, the trace was accepted and all recorded channels were transferred to the acquisition computers memory. The automated procedure continued, and traces were summed with their respective counterparts in memory, until each trace in memory consisted of 50 stacked or summed traces. These summed traces were then written to the computer's hard drive. The acous-

a

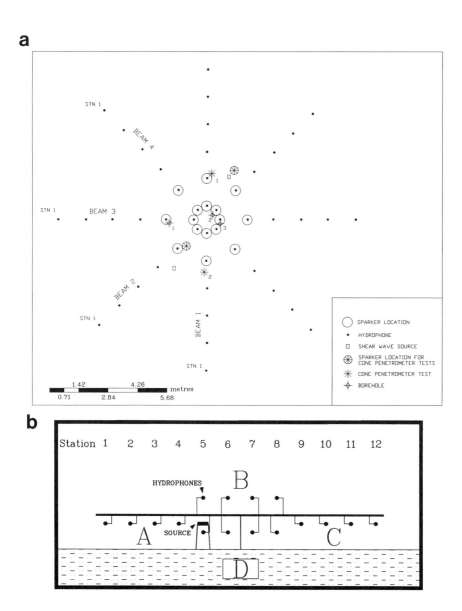

b

Figure 4. (a) Site plan for experimental field program and (b) cross-section of seabed recording geometry.

tic source was then moved to stations 6, 7 and 8 with the above acquisition procedure repeated for each source position. After the shot at station 8 the entire beam was rotated about its central point in 45 degree increments and the recording process described above was repeated. The complete source receiver coverage is shown in Figure 4a. The four recorded profiles are labelled Beams 1 to 4 and station number 1 is identified for each beam with a total of twelve stations per beam. The outer ring of sparker source point locations, at a radius of 2.13 m, defines the extent of the subsurface normal incidence coverage. The common depth point (CDP) coverage is defined by a circle of radius 4.87 m, passing through stations 3 and 10 on beams 1 through 4.

The entire procedure described above was carried out for two separate power settings, 480 J and 1080 J. Thus two complete surveys were acquired for this site, and are referred to as the low and high power surveys.

The physical-sampling program provided marine soils and bedrock cores for analysis and a geotechnical program provided some *in situ* measurements using a limited series of cone penetrometer tests (CPT's) as well as a series of seismic cone penetrometer tests (SCPT's) to acquire velocity information. Figure 4a shows the locations of the CPT's and boreholes.

The cone penetrometer was capable of recording shear (S) and compressional (P) wave data. This information was acquired when the cone penetrometer was stopped at 1 meter intervals as the cone was pushed into the seabed. The compressional wave arrival times were used to calculate average and pseudo interval velocity values for the soils. The horizontal offset between the three acoustic sources and the cone rods and the offset of the geophones from the cone tip were accounted for in all velocity calculations.

Three continuously-sampled boreholes were drilled by rotary techniques, using N-size casing. Standard penetration tests were conducted and representative but disturbed soil samples were obtained in the overburden, using a conventional 51 mm OD split-spoon (SS) sampler. Bedrock was cored in NQ size. All borehole depths reported are with respect to the sea floor. The exact hole locations were measured with respect to the underwater acoustic survey by divers.

§5. Data Analysis

Acoustic data analysis was divided into two phases with the first focused on determining subsurface stratigraphy by using the multichannel acoustic data, the subsurface samples and results from the penetrometer tests. The second phase involved analysis of the normal-incidence data and was concerned with measuring and quantifying changes in the waveform of the reflected energy as well as finding some means to visually compare the

16 normal-incidence acoustic signals. Use of the wavelet transform was investigated for this particular analysis.

5.1. Conventional data analysis and processing

Core samples extracted from the site confirmed the composition of the marine sediments for this area. The sediments consisted of a fine to medium-grained gravelly sand with some organic content overlying a sand-gravel deposit of between 4.2 and 4.6 m thick. Sandwiched between the fractured, layered bedrock and the sand-gravel deposit was a thin (< 1.0 meter) till. The depth to bedrock ranged between 4.7 and 5.1 m below the sea-bottom.

The cone penetrometer data were of limited value as this tool is designed for use with fine-grained soils. However, some useful data included the cone tip resistance which provided information on soil stiffness. The seismic data were used to estimate independently the compressional (P) wave velocity of the soil.

Data from the acoustic program were processed using conventional common depth point procedures for multichannel seismic analysis [38]. An example of a brute stack from the Beam 3 profile (Figure 4a) is given in Figure 5. Two principal acoustic horizons are easily observed; the seabed event, a positive (black) peak (trace excursions to the right), at about 2.0 ms and a second event, a trough at about 5.0 ms, both indicated by the arrow heads on either side of the plot. The seismic section is displayed with two-way vertical travel times versus horizontal position in the profile. There are many other weaker events evident but below about 6 ms the events become more disjointed and incoherent. Figure 6 shows (a) a brute stack for the Beam 3 profile, (b) filtered and gained normal incidence trace (repeated four times as a visual aid), (c) stratigraphic indicators from Borehole 1 and (d) the geotechnical field report of the samples showing stratigraphy. The reflection even at 5.0 ms in Figure 6a correlates with a boundary that was not inferred from the sampling program possibly due to low sample recovery rates (see Figure 6d). A boundary was evident in the geotechnical analyses and its existence corroborated from the force needed to push the cone penetrometer past this boundary. It is indicated by the triangle symbol (Figure 6c).

5.2. Wavelet transform analysis of an acoustic signal

Using a Morlet analyzing wavelet, the wavelet transform was applied to the normal incidence data from the field program. These data are recorded at the locations in Figure 4 where the sparker source and hydrophones are coincident. Figure 7a shows the normal incidence trace from profile Beam 3, station 5 with a geometrical spreading correction applied to the data. The geometrical spreading correction was estimated using the average velocity

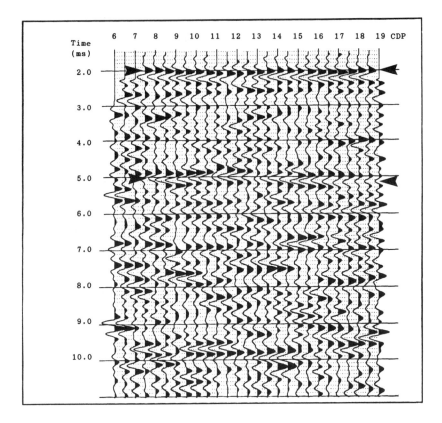

Figure 5. Brute stack for profile Beam 1 (refer Figure 4a). Bandpass filtered and gained for display.

function determined in phase I of the acoustic field program. The modulus of the wavelet transform is presented in Figure 7b and the spectral ratio is shown in Figure 7c.

The primary or direct arrival has a direct travel time of 0.0003 sec (Figure 7a). The next significant event is the reflection from the seabed, which arrives at 0.0021 sec, followed by another clear event arriving at about the 0.005 sec mark at 8.4 ln(Hz). There are no distinct events between the 0.0021 sec event and the event at 0.005 sec but there is high frequency energy in the signal which may be related to internal scattering of acoustic waves by cobbles and small boulders buried in the seabed. Following

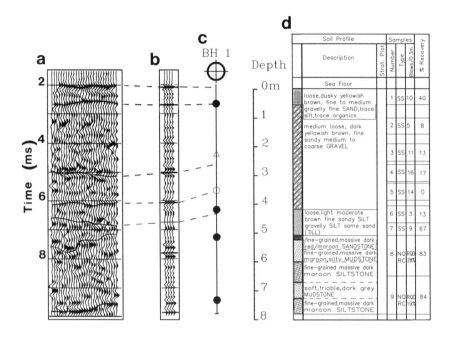

Figure 6. (a) Brute stack for profile Beam 3; (b) filtered normal incidence trace for station 5, Beam 3 (trace is replicated four times for visual aid); (c) borehole 1 showing identified stratigraphic boundaries; (d) soil profile description based on physical samples and cores.

the 0.005 sec event are several more pronounced excursions from the signal and these could be due to scattering from the fractured bedrock or internal multiple reflections.

The modulus of the wavelet transform for the acoustic signal (Figure 7b) exhibits bright areas (shown as white) where the signal strength is strong (note the direct arrival and the seabed event in Figure 7a). There is more detail represented in this display than would be found in a conventional power spectra display. The modulus of the wavelet transform associated with the direct pulse exhibits three lighter areas (see Figure 7b, between 7.5 and 10 ln(Hz) and 0.00025 and 0.001 sec). Comparison between the waveform of the direct pulse (Figure 7a) and the modulus of the wavelet transform for this pulse (Figure 7b) demonstrates the strength of the transform for signal decomposition. The direct pulse (Figure 7a) has an onset time of 0.0003 sec. Following this onset is the first positive peak, followed immediately by the first trough, and then a second smaller positive doublet. In Figure 7b the modulus of the wavelet transform of this pulse indicates that it consists of three primary frequency groupings. The large main energy burst at about 8.5 ln(Hz) and 0.0004 sec associated with the trough; the higher frequency elongated region between 8.7 and 9.2 ln(Hz) associated with the first peak and the high frequency event at 9.8 ln(Hz) associated with the doublet peak at 0.00075 sec. From this examination of the direct water-borne pulse it can be seen that the wavelet transform can resolve very high frequencies in a signal, and that interference patterns due to scattering are resolved well in time and space.

Recordings of this source in deeper water shows the possible cause of the interference in this pulse. In deeper water the bubble pulse separates completely from the initial pulse. In the water depths for these examples the bubble pulse is on top of the initial pulse making the pulse appear as a single burst. Detailed evaluation of the direct pulse allows for a better assessment of the subsequent reflected pulses and this assessment is possible with the wavelet transform.

The first echo is a reflection from the seabed, an event at 0.0021 sec. The source pulse has now undergone modification by the seabed as result of the reflection and subsequent partitioning of the energy at the water-seabed boundary. The two lower-frequency features are still present, but the region that was at the higher frequency appears brighter relative to the other two lower-frequency events (Figure 7b. Possible explanations for these differences are: the lower frequency energy transmits more of its energy into the seabed whereas more of the higher-frequency energy is reflected, or more of the higher-frequency energy is scattered generating a high degree of interference both constructively and destructively.

At 9.1 ln(Hz) following the seabed event there is a line of bright, almost merged regions extending for the length of the display. It is not very likely that 8100 Hz energy penetrated the seabed to any significant depth so this energy probably represents energy backscattered from the sub-seabed. Frequencies below 8100 Hz, for example the event at 0.005 sec, have values of about 3600 Hz, which are of similar intensity in frequency content to the direct pulse. There is not much energy evident between the seabed and 0.005 ms events for spectral values less than about 3000 Hz possibly indicating no major boundaries or acoustic impedances in this interval. This interpretation is corroborated by the stratigraphic information presented in Figure 6. Sediments at the test site included pebbles between 3 and 6 cm diameter which is approximately one quarter of the wavelength for the dominant scattered energy above the 8100 Hz (9 ln(Hz)) line. Larger

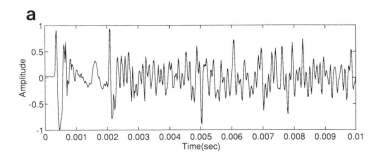

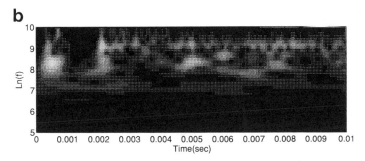

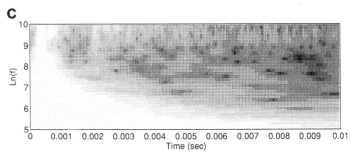

Figure 7. (a) Unfiltered, normal incident trace, from station 5, profile Beam 3 corrected for geometrical spreading losses. (b) Modulus of the wavelet transform of the trace in (a). (c) Spectral ratio based on modulus of wavelet transform of ungained trace.

scatterers were assumed to be present as the cone penetrometer test encountered large rocks on several occasions requiring the boreholes to be
drilled out in order to proceed past these points. During the process of
drilling through the sediments to obtain a bedrock core a large boulder
was encountered resulting in the recovery of a core through a portion of
this boulder.

The spectral ratio, calculated in the same way as the synthetic trace
example (Figure 2c), is displayed in Figure 7c. Small spectral ratios are
shown as dark or black regions and values for large spectral differences with
respect to the direct or source pulse are displayed in white.

The average velocity function determined from the multichannel processing of the common depth point acoustic data in phase I [23] was used to
convert the scales for the axes of the wavelet transform modulus plots from
time into depth beneath the seabed and a constant velocity of 1800 m/s
was used to convert frequency to wavelength. The modulus was replotted
with the new axes and is shown in Figure 8 as a function of depth below
the seabed against wavelength values. The scattered energy observed above
8100 Hz in Figure 7 now corresponds to wavelength values in the 0.10 to
0.30 m range. The strong event at a depth of 3 m (Figure 8) corresponds
to reflected energy with a dominant wavelength of 0.4 m. Another event
at about 4.5 m has a wavelength of 0.7 m. Figure 9 shows the modulus
of the wavelet transform for four normal incidence examples from Beam 3
plotted as a function of depth below seabed versus wavelength. The strong
energy event observed at 3 m depth in Figure 8 is evident on all of the
panels for this profile indicating a continuous boundary reflector (note the
5.0 ms event in Figure 5 and 7).

The preceding discussion highlights some of the potential features of
the wavelet transform display that offers a new diagnostic tool to the high-
resolution geophysicist. The analysis should lend itself well to two- dimensional image analysis procedures for feature extraction and correlation. By
utilizing an energy threshold condition coupled with a density measure
for the number of events per meter a classification system for subsurface
scatterers could be developed.

5.3. Estimation of attenuation

For the attenuation analysis a similar procedure was followed as described in section 3.2. Figure 10 shows a comparison between the variance
of the spectral ratio (black dots) and the value for the variance of the corrected data using the estimated spherical spreading correction (solid line).
The points cluster fairly closely about the solid line indicating that either
the spectral ratio or the corrected data could be taken for analysis. Attenuation estimates were derived from the modulus of the wavelet transform
of the data corrected for geometrical spreading losses. This approach was

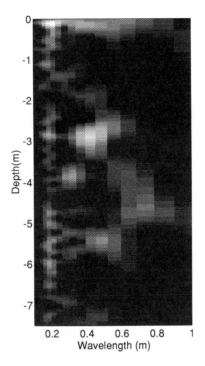

Figure 8. Modulus of the wavelet transform for the trace in Figure 7a plotted against depth and wavelength (units are meters).

the most successful in the synthetic example.

Curves, similar to the one in Figure 3a, were produced for all four stations (Figure 11) from profile Beam 3, but time axes were converted to depth axes using the average velocity function. Peaks near the seabed and at or near the 3.0 m depth were selected for each station. This procedure was repeated for profile Beam 1 and the values for the attenuation in the first 3 m are given in Table I.

The mean value is 4.6 nepers/m (40 dB/m) with a standard deviation of 1.9 nepers/m (17 dB/m). Data for coarse gravelly soils is sparse but values for shallow (3 m) depth are similar [37]. Lee and Malloy [19] report values for compressional wave attenuation that range from a low of 5.3 dB/m for metalliferous sediments to a high of 80 dB/m for a silty clay.

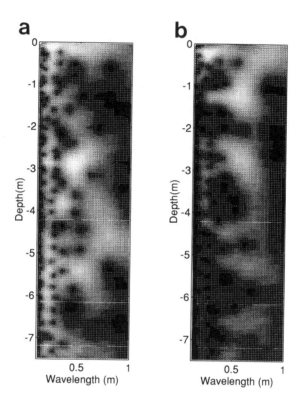

Figure 9. Modulus of the wavelet transform, for normal incidence traces along profile Beam 3 at stations 5 and 6 (a and b respectively) plotted as in Figure 8.

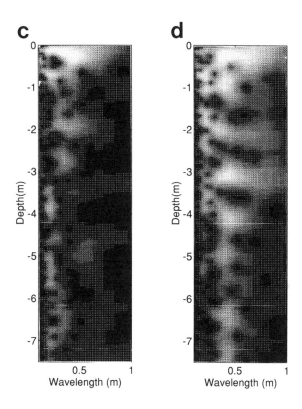

Figure 9. (contd.) Modulus of the wavelet transform, for normal incidence traces along profile Beam 3 at stations 7 and 8 (c and d respectively) plotted as in Figure 8.

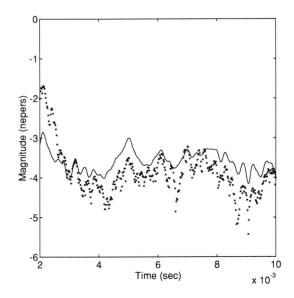

Figure 10. Attenuation curves for station 5, Beam 3 (see text for conversion between nepers and decibels).

Table I
Wavelet Transform Attenuation Values

	Beam 1 (nepers/m)	Beam 3 (nepers/m)
Station 5	4.4	8.3
Station 6	3.5	2.9
Station 7	3.5	2.8
Station 8	5.6	6.2

§6. Conclusion

Two analyses based upon the wavelet transform have been presented. The qualitative analysis discusses the features observed in the modulus of the wavelet transform in relation to the acoustic signal and the sub-seabed. It represents a starting point from which feature extraction and a quantitative analysis can begin with the goal of developing classifiers for the scatterers. It should be noted that only the Morlet analyzing wavelet has been utilized in this study. Examination of alternate wavelets that may be more diagnostic for classification of scattering should be pursued. The next stage of the study of wavelets for this application will employ other

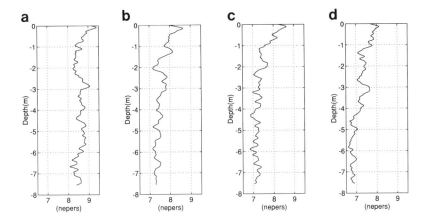

Figure 11. Attenuation curves for profile Beam 3, stations 5 to 8 (a–d respectively), plotted against depth.

analyzing wavelets in the evaluation of a finite difference model of wavefield scattering. Investigation into interference, tuning and detuning of acoustic energy should provide a model for assessing complex scattering phenomena which may answer questions concerning the size and nature of scatterers and their effect upon acoustic wavelengths of the same order.

The attenuation analyses suffers from the short data record length at the Terrenceville test site, and limits the full potential of the analysis. Digital sub-bottom profiler records with longer recording times of 20 ms and greater were recently acquired using the same sparker source as used at the field test site in Terrenceville. These data were collected as part of a U.S. navy seabed imaging program off Panama City, Florida. Extensive tests were carried out at this site and all data will be made available to the participants. The wavelet transform attenuation process will be refined and applied in an analysis of these data.

Based upon the review of the current literature on acoustic-geotechnical correlations for marine soils, there are several applications that should benefit from the wavelet transform analysis. Using C-CORE's 200-gravity centrifuge facility, very high frequency acoustic data will be collected in a study of sediment loading and transport. The attenuation of high frequency waves (> 100 kHz) propagated through centrifuged sediment will be analyzed using standard techniques that rely on windowed Fourier transforms or bandpassed signals which can then be compared to results obtained using wavelet analysis.

Some authors have questioned the validity of estimating attenuation using the method of spectral ratios. Sams and Goldberg [28] point out that the effect of windowing compressional waves to isolate them from slower waves is included in the calculation of the spectral ratio. They conclude that for borehole studies the spectral ratio technique will not give good results. This problem presents a good test for the application of the wavelet transform and the results should prove superior since the window functions can be avoided.

References

1. Buchan, S., D. M. McCann, and D. Taylor Smith, Relations between the acoustic and geotechnical properties of marine sediments, *Q. Jl. Eng. Geol.*, 5, 265–284, 1972.

2. Courtney, R. C. and L. A. Mayer, Calculation of acoustic parameters by a filter- correlation method, *J. Acoust. Soc. Am.*, 93, 1145–1154, 1993.

3. Dodds, D. J., Attenuation estimates from high resolution subbottom profiler echoes, *Presented at the Saclant Asw Research Conference on Ocean Acoustics Influenced by the Sea Floor*, La Spezia, Italy, June 9–12, NATO conference on bottom interacting ocean acoustics, 173–191, 1980.

4. English, J. M., S. T. Inkpen and J. Y. Guigné, A new, high-frequency, broadband seismic source. *Presented at the 23rd Annual Offshore Technology Conference in Houston*, Texas, OTC 6558, 1991.

5. Goupillaud, Pierre L., Three New Mathematical Developments on the Geophysical Horizon, *Geophysics: The Leading Edge of Exploration*, 40–42, June 1992.

6. Goupillaud, P., A. Grossmann and J. Morlet, Cycle-Octaves and Related Transforms in Seismic Signal Analysis, *Geoexploration*, 23, 85–102, 1984.

7. Goupillaud, P., A. Grossmann, J. Morlet, Cycle-Octave Representation for Instantaneous Frequency Spectra, *SEG Expanded Abstracts with Biographies*, 53rd Annual International SEG Meeting, 613–615, 1983.

8. Grant, F. S. and G. F. West, *Interpretation Theory in Applied Geophysics*, McGraw-Hill Book Company, New York, 1965.

9. Hamilton, E. L., Geoacoustic modelling of the sea floor, *Journal of the Acoustical Society of America*, 68, 1313–1340, 1980.

10. Hamilton, E. L., Compressional-wave attenuation in marine sediments, *Geophysics*, 37(4), 620–646, 1972.

11. Hamilton, E. L., Prediction of *in-situ* acoustic and elastic properties of marine sediments, *Geophysics*, 36, 266–284, 1971.

12. Hauge, P. S., Measurements of attenuation from vertical seismic profiles, *Geophysics*, 46(11), 1548–1558, 1981.

13. Haynes, R., A. M. Davis, J. M. Reynolds and D. I. Taylor-Smith, The extraction of geotechnical information from high-resolution seismic reflection data, In *Offshore Site Investigation and Foundation Behaviour*, 28, 215–228. Society of Underwater Technology, 1993.

14. Jacobson, R. S., G. G. Shor Jr., and L. M. Dorman, Linear inversion of body wave data - Part II: Attenuation versus depth using spectral ratios, *Geophysics*, 46(2), 152–162, 1981.

15. Jacobson, R. S., An investigation into the fundamental relationships between attenuation, phase dispersion, and frequency using seismic refraction profiles over structures, *Geophysics*, 52(1), 72–87, 1987.

16. Johnston, D. H. and M. N. Toksöz, Chapter 1, Definitions and terminology, In *Seismic Wave Attenuation*, Eds. M. N. Toksöz and D. H. Johnston, SEG Geophysical Reprint Series No. 2, Series Ed. F.K. Levin, Society of Exploration Geophysics, 1981.

17. Kronland-Martinet, R., J. Morlet, A. Grossmann, Analysis of sound patterns through wavelet transforms, *International Journal of Pattern Recognition and Artificial Intelligence*, 1, 273–301, 1987.

18. LeBlanc, L. R., L. Mayer, M. Rufino, S. G. Schock and J. King, Marine sediment classification using the chirp sonar, *J. Acoust. Soc. Am.*, 91, 107–115, 1992.

19. Lee, Homa and R. J. Malloy, Acoustic retrieval of seafloor Geotechnics: Technical Note, Civil Engineering Laboratory, Naval Construction Battalion Centre, Port Hueneme, California 93042, Program Nos. Yf52.556.091.01.109, Sponsor Naval Facilities Engineering Command (1977).

20. McCann, C. and D. M. McCann, The acoustic properties of marine sediments, *Defence Oceanology International Brighton*, 1990.

21. Morlet, J., G. Arens, E. Fourgeau and D. Giard, Wave propagation and sampling theory - Part I: complex signal and scattering in multilayered media, *Geophysics*, 47(2), 203–221, 1982.

22. Morlet, J., G. Arens, E. Fourgeau and D. Giard, Wave propagation and sampling theory - Part II: sampling theory and complex waves, *Geophysics*, 47(2), 222–236, 1982.

23. Pike, C. J., S. T. Inkpen, and J. Y. Guigné, Marine geophysical and geotechnical site characterization; Terrenceville, Newfoundland, Contract report for Energy Mines and Resources, Bedford Institute of Oceanography, March, C-CORE Contract No. 91–C5, 1991.

24. Portsmouth, I. R., M. H. Worthington and C. C. Kerner, A field study of seismic attenuation in layered sedimentary rocks - I VSP data, *Geophys. J. Int.*, 113, 124–134, 1992a.

25. Portsmouth, I. R., M. H. Worthington and J. P. Neep, A field study of attenuation in layered sedimentary rocks - II Crosshole data, *Geophys. J. Int.*, 113, 135–143, 1992b.

26. Ricker, Norman, The form and laws of seismic wavelets, *Geophysics*, 18(1), 10–40, 1953.

27. Rioul, Olivier and Marten Vetterli, Wavelets and signal processing, *IEEE Signal Processing Magazine*, 14–38, 1992.

28. Sams, M. and D. Goldberg, The validity of Q estimates from borehole data using spectral ratios, *Geophysics*, 55(1), 97–101, 1990.

29. Schock, S. G., L. R. LeBlanc and L. A. Mayer, Chirp subbottom profiler for quantitative sediment analysis, *Geophysics*, 54, 445–450, 1989.

30. Schoenberger, M. and F. K. Levin., Apparent attenuation due to intrabed multiples, II, *Geophysics*, 43(4), 730–737, 1978.

31. Stoll, R. D., Sediment acoustics, *Lecture notes in earth sciences*, Volume 26, Spring-Verlag, Berlin, 155pp, 1989.

32. Taylor Smith, D., Geotechnical characteristics of the sea bed related to seismo acoustics, In *Ocean Seismo-Acoustics*, Eds. T. Akal and J. M. Berkson, Plenum Press, New York, 483–500, 1986.

33. Telford, N. M., L. P. Geldart, R. E. Sheriff and D. A. Keys, *Applied Geophysics 2nd Ed.*, Cambridge University Press, New York, 1990.

34. Toksöz, M. N., D. H. Johnston and A. Tinur, Attenuation of seismic waves in dry and saturated rocks: 1 Laboratory Measurements, *Geophysics*, 44, 681–690, 1979.

35. Toksöz, M. N., and D. H. Johnston Chapter 2, Laboratory measurements of attenuation, In *Seismic Wave Attenuation*, Eds. M. N. Toksöz and D. H. Johnston, SEG Geophysical Reprint Series No. 2, Series Ed. F. K. Levin, Society of Exploration Geophysics, 1981.

36. Tyce, R. C., Estimating attenuation from a quantitative seismic profiler, *Geophysics*, 46, 1364–1378, 1981.

37. Tullos, F. N. and A. C. Reid, Seismic attenuation of gulf coast sediments, *Geophysics*, 34, 516–528, 1969.

38. Yilmaz, O., Seismic Data Processing, Doherty, S. M., Ed., Series: *Investigations in Geophysics* 2, Neitzel, E. B. Series Ed., Society of Exploration Geophysicists, Tulsa, OK, 1987.

This research has been partially supported by the Natural Sciences and Engineering Research Council of Canada (NSERC), ESSO Resources Canada Ltd., Gulf Canada Resources Ltd., Mobil Oil Canada Ltd. and Petro-Canada Resources Ltd.

Chris J. Pike
Centre for Cold Ocean Resources Engineering (C-CORE)
Memorial University of Newfoundland
St. John's, Newfoundland
Canada, A1B 3X5
e-mail: *cpike@kean.ucs.mun.ca*

Including Multi-Scale Information in the Characterization of Hydraulic Conductivity Distributions

Kevin E. Brewer and Stephen W. Wheatcraft

Abstract. Transport in heterogeneous porous media is highly dependent on the spatial variability of aquifer properties, particularly hydraulic conductivity. It is impractical, however, when modeling a real situation, to obtain aquifer properties at the scale of each grid block. Wavelet transforms are investigated as a tool to assess the movement of information between scales, and to incorporate the scale and location of hydraulic conductivity test data when interpolating to a fine grid. Wavelet transforms are particularly useful as they include parameters that define a spatial localization center that can be positioned in space, and changed in size. As a result, one can incorporate multiple measured hydraulic conductivity values in the wavelet transformation at a resolution matched to each measurement scale.

A multi-scale reconstruction method is developed here which uses forward and inverse wavelet transforms in conjunction with a pseudo-fractal distribution to fill in missing information around sparse data. This wavelet reconstruction method is compared to several more traditional interpolation schemes with respect to accuracy of solute transport prediction. The wavelet reconstruction algorithm is then used to examine the issue of optimum sample size and density for stationary and fractal random fields.

§1. Introduction

Various authors ([14] and [19]) have shown that the spatial variability of hydraulic conductivity, K, produces most of the transport dispersion observed at the field scale. If we could accurately reproduce the K-distribution at a fine scale in a numerical model, there would be no need for stochastic theories. However, it is impractical to collect fine scale K values for each grid block in a high resolution model. Moreover, even if we had the information, the model would likely contain enough grid blocks (and therefore enough unknowns) that it would take excessive computation time to

Wavelets in Geophysics   213
Efi Foufoula-Georgiou and Praveen Kumar (eds.), pp. 213–248.

be practical for routine use. Both of these factors (lack of information and large number of unknowns) lead to development of stochastic theories, upscaling methods, and concepts of effective parameters at coarser scales.

Our work is based on the premise that it would be useful to have a rigorous method of creating a K-distribution at a particular scale, using K information from larger and (perhaps) smaller scales. As a first step, we have restricted ourselves to the problem of reproducing (or reconstructing) a K-distribution at "the finest scale." By this, we mean that given a limited number of K test values from the smallest scale, and from several larger scales, how do we reconstruct a complete K-distribution (or equivalently, a complete K-grid) at the finest scale.

By test values, we mean the value of K obtained from an aquifer test, permeameter test, etc. It is important to realize that each K value has an associated scale determined by the volume of the porous media used to obtain the K value. For porous media properties, different K sampling techniques inherently average different sample volumes, which can result in different sample scales and values at the same physical location. We have focused our research on how to best use this scale information from K samples to enhance the characterization of porous media. Our results, with appropriate modification, could also be applied to any other scale affected property.

Summarizing, we know that field (coarse) scale contaminant dispersion is mostly caused by the fine scale variability of the K distribution. Ultimately, however, we cannot adequately sample at the fine scale to account for this variability, nor can we justify the excessive computation time for solving practical fine grid problems. Therefore, we have two possible options: (1) develop a method to adequately reconstruct the fine scale variability from coarser, multi-scale samples, and/or (2), develop a method to properly coarsen the fine scale variability to more accessible scales. In both cases, the central issue is how K variability moves, or transforms, in both directions between scales.

In this paper we will develop a method that uses the wavelet transform to include multi-scale sample information in the characterization of K-distributions. Figure 1 poses such a multi-scale reconstruction question for a hypothetical situation of only two scales of measurement. Although this figure shows the movement of coarse information to a fine scale, the problem of coarsening fine scale information is similar. These problems are somewhat analogous to, in electrical engineering terminology, a data fusion problem with incomplete information.

For our contaminant transport problems, the measure of reconstruction accuracy of the K-grid at the finest scale is best determined by comparison of the results of flow and transport models, since it is the variability of the finest scale K values that primarily controls transport. Although the

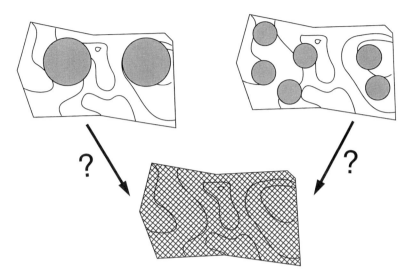

Figure 1. Schematic showing the problem of characterizing heterogeneous porous media on a fine grid utilizing information from different
scales.

accuracy of the individual reconstructed fine scale K-grid values at each
grid block is of some importance, it is of secondary interest. Better reconstruction techniques will result in higher accuracy of the reconstructed
breakthrough curves compared to the true (or known) breakthrough curves.
Reconstruction techniques should also honor the multi-scale sample values
exactly. Therefore, our primary concern in determining the success of a
reconstruction is not how individual finest scale element or node interpolated values match known values, but how accurately simulated transport
through each reconstructed media compares to the true transport results.

Our (or any) reconstruction method can only be said to be successful if it reconstructs actual field distributions. However, actual field data is
rarely (if ever) available in enough detail to properly test our reconstruction
method. Even in cases where high resolution field studies have been collected [5], there are no field scale tracer tests associated with such data sets.
Therefore, we have used various synthetic fine grid data sets as "truth,"
with comparisons made primarily with simulated transport results.

We will first show how wavelet transforms can be used for multi-scale
reconstruction by exploiting the inherent localization and frequency resolution of the wavelet transform. We have developed a wavelet reconstruction
method in which sample location, scale, and values are all used to recon-

struct a fine scale grid. For this initial work, we have assumed that there is no noise in our sample data; that is, our sample values perfectly represent the unknown information for the given sample location and scale. Inclusion of measurement uncertainty is delegated to future research, and might possibly follow along the lines of [2]. The wavelet reconstruction technique is compared with other traditional interpolation techniques by using a 256 × 256 node synthetically generated two-dimensional porous media and solving the flow and transport equations.

Additionally, we have used a modification of the wavelet reconstruction technique, dubbed wavelet extrapolation, to investigate what, if any, important scales of information exist in various synthetic representations of porous media K distributions. This extrapolation method creates estimates at each location for any scale finer than the given, completely sampled, scale. Here, we attempt to understand the relative importance of each scale to identify the inherently significant coarse scale representations of the fine scale variability.

§2. Wavelet Based Interpolation

2.1. Discrete transform introduction

To understand our wavelet based reconstruction and extrapolation techniques, we must first familiarize ourselves with some mathematical details of wavelet transforms. The wavelet transformation is similar to the Fourier transform in that a data set is transformed into the frequency (or scale) domain. The Fourier transform (via spectral analysis) has been used extensively to study the characterization of porous media [7], but a limitation of the Fourier analysis technique is the assumption that the information at each scale is homogeneous and can be essentially represented by a single "power" value. What makes the wavelet transform useful as a tool for multi-scale reconstruction of porous media is that this assumption is not necessary, resulting in the preservation of positional information (i.e. the heterogeneity) at each scale. Daubechies in [6] best expresses the wavelet transform as "a tool that cuts up data or functions or operators into different frequency components, and then studies each component with a resolution matched to its scale." Thus, with the wavelet transform, one can analyze data sets with heterogeneous information at each scale, which is commonly the case for porous media distributions.

The forward discrete wavelet transform consists of a convolution of a basis function (sometimes confusingly called the wavelet function) with a data set. Unlike the Fourier transform, which uses only sines and cosines as basis functions, the wavelet transform can use a variety of basis functions. Basis functions are discussed in detail in [6], among others. Here we discuss the important basis function characteristics that will concern our

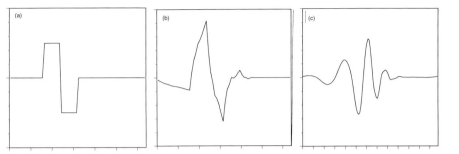

Figure 2. Graphs of the (a) (left) Haar basis function, (b) (middle) Daubechies4, and (c) (right) Daubechies12 basis functions.

reconstruction and extrapolation techniques.

Every basis function has localization (or positional) and frequency components which control the scale-to-scale relationships, and the in-scale positional relationships. By scale-to-scale relationships we mean the amount of coarsening (or conversely, the fining) from scale-to-scale. The basis function components are commonly set is such a way that the wavelet coefficients on each scale are related to the next coarser or finer scale by a factor of 2, and that the transform convolution begins with every odd-indexed element on each scale.

That is, if for a one-dimensional data set whose finest scale is 128 elements of 1 length unit, the next coarser scale will consist of 64 elements representing 2 original-scale length units. Similarly, the next coarser scale will consist of 32 elements representing 4 length units, and so on, until the coarsest possible scale, which will consist of only one element representing the entire 128 original-scale length. For our work, we have used the common Haar and Daubechies family basis functions [6] which have a scale-to-scale factor of 2. Each of these basis functions are also orthogonal, and so can be conveniently used on two-dimensional data sets by first performing each transform on the row data, then the column data.

Figure 2 shows graphs of the basis functions themselves. The Haar basis function is highly discontinuous, and as such, is often used to analyze discontinuous data sets. The Daubechies family basis functions are less discontinuous, with the higher order functions becoming more smooth. The basis function coefficients we used in the discrete transform equations are listed in Table I.

The discrete wavelet transform is a transformation of information from a fine scale to a coarser scale by extracting information that describes the

Table I
Basis function coefficients for the Haar and Daubechies4 functions.

Haar	
c_1	0.70710678
c_2	0.70710678
Daubechies4	
c_1	0.48296291
c_2	0.83651630
c_3	0.22414387
c_4	-0.12940952

fine scale variability (the detail coefficients) and the coarser scale smoothness (the smooth coefficients) according to:

$$\{s_m\} = [H]\{s_{m+1}\}; \quad \{d_m\} = [G]\{s_{m+1}\} \tag{1}$$

where s represent smooth coefficients, d represent detail coefficients, m is the level, and H and G are the convolution matrices based on the wavelet basis function. Higher values of m signify finer scales of information. The complete wavelet transform is a process that recursively applies Equation (1) from the finest to the coarsest level (scale). This describes a scale by scale extraction of the variability information at each scale. The smooth coefficients generated at each scale are used for the extraction in the next coarser scale.

The matrices H and G are created from the coefficients of the basis functions, and represent the convolution of the basis function with the data. The following is the general makeup for the H and G matrices for a four coefficient basis function that has a scale-to-scale relationship of 2 (basis functions that have more or less coefficients would be similar):

$$H = \begin{bmatrix} c_1 & c_2 & c_3 & c_4 & 0 & 0 & 0 & 0 & \cdots & 0 & 0 \\ 0 & 0 & c_1 & c_2 & c_3 & c_4 & 0 & 0 & \cdots & 0 & 0 \\ 0 & 0 & 0 & 0 & c_1 & c_2 & c_3 & c_4 & \cdots & 0 & 0 \\ \cdot & \cdot & \cdot & \cdot & \cdot & \cdot & \cdot & \cdot & \cdots & \cdot & \cdot \\ c_3 & c_4 & 0 & 0 & 0 & 0 & 0 & 0 & \cdots & c_1 & c_2 \end{bmatrix} \tag{2}$$

$$G = \begin{bmatrix} c_4 & -c_3 & c_2 & -c_1 & 0 & 0 & 0 & 0 & \cdots & 0 & 0 \\ 0 & 0 & c_4 & -c_3 & c_2 & -c_1 & 0 & 0 & \cdots & 0 & 0 \\ 0 & 0 & 0 & 0 & c_4 & -c_3 & c_2 & -c_1 & \cdots & 0 & 0 \\ \cdot & \cdot & \cdot & \cdot & \cdot & \cdot & \cdot & \cdot & \cdots & \cdot & \cdot \\ c_2 & -c_1 & 0 & 0 & 0 & 0 & 0 & 0 & \cdots & c_4 & -c_3 \end{bmatrix} \tag{3}$$

As a forward wavelet transform example, Figure 3a shows a simple 4 element one-dimensional data set $\{6,4,5,3\}$ and the first convolution application using the simple Haar basis function resulting in two $m = 2$ level detail coefficients and two $m = 2$ level smooth coefficients. Continuing in Figure 3b, the transform convolution is applied to the two $m = 2$ level smooth coefficients resulting in a single $m = 1$ level smooth coefficient and detail coefficient. Note that the $m = 2$ level detail coefficients have not been altered.

Alternatively, we can represent Figure 3a following the form of Equation (1) as:

$$\left\{ \begin{array}{c} 10\frac{1}{\sqrt{2}} \\ 8\frac{1}{\sqrt{2}} \end{array} \right\} = \left[\begin{array}{cccc} \frac{1}{\sqrt{2}} & \frac{1}{\sqrt{2}} & 0 & 0 \\ 0 & 0 & \frac{1}{\sqrt{2}} & \frac{1}{\sqrt{2}} \end{array} \right] \left\{ \begin{array}{c} 6 \\ 4 \\ 5 \\ 3 \end{array} \right\} \tag{4}$$

for the smooth coefficient upscaling, and:

$$\left\{ \begin{array}{c} 2\frac{1}{\sqrt{2}} \\ 2\frac{1}{\sqrt{2}} \end{array} \right\} = \left[\begin{array}{cccc} \frac{1}{\sqrt{2}} & -\frac{1}{\sqrt{2}} & 0 & 0 \\ 0 & 0 & \frac{1}{\sqrt{2}} & -\frac{1}{\sqrt{2}} \end{array} \right] \left\{ \begin{array}{c} 6 \\ 4 \\ 5 \\ 3 \end{array} \right\} \tag{5}$$

for the detail coefficient calculation. Continuing, we can represent the final calculation in Figure 3b in matrix notation for the smooth coefficient upscaling, and the detail coefficient calculation as, respectively:

$$9 = \left\{ \begin{array}{cc} \frac{1}{\sqrt{2}} & \frac{1}{\sqrt{2}} \end{array} \right\} \left\{ \begin{array}{c} 10\frac{1}{\sqrt{2}} \\ 8\frac{1}{\sqrt{2}} \end{array} \right\} \tag{6}$$

$$1 = \left\{ \begin{array}{cc} \frac{1}{\sqrt{2}} & -\frac{1}{\sqrt{2}} \end{array} \right\} \left\{ \begin{array}{c} 10\frac{1}{\sqrt{2}} \\ 8\frac{1}{\sqrt{2}} \end{array} \right\}. \tag{7}$$

Traditionally, only the detail coefficients at every scale, and the coarsest level smooth coefficients, are considered a complete set of wavelet coefficients. In our example of Figure 3, the complete set is {9, 1, 1.414, 1.414 }. In our work however, we are not so much interested with just the traditional wavelet coefficients, but in the entire suite of smooth and detail coefficients at each scale (or level) since we will show in the next section that our multi-scale samples can be related to smooth coefficients at various scales and locations.

The inverse discrete wavelet transform is similarly implemented via a recursive recombination of the smooth and detail information from the coarsest to finest level (scale):

$$\{s_{m+1}\} = [H]'\{s_m\} + [G]'\{d_m\} \tag{8}$$

where H' and G' indicate the transpose of the H and G matrices.

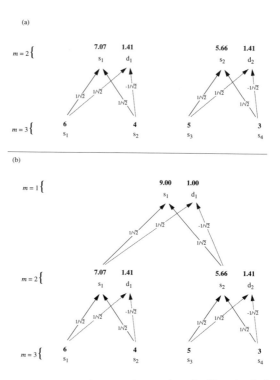

Figure 3. Discrete wavelet transform using the Haar basis function on an example data set {6,4,5,3}: (a) the transform of the original data, (b) the completed transform resulting in a single $m = 1$ smooth coefficient, and a single $m = 1$ detail coefficient, and two $m = 2$ detail coefficients.

Using the generated wavelet coefficients in our previous example of Figure 3, the inverse transform using the Haar basis function initially proceeds to generate the $m = 2$ smooth coefficients from the $m = 1$ smooth and detail coefficients, as shown in Figure 4a. Continuing, the finest scale (third-level) coefficients are generated using the $m = 2$ level information, as shown in Figure 4b.

In matrix notation, the Figure 4a transform is:

$$\left\{ \begin{array}{c} 10\frac{1}{\sqrt{2}} \\ 8\frac{1}{\sqrt{2}} \end{array} \right\} = \left\{ \begin{array}{c} \frac{1}{\sqrt{2}} \\ \frac{1}{\sqrt{2}} \end{array} \right\} 9 + \left\{ \begin{array}{c} \frac{1}{\sqrt{2}} \\ -\frac{1}{\sqrt{2}} \end{array} \right\} 1 \tag{9}$$

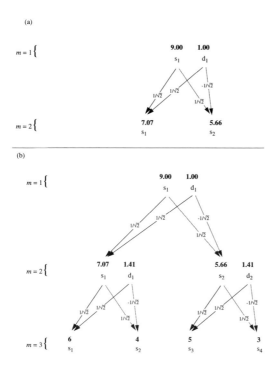

Figure 4. Discrete inverse wavelet transform using the Haar basis function on the wavelet coefficients from Figure 3: (a) the inverse transform from the first-level, (b) the completed inverse transform resulting in the original fine scale data values.

and the Figure 4b transform is:

$$
\left\{ \begin{array}{c} 6 \\ 4 \\ 5 \\ 3 \end{array} \right\} = \left[\begin{array}{cc} \frac{1}{\sqrt{2}} & 0 \\ \frac{1}{\sqrt{2}} & 0 \\ 0 & \frac{1}{\sqrt{2}} \\ 0 & \frac{1}{\sqrt{2}} \end{array} \right] \left\{ \begin{array}{c} 10\frac{1}{\sqrt{2}} \\ 8\frac{1}{\sqrt{2}} \end{array} \right\} + \left[\begin{array}{cc} \frac{1}{\sqrt{2}} & 0 \\ -\frac{1}{\sqrt{2}} & 0 \\ 0 & \frac{1}{\sqrt{2}} \\ 0 & -\frac{1}{\sqrt{2}} \end{array} \right] \left\{ \begin{array}{c} 2\frac{1}{\sqrt{2}} \\ 2\frac{1}{\sqrt{2}} \end{array} \right\} . \quad (10)
$$

Note that the resulting finest level smooth information is exactly the same as the original data set. Overall, both the forward and inverse wavelet transformations, when applied over many scales (levels), can be pictorially represented as in Figure 5.

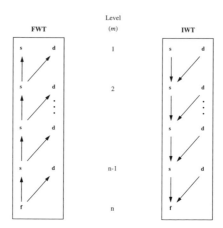

Figure 5. Exploded view of an one-dimensional wavelet transform. Reprinted with permission from the *SIAM Review*, 31(4), pp. 614–627. Copyright 1989 by the Society for Industrial and Applied Mathematics, Philadelphia, Pennsylvania. All rights reserved.

2.2. Multi-scale reconstruction

The objective of multi-scale reconstruction is to completely determine all finest scale values given a limited number of sample values from the finest and several coarser scales, and possibly other necessary information. This section introduces a multi-scale reconstruction algorithm based on the wavelet transform.

As previously shown (see Figure 3 and Figure 4), the wavelet transform can be represented by a linked pyramid (or lattice) of smooth and detail coefficients. Because hydraulic conductivity (K) samples contain scale information and are an inherent average over some area or volume, they are multi-scale samples and should be amenable to this reconstruction algorithm. Therefore, our K samples at different scales and locations, for input to our reconstruction algorithm, are equivalent to the smooth coefficients on the lattice, both representing an average of finer scale information.

To generate the desired finest scale values from coarser scale smooth coefficients, the lattice relationships necessitate that all other smooth and detail coefficients, if unknown, also be generated. That is, to generate all the unknown finest scale values via application of the discrete inverse wavelet transform, all the next coarsest scale smooth and detail coefficients must be known or calculated. Since the discrete forward and inverse wavelet transforms create relationships between adjacent levels (scales) of smooth and detail coefficients, these coarser smooth coefficients can be calculated from even coarser scale (via interpolation using the inverse transform) and from the finer scale (via upscaling using the forward transform) coefficients. The same will be true for all unknown coefficients at any coarser scale, except the coarsest possible scale, where only finer scale information can be used. The problem is then of solving for all the unknown smooth and detail coefficients on all levels with a limited number of known coefficients.

There are different potential solution techniques to solve this wavelet transform based multi-scale reconstruction problem. Since our known sample data is sparse (i.e., a large number of unknown smooth coefficients on a number of different scales) and we are limiting our application to two-dimensions, we have elected to use a simple cyclic iteration and averaging technique which is in some ways similar to multigrid cycling. Other possible techniques have been proposed [3], and may be utilized in future work.

Cyclic iteration essentially consists of starting at the coarsest scale, proceeds to the finest scale, and then cycles to the coarsest and back down to the finest until convergence is reached for each finest scale value. At each scale, each unknown coefficient is calculated. For all, but the finest and coarsest scales of interest, this means that the value interpolated from coarser information and the value upscaled from finer information is weight averaged to the final estimate. In our work, we used equal weights for the interpolated and upscaled values. The cyclic iteration (coarsest-to-finest-to-coarsest, etc.) is necessary to "move" the limited information around the lattice, as appropriate. Algorithm convergence is determined when the mean, maximum, and variance of change in fine scale values between iterations is below predetermined criteria, typically equal to two or three orders of magnitude less than the precision of the sample values.

The scales of interest, i.e. the scales inclusive of the finest and coarsest in our cyclic iterations, are bounded on the fine-end and the coarsest-end by the scales in the sample set. Our implementation requires complete test value coverage at the coarsest scale to ensure each fine scale generated value is supported by at least one known test value, and to bring closure to the algorithm. It is important to note that each smooth coefficient (i.e., sample value) contains all detail and coarse information from all higher (coarser) scales.

In general, the requirement for complete test value coverage at the coarsest scale is not necessary. For site characterization problems that have a statistically significant number of test values on any one particular scale, a statistical relationship could be derived to assign stochastic sample values to the remaining unknown sample locations at that level. That level would become fully populated with real and stochastic values, and any coarser scale test values would not be necessary.

Essentially all detail coefficient values at any scale cannot typically be calculated, since enough spatial coverage of known smooth coefficients to directly calculate detail coefficients will rarely occur. Therefore, based on a pseudo-fractal relationship, we assign stochastic values for all detail coefficients assuming an underlying infinite correlation at the finest scale. The choice of a pseudo-fractal relationship for detail coefficients is arbitrary, but reasonable, based on recent work such as [17] and [20]. It is possible to use log-normally-distributed stationary random fields or a variety of other statistical distributions. There is no inherent algorithmic limitation for use of any other desired detail coefficient relationship.

Wornell and Oppenheim in [21] showed the fractal relationship between expected values of the detail coefficients as:

$$E[(d_m)^2] = \sigma^2 2^{-\gamma m} \tag{11}$$

where σ^2 is a positive constant that depends on the choice of the wavelet basis function, γ is related to the desired fractal dimension, and m is the scale. Adopting Equation (11) for the assignment of specific detail coefficients, we have for our two-dimensional algorithm:

$$d_m = fg2^{-\frac{1}{2}b(m-1)} \tag{12}$$

$$b = 2(3 - D) + 1 \tag{13}$$

where f is a fitting parameter, D is the desired fractal dimension ($3.0 > D > 2.0$), and g is a Gaussian random number. We found $f = 100$ to be the best fitting parameter value based on visual analysis of the reconstructed data sets. The random component in our work was generated via the RAN1 and GASDEV random number generators given in [13]. The choice of fractal dimension is discussed below.

The reconstruction algorithm begins with the assignment of the sample values to smooth coefficients at the appropriate location and level (based on the location and scale of the sample). Since the wavelet transform smooth coefficients are not normalized, the sample values are adjusted prior to their assignment. For example, given a completely smooth data set {1,1,1,1}, the $m = 2$ level smooth coefficients would not be 1 as expected from a statistical mean, but would be 1.414 (for the Haar basis). In this case, a $m = 2$ level sample value of 1 would be adjusted to 1.414 before assignment to the smooth coefficient. This adjustment assumes that the sample values are representative of a statistical mean with the amount of adjustment based on the basis function used. Futher research is needed to determine the proper adjustment for actual hydraulic conductivity test values.

After the assignment of the sample values, the detail coefficients are assigned during the initial downward (coarsest to finest scale) leg. After calculating the finest scale estimates, the cyclic iterations begin, with convergence calculations after each finest scale estimation. The algorithm ends when convergence is reached. Pseudo-code for our wavelet reconstruction algorithm is given in Table II.

Since the calculation for each coefficient is independent on each level, the algorithm is inherently parallelizable. Using either distributed and/or parallel computing capabilities, a significant decrease in the execution time for large problems would be experienced. Although not necessary for our work, this capability may be highly desirable for large problems (especially three-dimensional data sets).

Wavelet reconstruction can also accommodate use of multiple detail coefficient relationships and/or wavelet basis functions. This allows incorporation of soft-information, such as known geologic relationships of values which are different between scales. For example, large scale information might be best modeled by a smoothing basis function, whereas the finer scales could be better modeled by a discontinuous (erratic) basis function. This scenario could represent an interbedded sand and gravel river deposit influenced by an overall hydraulic downstream particle sorting. In our initial work, however, we have not implemented these extensions and only use one wavelet basis function and one stochastic detail coefficient distribution for all scales.

2.3. Multi-scale extrapolation

We have also developed a pure stochastic extrapolation algorithm based on the wavelet transform to study important scales of interest. This extrapolation method could be combined with the multi-scale reconstruction algorithm for situations where the desired grid level is smaller than the attainable sample values, or when it is desirable to maintain the similarity of the variable distribution upon closer inspection of a subset area of interest.

The algorithm requires complete sample coverage at only one scale, and outputs all values at the desired finest scale. Detail coefficients are stochastically generated, as discussed above, for all scales down to the finest scale desired. Then, level-by-level, the smooth and detail coefficients are used to interpolate the finer scale smooth coefficients via the inverse wavelet transform. The algorithm stops when all the finest desired level smooth values are generated. Pseudo-code for the extrapolation algorithm is given in Table III .

§3. Traditional Interpolation Methods

The wavelet reconstruction technique was compared to three traditional interpolation methods: simple assignment, kriging, and conditional simulation. These methods are briefly reviewed below. Kriging and conditional simulation are classical geostatistical methods that have been in use in mining engineering and geological fields for over a decade. Journel

Table II
Pseudo-code of the wavelet multi-scale reconstruction algorithm.

```
C     SETUP
READ D    -- FRACTAL DIMENSION
BETA = 2*(3-D)+1
C     READ IN SAMPLE INFORMATION
FOR ALL SAMPLES
   READ SAMPLE => LOCATION, VALUE, LEVEL
   SMOOTH_COEF(LOCATION, LEVEL) = ADJUST_VALUE(SAMPLE VALUE)
   KNOWN_FLAG(LOCATION, LEVEL) = 1
END FOR
C     ASSIGN DETAIL COEFFICIENTS
FOR EVERY LEVEL
   FOR EVERY LOCATION
      DETAIL_COEF(LOCATION, LEVEL) = GAUSS*2**(-0.5*(LEVEL-1)*BETA)
   END FOR
END FOR
C     PERFORM INITIAL DOWN SWEEP
FOR LEVEL = MINLEVEL+1 TO MAXLEVEL-1
   FOR EVERY LOCATION
      IF KNOWN_FLAG(LOCATION, LEVEL) IS NOT 1 THEN
         XWT = WAVELET_TRANSFORM(LEVEL, LOCATION, FORWARD_FLAG)
         XIWT = WAVELET_TRANSFORM(LEVEL, LOCATION, INVERSE_FLAG)
         SMOOTH_COEF(LOCATION, LEVEL) = AVERAGE(XWT, XIWT)
      END IF
   END FOR
END FOR
C     CALCULATE THE FINEST SCALE (NO FORWARD TRANSFORM POSSIBLE)
LEVEL = MAXLEVEL
FOR EVERY LOCATION
   IF KNOWN_FLAG(LOCATION, LEVEL) IS NOT 1 THEN
      XIWT = WAVELET_TRANSFORM(LEVEL, LOCATION, INVERSE_FLAG)
      SMOOTH_COEF(LOCATION, LEVEL) = XIWT
   END IF
END FOR
C     GO UP AND DOWN UNTIL CONVERGENCE IS ACHIEVED
CONVERGENCE_FLAG IS FALSE
DO UNTIL CONVERGENCE_FLAG IS TRUE
C     GO UP
FOR LEVEL = MAXLEVEL-1 TO MINLEVEL+1 BY -1
   FOR EVERY LOCATION
      IF KNOWN_FLAG(LOCATION, LEVEL) IS NOT 1 THEN
         XWT = WAVELET_TRANSFORM(LEVEL, LOCATION, FORWARD_FLAG)
         XIWT = WAVELET_TRANSFORM(LEVEL, LOCATION, INVERSE_FLAG)
         SMOOTH_COEF(LOCATION, LEVEL) = AVERAGE(XWT, XIWT)
      END IF
   END FOR
END FOR
C     GO DOWN
FOR LEVEL = MINLEVEL+2 TO MAXLEVEL-1
   FOR EVERY LOCATION
      IF KNOWN_FLAG(LOCATION, LEVEL) IS NOT 1 THEN
         XWT = WAVELET_TRANSFORM(LEVEL, LOCATION, FORWARD_FLAG)
         XIWT = WAVELET_TRANSFORM(LEVEL, LOCATION, INVERSE_FLAG)
         SMOOTH_COEF(LOCATION, LEVEL) = AVERAGE(XWT, XIWT)
      END IF
   END FOR
END FOR
C     DO FINEST LEVEL
LEVEL = MAXLEVEL
FOR EVERY LOCATION
   IF KNOWN_FLAG(LOCATION, LEVEL) IS NOT 1 THEN
      XIWT = WAVELET_TRANSFORM(LEVEL, LOCATION, INVERSE_FLAG)
      SMOOTH_COEF(LOCATION, LEVEL) = XIWT
      UPDATE MEAN_CHANGE AND MAX_CHANGE
   END IF
END FOR
IF MEAN_CHANGE < DESIRED_MEAN_CONVERGENCE _AND_
      MAX_CHANGE < DESIRED_MAX_CONVERGENCE THEN
   CONVERGENCE_FLAG IS TRUE
END IF
END DO
C     WRITE THE FINEST LEVEL VALUES
FOR EVERY LOCATION
   WRITE SMOOTH_COEF(LOCATION, MAXLEVEL)
END FOR
STOP END
```

Table III
Pseudo-code of the wavelet multi-scale extrapolation algorithm.

```
C    SETUP
READ D    -- FRACTAL DIMENSION
BETA = 2*(3-D)+1
READ MAXLEVEL
READ OTHER VARIABLES
C    READ IN SAMPLE INFORMATION -- COMPLETE ON ONE SCALE
FOR ALL SAMPLES
   READ SAMPLE => LOCATION, VALUE, LEVEL
   SMOOTH_COEF(LOCATION, LEVEL) = SAMPLE VALUE
   MINLEVEL = LEVEL
END FOR
C    ASSIGN DETAIL COEFFICIENTS WHEN SWEEPING DOWN
FOR LEVEL = MINLEVEL+1 TO MAXLEVEL
   FOR EVERY LOCATION
      DETAIL_COEF(LOCATION, LEVEL) = GAUSS*2**(-0.5*(LEVEL-1)*BETA)
      XIWT = WAVELET_TRANSFORM(LEVEL, LOCATION, INVERSE_FLAG)
      SMOOTH_COEF(LOCATION, LEVEL) = AVERAGE(XWT, XIWT)
   END FOR
END FOR
C    WRITE THE FINEST LEVEL VALUES
FOR EVERY LOCATION
   WRITE SMOOTH_COEF(LOCATION, MAXLEVEL)
END FOR
STOP END
```

and Huijbregts [9] and Hohn [8] detail the development and mathematical justification behind these and other geostatistical methods.

3.1. Simple assignment

The first and simplest of the traditional methods is actually an assignment method, rather than an interpolation one, and consists of assigning a sample value to each fine scale grid location based on the finest scale sample encompassing that fine scale grid location. For example, if an unknown fine scale grid location was contained in the sampling area of a medium scale sample and a coarse scale sample, the medium scale sample value would be assigned to this fine grid location. In other words, the medium scale sample is the finest scale sample encompassing the unknown fine grid location. In another example, if a different unknown fine scale grid location is only encompassed by a coarse scale sample, that coarse scale sample value would be assigned to the fine scale target location. Assignments are determined in a like manner for all unknown fine grid locations. We have included the simple assignment method in our analysis primarily because it is expected to represent the worst case in our reconstruction accuracy comparison.

3.2. Kriging

Kriging is a best linear unbiased estimator (BLUE) and uses the spatial covariance of the samples to determine the best estimate at a particular point. All samples are assumed to have point support and thus to be from only one scale. As developed in [9], the kriging system consists of solving for sample weights, λ:

$$[W]\{\lambda\} = \{B\} \tag{14}$$

where W is the sample-point to sample-point covariance matrix and B is the sample-point to target-point covariance vector. The weights are then used on the sample values to calculate an estimate at the target point. This procedure is applied at each unknown grid point. The kriging algorithm in [1] was used with a slight modification to incorporate the simple assignment technique, described above, for those fine grid locations which were beyond the algorithm range.

3.3. Conditional simulation

Conditional simulation is a method by which a correlated stochastic random field on a fine scale, known as a simulation field, is conditioned to limited sample information. The result is a correlated stochastic realization that exactly matches sample values at the point these samples occur. The unconditioned correlated random field is generated from the second order statistics (correlation and mean) of the samples. The conditioning is performed with a kriging-like algorithm. Like kriging, sample values are expected to be from a single scale (the scale of the unconditioned field).

Proper generation of the unconditioned random field is very important as an incorrect field will result in poor conditional simulation results. Turning bands methods are typically used to generate the correlated random fields, and as pointed out by [16], using an inadequate number of bands can result in striations or banding in the simulation field. For our work we have incorporated the turning bands generator in [16] with a conditioning algorithm based in part on the algorithm given in [12].

§4. Evaluation of Methods

4.1. Reconstruction comparison

The applicability and potential accuracy of the wavelet reconstruction technique was investigated by comparing various reconstruction techniques using a single generated sample set from a known synthetic two-dimensional hydraulic conductivity distribution. A synthetic fine scale grid was used to know, with certainty, the actual fine scale values and the resulting "true" particle breakthrough curve. An equitable comparison could therefore be made for each reconstruction technique.

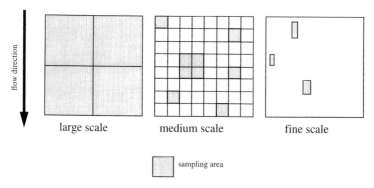

large scale   medium scale   fine scale

sampling area

Figure 6. Sampling schematic.

The known fine scale grid was a two-dimensional set (128×128) of K values ranging four orders of magnitude. The $\ln(K)$ values were generated as a pseudo-fractal from a spectral fractional Brownian motion (fBm) algorithm [18]. The input fractal dimension of 2.4 was representative of geologic media scaling relationships ([10] and [17]).

The known grid was sampled at three different scales, as schematically shown in Figure 6. There were four sample values (representing complete area coverage) at the largest scale, nine sample values at a medium scale, and 227 fine scale samples values grouped in three primary areas. All the sample locations were chosen to mimic a typical site characterization sampling plan using "best professional judgement," without regard to whether it would favor one method over another.

Obviously, "best professional judgement" is a limited method for sample selection as it is based entirely on so called best guesses. A more rigorous approach would have been to evaluate multiple sample sets in a type of Monte Carlo process, in an effort to obtain the best sample suite for each method. These issues will be discussed in future work.

In general, the sample locating philosophy was: (1) to spread a limited number of samples over the entire study area in an attempt to understand the overall variability on each sampled scale, and (2), to group some of the samples on each scale in areas of high variability. The nine medium scale samples were therefore located throughout the domain, with a subset of four samples grouped near the domain center. The three areas of dense fine scale samples were located in areas of interesting medium and large scale sample value variability. These fine scale sample groups were oriented in line with the flow direction to maximize information that directly affects dispersion.

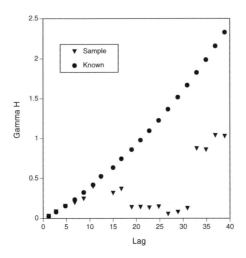

Figure 7. Semivariograms for the entire, known pseudo-fractal synthetic fine scale grid and the resulting sample set.

The medium and large scale sample values were calculated using a simple geometric mean of all known K values in each sample area [11]. This sample set (sample value, location, and scale) was used as input for the simple assignment and wavelet interpolation techniques. For the kriging and conditional simulation interpolation techniques, the sample set was modified by locating the medium and large scale samples at each sample centroid, and discarding the sample scale information. The sample semivariogram, Figure 7, was used for determining the covariance structure for the kriging and conditional simulation reconstruction techniques. A range of 12, sill of 0.5, and a nugget of 0.0 was determined from the sample semivariogram. The sample semivariogram is not similar to the known semivariogram, and this difference is a likely cause for part of any resulting error for the kriging and conditional simulation techniques.

Only one fine scale reconstruction realization from the sample set was required for the deterministic simple assignment and kriging techniques. The wavelet interpolation and the conditional simulation techniques, however, used a Monte Carlo ensemble average. The Daubechies4 wavelet basis function [6] was used for the wavelet reconstruction technique.

For each of the fine scale reconstructions, as well as for the known grid, the flow equation was solved with a multigrid approach [4] with no-flow side

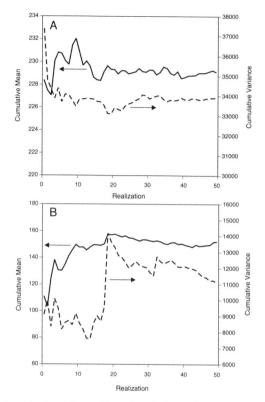

Figure 8. Cumulative Monte Carlo breakthrough curve statistics for (a) the wavelet reconstruction, and (b) the conditional simulation techniques.

boundaries and constant head up- and down-gradient boundaries. Particle tracking was used on each solved flow domain to simulate transport. Particles were released at the up-gradient boundary and tracked through the flow domain. Since the sides were no-flow boundaries, all particles exited at the down-gradient boundary. The individual particle travel times were used to create breakthrough curves. For the wavelet and conditional simulation techniques, 50 Monte-Carlo realizations were found to be sufficient to generate stable ensemble breakthrough curve statistics, as shown in Figure 8.

The shaded contour maps shown in Figure 9 are the fine scale $\ln(K)$ values from the known grid, and the simple, kriged, conditional simulation, and wavelet reconstructions. As shown, the fractal characteristic of the known $\ln(K)$ grid, the blocks of assigned values for the simple reconstruction, as well as the expected smoothing of the point samples on the kriged $\ln(K)$ reconstruction are readily observable. The conditional simulation realization exhibits the correlated stochastic nature underlying the technique. Although the wavelet $\ln(K)$ reconstruction is not exact, it does show the influence of the test scales (evidenced by the generally high grid values in the southwest corner, consistent with large scale sample values), indicating the coarser level information is indeed being transmitted to finer scales. Resampling the wavelet reconstructions gives exact reproduction of the original sample set values.

The simple and kriged particle breakthrough curves (Figure 10) exhibit a "bench" which is not evident in the known breakthrough curve. This bench is apparently due to the large difference in K values at the coarsest sample scale directly influencing the reconstructions, since at that scale, a sample covers 50% of the domain perpendicular to the flow direction. The wavelet breakthrough curve appears to exhibit some of the characteristics of the known curve. The conditional simulation ensemble curve is very smooth and generally indicates early breakthrough at most concentrations.

The mean error of each breakthrough curve can be calculated by:

$$\overline{E} = \frac{1}{n}\left\{\sum_i^n \frac{|r_i - k_i|}{k_i}\right\} \tag{15}$$

where r_i is the time for the ith particle to travel through the reconstructed domain, k_i is the time for the ith particle to travel through the known domain, and for each of the n particles. These resulting mean error values are shown in Table IV. As the percentages show, the wavelet and conditional simulation techniques recreate breakthrough curves better than the simple and kriging techniques. For this particular sample set, it appears that conditional simulation technique is slightly better than wavelet reconstruction. This result may be an artifact of the sample suite or the basis function used in the wavelet method.

4.2. Optimum scale

Addressing the question of whether there are optimum sample scales for various synthetic two-dimensional hydraulic conductivity distributions, the wavelet extrapolation technique was used with complete sample coverage at various scales. An optimum sample scale is defined as one where the greatest reduction in breakthrough curve error occurs, which is indicative of important scales representing the coarsening process of the fine scale variability. Identification of optimum sample scales are also important in developing an effective and efficient site characterization procedure. The experiments consisted of using the wavelet extrapolation technique to extrapolate complete sample information at each scale down to the original finest scale. The Daubechies4 wavelet basis function was again used for all

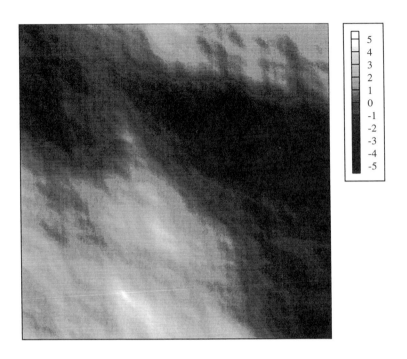

Figure 9. Fine scale ln(K) shaded contour maps for: (a) The known grid.

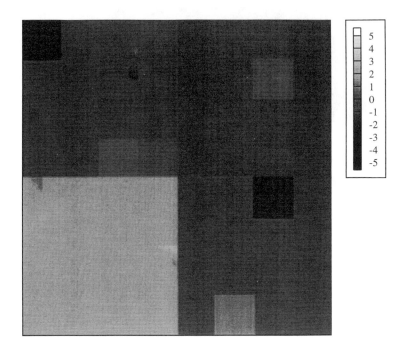

Figure 9. **(contd.)** (b) The simple assignment reconstruction.

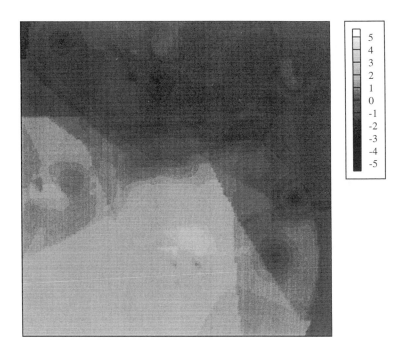

Figure 9. (contd.) (c) The kriging reconstruction.

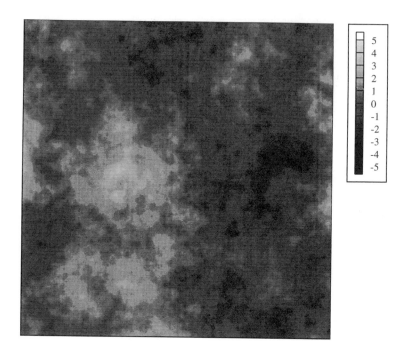

Figure 9. **(contd.)** (d) One realization of the conditional simulation reconstruction.

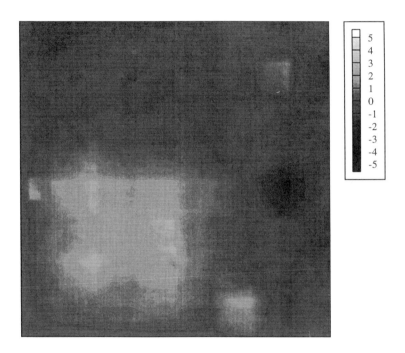

Figure 9. **(contd.)** (e) One realization of the wavelet reconstruction.

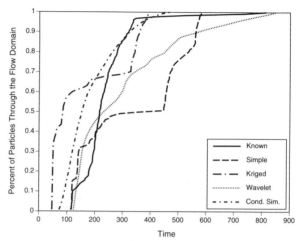

Figure 10.   Particle breakthrough curves.

Table IV
Mean breakthrough curve error for reconstruction techniques.

Reconstruction Technique	Mean BTC error
Simple Assignment	50%
Kriging	48%
Wavelet Reconstruction Ensemble	31%
Conditional Simulation Ensemble	25%

extrapolations.

With a 128×128 regular grid, 6 separate levels can be completely sampled. Samples at each level were generated using the geometric mean of the known fine scale values. Each complete level sample set is then used, separately, to reconstruct the fine grid via the wavelet extrapolation technique. The flow equations are solved and particle breakthrough curves are generated. Since the extrapolation technique is stochastic, a Monte Carlo suite of thirty reconstruction realizations was necessary to ensure stability of the first and second order moments of the ensemble breakthrough curve. These techniques are similar to those described in Subsection 4.1. An example of a breakthrough curve suite for all levels is given in Figure 11. This fig-

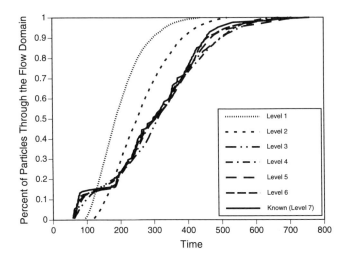

Figure 11. Suite of breakthrough curves showing the increasing accuracy of the reconstructions as finer levels of information are used.

ure visually shows the decrease in error (or equivalently, an increase in the incorporation of the fine scale variability) as increasingly detailed sample information is used.

The first group of numerical experiments used a synthetically generated 12 grid length (approximately 9% of the domain) correlated stochastic fine scale $\ln(K)$ grid as the known values. The known field was generated via the turning bands algorithm used previously. Figure 12 shows the resulting mean and variance of the ensemble breakthrough curve mean error for each level. The horizontal axis shows the number of samples in each level's sample set. At the known fine grid level, 16,384 "samples," shows zero error. This plot shows the expected continual decrease in mean error and variance with finer sampling.

Using the mean values from the high slope part of Figure 12, and differencing the error values between levels, we obtain Figure 13 which shows the decrease in error from each previous level. This plot shows that the greatest decrease in error occurs at 64 samples, which is a sample size of 16 × 16 grid units. This is very close to the correlation length of 12 and appears to indicate that for correlated random fields, the optimum scale is equivalent to the correlation length.

To confirm that this result was not due to the particular grid size used, the experiment was re-run with a 512 × 512 correlated stochastic

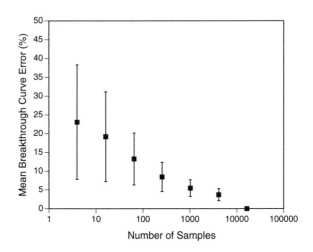

Figure 12.   Mean and variance of ensemble breakthrough curve errors.

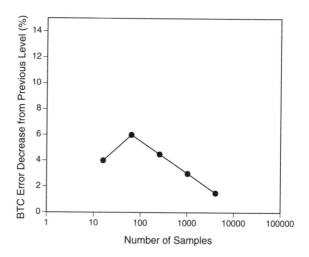

Figure 13.   Mean error change for increasing finer samples levels, $128 \times$ 128 correlated stochastic field with a correlation length of 12 units.

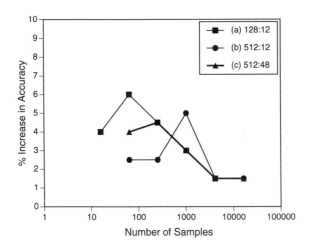

Figure 14. Mean error change for (a) 128 × 128:12, (b) 512 × 512:12, and (c) 512 × 512:48 correlated stochastic fields.

two-dimensional field with a correlation length of 12 units, and also with a correlation length of 48 units. The first field is effectively a reduction in relative correlation length from 9% (12/128) of the overall domain length, to 2% (12/512). The second field has the same percentage correlation length as the first experiment (9%), except that a finer overall grid is used. The results for these two new experiments and the original experiments are shown in Figure 14.

The results show that the peak has shifted in the 512 × 512:12 case to 1024 samples. This is equivalent to a 16 × 16 unit sample size, which is again nearly equivalent to the correlation length (12) of the field. The peak for the 512 × 512:48 case is approximately equivalent to the first experiment, and is again approximately equivalent to the correlation length of the field. Although results from only one realization for each experiment is shown, numerous other experimental replications of the stochastic fine grid field confirm these results.

The second group of numerical experiments used a pseudo-fractal (i.e., infinite correlation) fine scale $\ln(K)$ grid for the known fine grid. The pseudo-fractal field was generated with the spectral fBm algorithm described before, with a representative fractal dimension (D=2.4). Figure 15 shows the resulting plot for a 128 × 128 grid, where 64 samples (16 × 16 grid units) appears optimum. Rerunning the experiment with a known fine

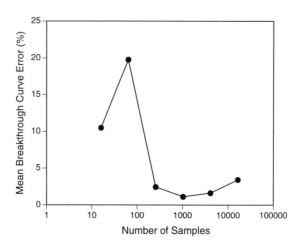

Figure 15.   Mean error change for increasing finer samples, 128×128 pseudo-fractal field.

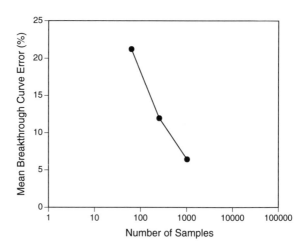

Figure 16.   Mean error change for increasing finer samples, 512×512 pseudo-fractal field.

grid of 512×512 and the same fractal dimension, no change in the sample optimum (64 samples) was evident, as shown in Figure 16. In this case, however, the sample size is 64×64 grid units. The cause for the difference in optimum sample size, yet consistency in sample quantity is not clear.

4.3. Wavelet basis effects

As noted in Subsection 2.1, many different wavelet basis functions can be used in the wavelet transform, and consequently, also in the wavelet reconstruction and extrapolation techniques. Different wavelet basis functions will preferentially move, between scales, different characteristics of the target dataset. For example, use of the Haar basis function will emphasize discontinuities in the target dataset, and similarly, would enhance reconstruction of a "discontinuous" fine grid field from a sample set. Other wavelet basis functions, such as the Daubechies family used throughout this research, emphasize the smoothness of the examined data.

The wavelet reconstruction experiments for fine scale $\ln(K)$ grids proved to be highly dependent on the wavelet basis chosen. Figure 17 shows a Haar basis wavelet reconstruction realization $\ln(K)$ shaded contour map. The resulting discontinuous field is evident, especially compared with the Daubechies4 reconstruction in Figure 9e. These discontinuities in the Haar reconstruction, which are not inherent in the known grid, made it unsuitable as a basis function for reconstructing our known fractal distributions.

An example of the effect on breakthrough curve error by utilizing different wavelet basis functions is shown in Figure 18. The Haar wavelet basis function has greater breakthrough curve error than the Daubechies4 wavelet function at all sample scales. The error decrease between scales is also generally slower for the Haar basis. This indicates that the Haar wavelet basis function does not interpolate the finer scale variability as well as the Daubechies family. This effect is again primarily due to the difference in the smoothing properties between the Haar and Daubechies wavelet basis functions. Since our known field (a pseudo-fractal) is relatively smooth, the Daubechies4 wavelet function performs better. Similar results should be expected for correlated stochastic grids.

§5. Conclusions and Recommendations

A primary purpose of this initial work was to investigate the usefulness of the wavelet technique for reconstructing fine scale variability and moving multi-scale information. A secondary purpose was to use the wavelet extrapolation technique to begin to understand what, if any, optimal scales may exist. Based on the results discussed in Section 4, a number of conclusions can be stated:

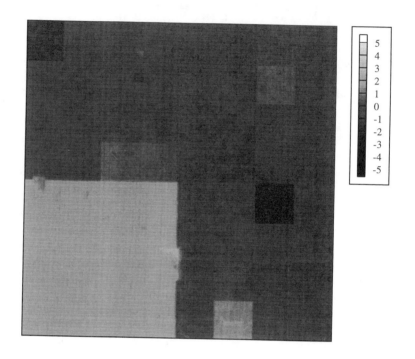

Figure 17. Shaded contour map of fine scale ln(K) field reconstructed using Haar basis function wavelet.

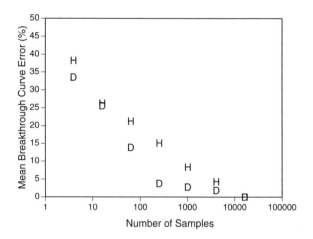

Figure 18. Effect of different wavelet functions on the mean breakthrough curve error for the 128 × 128 pseudo-fractal field.

- First, we have demonstrated that the wavelet reconstruction technique is able to incorporate the location and scale of each sample in reconstructing a fine grid as evidenced visually and by the exact resampling. The wavelet reconstruction provides a subjectively better representation of the known values than the other compared methods as shown by Figure 9. The resulting breakthrough curve mean error for the wavelet technique was similar to the conditional simulation technique error, and better than the others. Thus, we are confident in stating that the wavelet reconstruction technique is a viable method to incorporate scale information in fine scale reconstructions. Further work is needed to understand the sensitivity of this method to the sample suite, and to determine optimum sample suite characteristics.

- Second, our present results indicate that choice of wavelet function is important. Smoother wavelet functions, like the Daubechies family, recreated the synthetic fractal and correlated stochastic distributions better, as well as enabled a more rapid reduction in breakthrough curve error. Future work is needed to determine what qualities or properties of a wavelet basis function best aid reconstruction of various porous media properties. It is hoped that these wavelet basis qualities might be directly related to geologic properties of porous media.

- Third, an optimum sample scale appears to be related to the correlation length of a correlated stochastic fine scale field. This implies that the maximum sampling benefit would be from samples at the scale of the correlation length. Practically, this implies that a site characterization procedure should first be designed to determine the correlation length of the hydraulic conductivity field, possibly via sampling in one area at different scales. The procedure would be to then cover the remaining site with samples at the correlation scale. This type of site characterization procedure is quite different than what is currently considered "best professional judgement."

§6. Concluding Remarks

Since fractals have correlation at all scales, the consistency of the optimum sample quantity (not scale) for the fractal field experiments is somewhat puzzling. One possible explanation might be that there is a similarity in the way wavelet techniques function compared to statistical relationships of the number of required samples necessary for adequate representation of a normal distribution. This may be an indication that fractal distributions are best modeled by the wavelet method. Additional experiments are needed.

Besides the areas of future research described above, part of a future research plan should include increasing the robustness of the wavelet technique to handle irregular sample locations and overlapping sample volumes. In addition, applying the reconstruction technique to a tightly controlled field-generated multi-scale data set is needed. Comparison of the reconstruction of a fine scale grid, and various contaminant plumes should further clarify the limitations and expectations of the wavelet reconstruction application.

References

1. Carr, J. R., UVKRIG: A FORTRAN-77 program for universal Kriging, *Comp. and Geosci.*, 16, 211–236, 1990.
2. Chou, Kenneth C., A stochastic modeling approach to multiscale signal processing, Ph.D. Thesis, Massachusetts Institute of Technology, May 1991.
3. Chou, K. C., A. S. Willsky, A. Benveniste, and M. Basseville, Recursive and iterative estimation algorithms for multi-resolution stochastic processes, *Proceedings of the 28th Conference of Decision and Control, Tampa, Florida, IEEE*, December 1989.

4. Cole, C. R. and H. P. Foote, Multigrid methods for solving multiscale transport problems, in *Dynamics of Fluids in Hierarchical Porous Media*, Academic Press, 273–303, 1990.

5. Davis, J. M., R. C. Lohmann, F. M. Phillips, and J. L. Wilson, Architecture of the Sierra Landrones Formation, central New Mexico; depositional controls on the permeability correlation structure, *GSA Bulletin*, 105, 998–1007, 1993.

6. Daubechies, I., *Ten lectures on wavelets*, p. 1, SIAM, Philadelphia, 1992.

7. Dykaar, B. B. and P. K. Kitanidis, Determination of the effective hydraulic conductivity for heterogeneous porous media using a numerical spectral approach; 1, Method, *Water Resour. Res.*, 28, 1155–1166, 1992.

8. Hohn, Michael E., *Geostatistics and Petroleum Geology*, Van Nostrand Reinhold, New York, 1988.

9. Journel, A. G., and C. J. Huijbregts, *Mining Geostatistics*, Academic Press, San Diego, 1978.

10. Mandelbrot, B. B., *The fractal geometry of nature*, W.H. Freeman and Co., San Francisco, 1983.

11. Neuman, S. P., Statistical characterization of aquifer heterogeneities: An overview, *G.S.A.*, Special Paper 189, 81–102, 1982.

12. Pardo-Iguzquiza, E., M. Chica-Olmo, and J. Delgado-Garcia, SICON1D, A FORTRAN-77 program for conditional simulation in one dimension, *Computers & Geosciences*, 18, 665–688, 1992.

13. Press, W. H., B. P. Flanners, S. A. Teukolsky, and W. T. Vetterling, *Numerical Recipes: The art of scientific computing*, 2nd ed., Cambridge Univ. Press, Cambridge, 1992.

14. Smith, L., and F. W. Schwartz, Mass transport, 1, A stochastic analysis of macroscopic dispersion, *Water Resour. Res.*, 16, 303–313, 1980.

15. Strang, G., Wavelets and dilation equations,: a brief introduction, *SIAM Review*, 31, 614–627, 1989.

16. Tompson, A. F. B., R. Ababou, and L. Gelhar, Application and use of the three dimensional turning bands random field generator: single realization problems, MIT Department of Civil Engineering, Report Number 313, 1987.

17. Turcotte, D. L., *Fractals and chaos in geology and geophysics*, Cambridge University Press, Cambridge, 1992.

18. Voss, R. F., Fractals in nature: From characterization to simulation, in *The Science of Fractal Images*, H.-O. Peitgen, and D. Saupe, Springer-Verlag, New York, 1988.

19. Warren, J. E., and F. F. Skiba, Macroscopic dispersion, *Trans. S.P.E.*, 231, 215–230, 1964.

20. Wheatcraft, S. W. and S. W. Tyler, An explanation of scale-dependent

dispersivity in heterogeneous aquifers using concepts of fractal geometry, *Water Resour. Res.*, 24, 566–578, 1988.

21. Wornell, G. W., and A. V. Oppenheim, Estimation of fractal signals from noisy measurements using wavelets, *IEEE Trans. Sig. Proc.*, 40, 611–623, 1992.

This work was primarily funded by U.S. E.P.A. Grant CR817379, "Site Characterization for Hazardous Waste Sites." This work has not been reviewed by EPA, and does not necessarily represent the position or opinions of EPA. This work was also partially funded by the Desert Research Institute, Water Resources Center, Reno, Nevada.

Kevin E. Brewer
Department of Geological Sciences/ 172
University of Nevada, Reno
Reno, Nevada 89557
USA
e-mail: *kevinb@hydro.unr.edu*

Stephen W. Wheatcraft
Department of Geological Sciences/ 172
University of Nevada, Reno
Reno, Nevada 89557
USA
e-mail: *steve@hydro.unr.edu*

Wavelet-Based Multifractal Analysis of Non-Stationary and/or Intermittent Geophysical Signals

Anthony Davis, Alexander Marshak, and Warren Wiscombe

Abstract. We show how some of the most popular wavelet transforms can be used to compute simple yet dynamically meaningful statistical properties of a one-dimensional dataset representative of a geophysical field or time-series. The observed properties of turbulent velocity fields and of liquid water density in clouds are used throughout as examples along with theoretical models drawn from the literature. We introduce a diagnostic tool called the "mean multifractal plane," simpler than (but representative of) the two well-known multifractal hierarchies of exponents $H(p)$ and $D(q)$, associated with structure functions and singularity analysis respectively. It is used to demonstrate the pressing need for a new class of stochastic models having both additive/non-stationary and multiplicative/intermittent features.

§1. Introduction and Overview

Geophysical systems are notoriously complex, driven by external forcing and ridden with internal instability due to their generically nonlinear dynamics. Whether traceable to the break-up of larger ones or the amplification of smaller ones, structures are present on all observable scales. The signals we depend on to probe these systems are naturally "rough" and/or "jumpy" (i.e., non-differentiable and/or discontinuous) and little insight is gained by statistical analysis unless the methods we adopt are fully adapted to this situation. Traditional approaches include one-point histograms (typically searching for Gaussianity), two-point auto-correlations (typically searching for an exponential decay) and energy spectra (typically searching for dominant frequencies). They are better suited for electrical engineering applications than for geophysical research, although all of the above listed statistics can be used in ways and circumstances we will describe in various parts of this paper. In the last couple of decades, we have

Wavelets in Geophysics
Efi Foufoula-Georgiou and Praveen Kumar (eds.), pp. 249–298.

seen Fourier analysis complemented by wavelet analysis, an increasing interest in lognormal and power-law distributions and an explosion of ideas centered on scale-invariance and fractals.

Scale-invariance is the natural framework for developing statistical tools that account for all scales in presence at once. We focus on random quantities that depend parameterically on scale – an idea idiosyncratic to wavelet theory as well – and seek power-law behavior with respect to the scale parameter for their statistical properties. If a single power-law exponent is sufficient to describe all the statistics within a whole family, we talk about monoscaling and a "monofractal" model is called for; otherwise (the most general case), we are dealing with multiscaling and "multifractal" models are in order. Multifractal statistics have proven useful in almost every reach of nonlinear science, in the laboratory as well as in computational experiments (including the exploration of deterministic chaos). It is hardly surprising that geophysical applications of multifractal and/or chaos theory have been so successful, judging by the growing number of researchers involved. However this says more about the geophysical community's data analysis needs and about the expectations raised by the scaling paradigm than about the degree of maturity reached by multifractal statistical methodology. Many questions remain open on issues ranging from the preprocessing required by certain procedures (cf. introductory remarks to section 4) to the validation of multiscaling studies (e.g., "Is the observed multiplicity simply due to finite size effects?") but also on more fundamental issues, for instance, the role of stationarity.

Because we depend on the spatial coordinate to perform all averaging operations when analyzing data, our first task is to determine which aspects of the signal are more likely to be stationary. In the upcoming section (and the Appendix) we show that, within the framework of scale-invariance, this can be done using spectral analysis. Assuming thereafter that the geophysical process of interest is non-stationary but with stationary increments (a frequent occurrence in nature), we turn to p-th order structure functions in section 3. How can they be obtained from the wavelet transform of the data? What are the general properties of the associated family of exponents $\zeta(p)$? How do they relate to geometrical, spectral and dynamical concepts? Another option altogether is to derive a stationary "measure" from the non-stationary signal, e.g., by taking absolute differences between (appropriately separated) points; one can then apply some form of singularity analysis to it (we simply consider here the q-th order moments of the coarse-grained measure). The same questions as above are asked about this approach and associated family of exponents $K(q)$ in section 4. The most important outcome is the mapping of structure functions to non-stationarity and of singularity analysis to intermittency, two challenging aspects of nonlinear dynamics that happen to have well-defined statistical meanings. Often however, two large families of exponents is too much to handle on practical grounds. In section 5 we briefly describe a way of relating them in an (almost) one-to-one fashion and go on to select a single exponent from each approach. This enables us to devise a plot where one axis, $H_1 = \zeta(1)$, measures the non-stationarity of the system and the other, $C_1 = K'(1)$, its intermittency; we call this the "mean multifractal plane" and discuss some possible applications, most importantly to the relatively new field of multi-affine [57] stochastic modeling. Finally we summarize and offer some concluding remarks in section 6.

We have surveyed the literature on a per topic basis: stationarity and spectral analysis (section 2, Appendix), structure functions and multi-affinity (section 3), singularity analysis and multiplicative cascade models (section 4). The wavelet literature *per se* is already enormous but by-and-large wavelet practitioners rely heavily on the existence of the inverse wavelet transform [14]. Signal reconstruction is exact (no information is lost) and, when made approximate, it is generally in the spirit of data compression. Computational efficiency is often sought using the orthonormality of many well-known wavelet bases [15]. These are not our concerns. In statistical analysis, a large amount of information is thrown away in the process of extracting a small number of sure quantities (i.e., averages) from a large number of random ones (i.e., the data) and furthermore the computational burden is slight in comparison to the overhead in data production. There is also a large body of multifractal literature, reviewed here as needed. We will borrow heavily from the relatively small (but steadily growing) number of papers that use wavelets in connection with multifractal models and/or data structures (generally pertaining to turbulence). Most of these publications appeared in physics journals, emanating from the group spearheaded by Alain Arnéodo of the Centre de Recherche Paul Pascal (Pessac, France). His lectures and the papers he co-authored gave us the impetus for this study.

§2. Preliminary Considerations on Stationarity, Ergodicity and Scale-Invariance

2.1. Data specification and spectral analysis

We consider given a discrete set of $N + 1$ data points:

$$f_i = f(x_i), \quad x_i = il \quad (i = 0, 1, \ldots, N), \tag{1}$$

sampled from $f(x), x \in [0, L]$, some scalar geophysical field (x is position) or time-series (x is time). The overall length of the one-dimensional dataset is L and l is the grid constant (or $1/l$, the sampling rate). We will furthermore assume that

$$N = \frac{L}{l} \gg 1. \tag{2}$$

In many cases, a time-series is really a picture of a geophysical field. Consider for instance the time-series $f(x)$ in Figure 1a. It represents liquid water content (LWC) captured from an airborne platform during FIRE (First ISCCP Regional Experiment) while flying through an extended marine stratocumulus (StCu) deck off the southern coast of California on June 30, 1987. (This type of cloud is of particular interest to climate modelers due to its frequent occurrence over vast regions, hence a first order effect on the planetary radiation budget.) Since the aircraft was flying at roughly constant speed and altitude, we can view this as a snap-shot of the local atmospheric LWC field, sampled along a linear transect (Taylor's frozen turbulence hypothesis). The parameters of this dataset are $l \approx 5$ m and $N = 2^{14} = 16384$, hence $L \approx 82$ km. We note the "roughness" of this graph which we will quantify in section 3. Figure 1b presents a related dataset $\varepsilon(x)$ based on absolute differences over a certain scale (to be discussed shortly). We note the inhomogeneous "spikiness" of this field which, in the tradition of turbulence studies, we view as a direct manifestation of intermittency (sudden bursts of high frequency activity) in the original signal. We will quantify intermittency in this sense in section 4.

The information in Equation (1) can be represented by a discrete Fourier transform: $\widehat{f}_{i-N/2} = \widehat{f}(k_{i-N/2})$, $k_{i-N/2} = (i - N/2)/L$ ($i = 0, 1, \ldots, N$). Being interested in finding the statistically robust features of $f(x)$, we start with the energy spectrum:

$$E_f(k) = \langle \sum_{\pm} |\hat{f}(\pm k)|^2 \rangle, \quad 1/L \leq k \leq 1/2l, \tag{3}$$

dropping subscripts for simplicity. This is probably the most popular statistic in time-series analysis, beyond 1-point p.d.f.'s (i.e., simple histograms of f-values as in Figure 1a). The energy spectrum $E_\varepsilon(k)$ of the ε-field in Figure 1b is computed in the same way and both spectra for the data in Figure 1 are plotted in Figure 2.

In Equation (3) we use $\langle \cdot \rangle$ to denote an "ensemble" average which involves in principle every possible realization of the random process $f(x)$. In geophysical practice however, we are generally provided with the outcome of a small number of experiments – possibly even a single one – and we are forced to sacrifice spatial information to obtain spatial averages, as estimates of their ensemble counterparts. Processes for which spatial averages converge to the corresponding ensemble statistics in the limit of large averaging sets are called "ergodic." Estimating an energy spectrum with a single realization, dropping the triangular brackets in Equation (3), amounts to making an ergodicity assumption. Throughout this paper, we use the same notation for ensemble averages and their spatially obtained estimates. The interested reader is referred to [32] and [29] for a detailed analytical study of multiplicative cascade models where cases of ergodicity violations can be found; this important class of models will be briefly discussed in section 4 and in the Appendix.

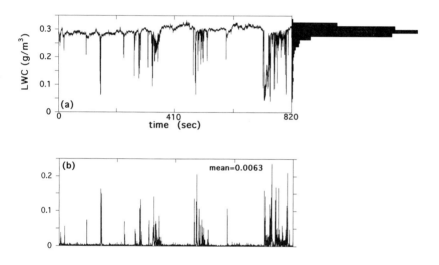

Figure 1. *A horizontal probing of atmospheric liquid water content (LWC), showing the internal structure of a marine stratocumulus deck during FIRE.* (a) The raw LWC data $f(x)$, sampled at 20 Hz (every $l \approx 5$ m, aircraft speed ≈ 100 m/s) for 820 s ($L \approx 82$ km). Also indicated is the negatively skewed 1-point p.d.f. (the downwards spikes responsible for this skewness are probably related to instabilities that entrain of dry air from cloud top). (b) An absolute gradient field $\varepsilon(x)$ associated with $f(x)$ in panel (a). The differences are taken at the lower end (η) of the non-stationary scaling regime defined by the structure functions (cf. Figure 5a) or the energy spectrum (cf. Figure 2); in this case, we use a distance of 4 pixels ($\eta \approx 20$ m).

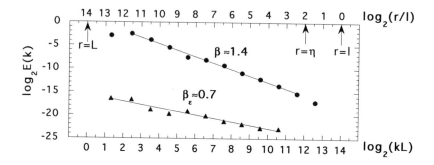

Figure 2. *Scale-invariance of the spectral densities for the datasets in Figure 1.* Log (base 2) of the energy $E(k)$ versus $\log_2(kL)$ for $2/L \leq k \leq 1/2l$. These bounds are set upwardly by the Nyquist frequency and downwardly by the minimal noise reduction option using routine SPCTRM (two non-overlapping partial spectra are averaged) in [50]. We also give a corresponding log-scale for r/l (using $r \approx 1/k$ with $L/l = 2^{14}$) with several remarkable values highlighted. We notice the lack of small scale (high k) information for the ε-field which is formed by taking differences over 4 grid points (corresponding to a distance $\eta \approx$ 20 m). As advocated by Marshak *et al.* [40], the energy as well as the (linearly discretized) wavenumbers are averaged inside octave-sized bins. Absolute slopes (power-law exponents) are indicated in both cases, showing this LWC dataset to be non-stationary ($\beta > 1$) over scales from 20 m to at least 5 km (50 s in Figure 1) while its associated ε-field is stationary ($\beta_\varepsilon < 1$). Davis *et al.* [18] show that the transition to stationary behavior ($E(k) \approx$ constant) at $r > R \approx$ 3–5 km ($\log_2(kL) <$ 2.5) is traceable to a single event (namely, the intense $\approx R$-sized "dip" at $\approx$ 700 s in Figure 1) and therefore is not robust with respect to the addition of new data. (An improved estimate of this "integral" scale for LWC in marine StCu is found to be $\approx$ 20 km.)

2.2. Stationarity and the spectral exponent

Spectral analysis is often used to find dominant frequencies (hence time-scales) in a time-series, viewed as a harmonic (and possibly deterministic) signal contaminated with random noise, the latter having a characteristically continuous spectrum. However, in natural systems it is usually futile to try to separate a smooth signal from the noise. These systems are turbulent and the signals they produce are noise. The prevailing type of noise in Nature is furthermore "scale-invariant," meaning that it is free of characteristic scales, at least over some large "scaling" range:

$$\eta \leq r \leq R \tag{4}$$

which will depend on the type of geophysical signal. The corresponding spectral densities reflect this by showing power-law distributions of energy with respect to scale ($r \sim 1/k$):

$$E_f(k) \propto k^{-\beta}, \quad E_\varepsilon(k) \propto k^{-\beta_\varepsilon} \tag{5}$$

over the associated range of wavenumbers. In the following, we will assume for simplicity that the dataset is scale-invariant in the full instrumentally accessible range of scales ($l \leq r \leq L$) in our general discussions. In our data-specific discussions (Section 3.4 and Section 4.4), we relax this assumption. It is noteworthy that the exponent β can be retrieved from a "wavelet spectrum" [21] under conditions spelled out by Fantodji *et al.* [20].

As shown in Figure 2, the LWC datasets scale over most of the accessible range of scales with $\beta \approx 1.4$ for the LWC data in Figure 1a and $\beta_\varepsilon \approx 0.7$ for the associated absolute gradient field in Figure 1b; details on these and several other related LWC time-series are given by Marshak *et al.* [40] and Davis *et al.* [16]. Duroure and Fantodji [19] compare the scaling of Fourier- and wavelet spectra for droplet density (rather than LWC) in StCu as well as for several stochastic models, similarly for Gollmer *et al.* [27] with liquid water path (the vertical integral of LWC retrieved from passive microwave radiometry).

One-point p.d.f.'s make no use of the ordering of the data, leaving wide open the possibility of correlations between f-values at different points (more loosely, "structures" or "patterns" in the data). The energy spectrum partially describes these correlations and has one advantage over many other popular statistics, including auto-correlation functions: it remains well-defined whether or not the process is "stationary," i.e., statistically invariant under translation. In the Appendix, we discuss stationarity in some detail because, being a prerequisite for ergodicity, it is necessarily invoked in every data analysis task where spatial averaging is performed as a surrogate for ensemble averaging. In the same Appendix, we argue that the spectral exponent can in fact be used to distinguish between stationary ($\beta < 1$) and non-stationary ($\beta > 1$) scaling situations.

According to this criterion, Figure 1a illustrates non-stationarity, at least for scales in the range 20 m $\approx 4l \leq r \leq L/16 \approx 5$ km (cf. Figure 2). Although this may seem paradoxical to the casual observer, Figure 1b illustrates a stationary process. Indeed it is characterized visually by long stretches with little activity suddenly interrupted by spikes coming in a wide variety of heights and with different degrees of clustering. *A priori* this sounds rather non-stationary. Conventional wisdom wants stationarity to be the tendency that "things are sort of the same at all times;" a better way of envisioning it would be a tendency to "quickly come back to typical (i.e., most probable) values." The picture conveyed by the former description is too restrictive, apparently excluding the strong deviations from average values observed in Figure 1b. We will substantiate this view in the Appendix as well.

§3. Quantifying and Qualifying Non-Stationarity with Structure Functions

In this section we interest ourselves primarily in non-stationary processes with stationary increments $(1 < \beta < 3)$ which are ubiquitous in nature. So, following the ideas surveyed in the Appendix, we will focus on the field of increments of a random process $f(x)$ over a distance r: $\Delta f(r; x) = f(x + r/2) - f(x - r/2)$, $r/2 \leq x \leq L - r/2$ $(0 < r \leq L)$. The statistical moments of $|\Delta f(r; x)|$ are a two-point statistic known as "structure functions of order p" and they are independent of x: $\langle |\Delta f(r; x)|^p \rangle \equiv \langle |\Delta f(r)|^p \rangle$; this incremental stationarity is a necessary condition to feel justified in estimating $\langle |\Delta f(r)|^p \rangle$ by averaging $|\Delta f(r; x)|^p$ over x. In most cases small increments are a frequent occurrence so one cannot take arbitrarily large negative values for p.

Arnéodo, Bacry and Muzy have shown in a series of papers ([5], [7], [9] and [45]) how wavelets can be explicitly used to perform structure function analysis in its standard and more general forms. We summarize their ideas in the upcoming subsection then review the implications of scale-invariance; finally, we discuss turbulence studies and apply the technique to the LWC data presented above, underscoring the dynamical meaning of structure function exponents in connection with non-stationarity.

3.1. Wavelets for structure function analysis

Define the "indicator" function of the open r-sized interval $(-r/2, +r/2)$, $r > 0$:

$$I_r(x) = \begin{cases} 1 & |x| < r/2 \\ 0 & |x| \geq r/2 \end{cases}. \tag{6a}$$

In the following, we will also need

$$\varphi_\sigma(x) = \frac{1}{\sigma\sqrt{2\pi}} \exp(-x^2/2\sigma^2), \tag{6b}$$

a Gaussian with vanishing mean and standard deviation σ.

The field $\Delta f(r; x)$ can be viewed as the convolution product of $f(x)$ with $\partial I_r(x) = \delta(x + r/2) - \delta(x - r/2)$ where $\delta(\cdot)$ designates Dirac's generalized function. So we have

$$\Delta f(r; x) = [f * \partial I_r](x) = \int f(x')[\delta(x + r/2 - x') - \delta(x - r/2 - x')] \, dx', \tag{7}$$

where boundary complications are dealt with by making $f(x)$ periodic with period L. Since the kernel ∂I_r of the integral operator in Equation (7) is parameterized by a variable scale r, we are reminded of a wavelet transform. Indeed, it is easy to show that

$$\Delta f(r; x) = \frac{1}{r} \int f(x')[-\partial I_1(\frac{x' - x}{r})] \, dx'. \tag{8}$$

Muzy *et al.* [45] call $-\partial I_1(x) = \delta(x - 1/2) - \delta(x + 1/2)$ a "poor-man's wavelet" and suggest a more general approach to increment estimation based *a priori* on a restricted class of anti-symmetric wavelets $\Psi(x)$, orthogonal in particular to constant functions but not to linear ones, that furthermore contain a single oscillation. Along with $-\partial I_1(x)$, Figure 3 shows classical examples:

$$\Psi_G(x) = -\varphi'_{1/2}(x) = 4x\varphi_{1/2}(x), \tag{9a}$$

on the one hand, and

$$\Psi_H(x) = \begin{cases} \text{sign}(x) & |x| \le 1 \\ 0 & |x| > 1 \end{cases} \tag{9b}$$

on the other. The latter and simpler (piece-wise constant) case is the well-known Haar wavelet. In analogy with Equation (8), we consider the continuous wavelet transform

$$T_\Psi[f](a, b) = \frac{1}{a} \int f(x)\Psi(\frac{x - b}{a}) \, dx, \quad a > 0, \ b \in \Re, \tag{10}$$

where the wavelet is real (so Ψ is used instead of its conjugate Ψ^*). Notice the non-standard normalization in Equation (10): $1/a$ appears, rather than $1/\sqrt{a}$. This inhibits the strict orthonormality of the wavelet basis (not required in the following) but ensures the constancy with respect to a of the integral of the positive part of the rescaled wavelet, $a^{-1}\Theta[\Psi(x/a)]$ where $\Theta(\cdot)$ is the Heaviside step function. This facilitates scaling comparisons.

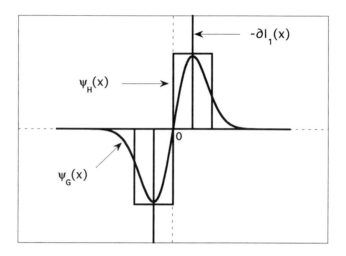

Figure 3. *Acceptable wavelets for structure function analysis.* The first derivative $\Psi_G(x)$ of $(2/\pi)^{-1/2}\exp(-2x^2)$; Haar's piecewise-constant wavelet $\Psi_H(x)$ and $-\partial I_1(x)$, the "poor-man's wavelet" of Muzy *et al.* [45], i.e., two δ-functions of opposite sign. Notice that all three are anti-symmetric but have ever smaller supports: all of $\Re$, a compact subset, two discrete points.

The premise of Arnéodo and his co-workers is that, being increment-like, the random wavelet coefficients in Equation (10) have statistical properties similar to those of $\Delta f(r;x)$ with scale r playing the role of the dilation parameter a and position x that of the translation parameter b. In particular, $\langle |T_\Psi[f](a,b)|^p \rangle$ will be independent of b and we are justified in estimating this (absolute) moment by averaging over the location. In summary, we expect that

$$\langle |\Delta f(r)|^p \rangle \approx \frac{1}{L}\int_0^L |T_\Psi[f](r,b)|^p\,db, \quad 0 < r \le L,\; p > -1, \quad (11)$$

independently of the choice of Ψ (say) in Equations (9a) and (9b). Using a variety of numerical and analytical methods, Arnéodo *et al.* ([5], [7]), Bacry *et al.* [9] and Muzy *et al.* [45] have shown this conjecture to hold (in the sense of scaling described below) for a wide class of random processes. We refer to [45] (and references therein) for the restriction on p

in Equation (11). These authors also show how wavelets with more oscillations (e.g., successive derivatives of $\varphi_\sigma(x)$, orthogonal to higher order polynomials) can be used to find the scale-invariant properties of a structure function-like statistic in Equation (11) for $p \leq -1$, which is dominated by regular points where the signal $f(x)$ is differentiable.

3.2. Scale-invariant structure functions, generalities

Random scale invariant fields/signals $f(x)$ are characterized by scale-conditioned statistics that follow power-laws with respect to the scale parameter. Equation (11) describes a way of estimating such a statistic (r is held constant during the spatial averaging), we thus anticipate

$$\langle |\Delta f(r)|^p \rangle \propto r^{\zeta(p)}, \tag{12}$$

over some large range of scales $[\eta, R]$ which may partially or entirely overlap with the instrumentally accessible range $[l, L]$.

Some general statements can be made about the family of exponents $\zeta(p)$. Firstly, one exponent is known *a priori*, from definitions: $\zeta(0) = 0$. Secondly, $\zeta(p)$ will be a smooth differentiable function of p no matter how rough the data $f(x)$, being essentially a sum of many exponential functions of p. Thirdly, it can be shown that, if the signal $f(x)$ has absolute bounds, then $\zeta(p)$ is monotonically non-decreasing ([23] and [39]). Finally, it can be shown on more general grounds that $\zeta(p)$ is concave ($\zeta''(p) < 0$).

To see this, choose units of length where $L = 1$ and f-units where $\langle |\Delta f(1; x)|^p \rangle \approx O(1)$; we then have $\langle |\Delta f(r; x)|^p \rangle \approx r^{\zeta(p)}$, making explicit the weak dependence of the prefactors on p. It follows that $-\zeta(p) \ln r$ is the second characteristic (or cumulant generating) function $\ln\langle \exp(-p\xi_r) \rangle$ of the random variables $\xi_r = -\ln |\Delta f(r; x)|$ and is therefore convex [22].

Stationary processes of course have stationary increments but these will scale trivially due to the invariance under translation: $\zeta(p) \equiv 0$ (scale-independent increments). However, due to the effects of finite spatial resolution, even theoretical models lead to numerically small $\zeta(p)$'s (see [39] for an example). At the other end of the non-stationarity scale, we have continuous functions with bounded non-vanishing first derivatives, yielding $|\Delta f(r; x)| \propto r$ (for almost every x) hence $\zeta(p) = p$.

Any concave $\zeta(p)$ with $\zeta(0) = 0$ allows the definition of a hierarchy of exponents:

$$H(p) = \frac{\zeta(p)}{p}. \tag{13}$$

By "hierarchy" we mean a monotonic function, in this case non-increasing. The reader is referred to Parisi and Frisch's [48] original paper on the multifractality of turbulent velocity signals/fields for the interpretation of the $\zeta(p)$ and $H(p)$ functions in terms of variable orders of singularity. ("Singularity" is used here in the sense of Hölder-Lipshitz heuristics: how does

$h(x)$ in $|\Delta f(r; x)| \propto r^{h(x)}$ vary, statistically speaking, with x?)

Obtaining $\zeta(p)$ and/or $H(p)$ is the goal of this whole approach, providing us with a means to characterize non-stationarity, in both quantitative and qualitative terms (examples to follow).

3.3. Geometrical, spectral and dynamical implications of structure function statistics

Processes with a variable $H(p)$ have recently been described as "multi-affine" [57] or "non-stationary multifractals" [39], leaving processes with a constant $H(p)$ or linear $\zeta(p)$ to be "mono-affine" or "non-stationary monofractals." This last class of random functions is indistinguishable from "fractional" Brownian motions (used as examples in Section 3.4). Such cases are often referred to as "self-affine" in the literature but this is an unnecessarily restrictive view of statistical self-affinity.

The expressions multi- and mono-affinity refer in fact to a well-known geometrical meaning of the $p = 1$ structure function. The quantity $\langle |\Delta f(r)| \rangle$ can indeed be related to the fractal structure of the graph $g(f)$ of $f(x)$, viewed as a statistically self-affine geometrical object in 2-dimensional Euclidean space. We have [38]

$$H_1 = H(1) = \zeta(1) = 2 - D_{g(f)} \geq 0, \qquad (14a)$$

where $D_{g(f)}$ is the fractal dimension of $g(f)$ or, in slightly different words, H_1 is the codimension of $g(f)$ and is also known as the "roughness" exponent. This immediately tells us that the largest possible value for H_1 is 1, attained for (almost everywhere) differentiable functions which have smooth graphs of non-fractal dimension $D_{g(f)} = 1$, like any line or otherwise rectifiable curve. At the opposite limit of $H_1 = 0$ (stationarity), we find graphs that fill space: $D_{g(f)} = 2$, a direct consequence of the necessary discontinuity of stationary scaling processes discussed in the Appendix.

An infinite number of exponents is needed to describe multi-affine processes completely. For instance, there is no general relation between H_1 and the spectral exponent which is however connected with the $p = 2$ case:

$$\beta = \zeta(2) + 1 = 2H(2) + 1 \geq 1, \qquad (14b)$$

the Wiener-Khintchine relation for non-stationary scaling processes (cf. Appendix). We will nevertheless retain H_1 as the single most important exponent in the whole $H(p)$ hierarchy. The geometrical significance given in Equation (14a) is not as important as the fact that H_1 defines as simply as possible the linear trend in $\zeta(p)$, cf. Figure 5b. In summary, we view H_1 as a direct quantifier of the system's non-stationarity and the complete $\zeta(p)$ or $H(p)$ functions as a means to qualify this non-stationarity.

We can take this statement one step further than its purely statistical sense. In the present circumstances, the expression "non-stationarity" is dynamically correct. In the Appendix we recall that non-stationary scale-invariant random processes, ubiquitous in nature, are characterized by large wanderings from mean (or rather most probable) values with relatively long return times. At very long times we expect a transition to stationary behavior to occur at least for geophysical fields which have absolute bounds. In nonlinear dynamical systems (of ODEs), stationary (time-independent) solutions generally occupy a very small portion of parameter space; an even smaller portion are stable and even these are generally unstable with respect to finite perturbations. In the chaotic regime time-dependent (non-stationary) solutions starting close to each other wander away exponentially fast at small times; at long times the outer size of the (usually strange) attractor puts bounds on their relative deviation. Shifting from computational to natural systems we know, at best, the governing nonlinear system of coupled PDEs (usually with forcing and dissipation terms); these are also generically unstable and their solutions are attracted into a certain state of "turbulence." At worst, we have no clue of the governing equations and rely primarily on observations for insight. The case of clouds and many other geophysical systems is intermediate; we advocate a more systematic use of structure functions, wavelet-based or not, to characterize the non-stationary aspects of their dynamics.

3.4. Examples from theory, turbulence and cloud structure

In Figure 4 we see five random processes, all theoretical, with different degrees of non-stationarity. At the top, we find a sample of "flicker" or $1/f$ noise ($\beta = 1$) which is marginally stationary: $\zeta(p) = H(p) \equiv 0$. At the bottom, we find a noiseless trend which represents everywhere differentiability: $\zeta(p) = p$ and $H(p) \equiv 1$. Between these two extremes, we find three samples of "fractional" Brownian motion (fBm), illustrating less-and-less stationarity: $\zeta(p) = pH_1$ or $H(p) \equiv H_1$ with $H_1 = 1/3, 1/2, 2/3$. The relation $\zeta(p) = pH_1$ has a simple probabilistic meaning. Using Equation (13), it implies $\langle |\Delta f(r)|^p \rangle \approx \langle |\Delta f(r)| \rangle^p$, i.e., that the p.d.f. of $|\Delta f(r; x)|$ is narrow enough to enable a simple dimensional argument to relate quantitatively all moments. The Gaussian p.d.f.'s used in Figure 4 for fBm are good examples of weakly variable increments. These models however are monoscaling (indeed mono-affine) whereas the next two examples argue that natural processes tend to be multiscaling (hence multi-affine).

The most notable structure function analysis in the turbulence literature is due to Anselmet *et al.* [1] who determined $\zeta(p)$ for wind tunnel flows at high Reynolds numbers. The authors find $H_1 \approx 1/3$ and, more importantly, confirm Kolmogorov's [34] prediction that $\zeta(3) = 1$ (cf. section 5). They also extend their averaging procedures up to $p = 18$; this is not a simple task because the larger the order p of the statistical moment in Equation (12), the more it is dominated by extreme events in $\Delta f(r; x)$ which

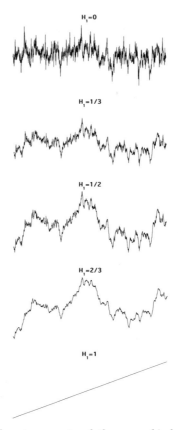

Figure 4. *Different amounts of the same kind of non-stationarity.*
Samples of Mandelbrot's [38] fractional Brownian motion, $B(x)$, $0 \leq x \leq L = 1024$, for $H_1 = 0, 1/3, 1/2, 2/3$ and 1, from top to bottom.
The mid-point displacement algorithm was used [49]: take $B(0) = 0$
and $B(L) = 1$; generate a zero-mean Gaussian deviate with standard
deviation $\sigma = \sqrt{2^{1-H_1} - 1}$ $(0 \leq H_1 \leq 1)$ and add it to $[B(0) + B(L)]/2$
to obtain $B(L/2)$; divide σ by 2^{H_1} and repeat between $x = 0, L/2$ for
$B(L/4)$, and $x = L/2, L$ for $B(3L/4)$; proceed similarly to smaller scales
(10 divisions in this case). The basic non-stationarity parameter H_1 increases from the marginally stationary case ($1/f$ noise at $H_1 = 0$) to the
extreme case of a noiseless linear trend ($B(x) = x/L$ at $H_1 = 1$). This
family of processes are all non-stationary ($\beta = 2H_1 + 1 > 1$) with stationary increments ($\beta < 3$) and can be said to have the same "kind" of non-
stationarity, namely, the monoscaling (or mono-affine) kind: $H(p) \equiv H_1$.
Other (multi-scaling) types of scale-invariant non-stationary models are
listed in the text.

are poorly sampled (by definition) in data analysis situations. The authors found not only a nonlinear $\zeta(p)$, thus establishing the multiscaling, but used it for discriminating between the competing so-called "intermittency" models. Anselmet and his co-workers used standard structure functions based on two discrete values in their time-series. Applications of *bone fide* wavelets to the analysis of turbulent signals abound and have been extensively reviewed by Farge [21]. Of particular interest here are results on velocity time-series generated in three-dimensional turbulent flows: Argoul *et al.* [2] use the symmetric wavelets discussed in Section 4.1 below to visualize the turbulent cascade process (rather than perform scaling analyses) while Muzy *et al.* [44] use the same symmetric wavelets as well as their anti-symmetric counterparts discussed above in scaling analyses.

Figure 5a shows $\langle|\Delta f(r)|^p\rangle$ for the LWC data in Figure 1a, using the traditional discrete value calculation of structure functions. The scaling range $[\eta, R]$ where the exponents are defined is clearly marked. However, Davis *et al.* [18] show that the break that defines the upper limit of $R \approx 5$ km can be traced to a single event (a "dip" in LWC of approximately that width that occurs *c.* 700 sec. into the flight on Figure 1a). In other words, this estimate of this transition point to large-scale stationary behavior, $\langle|\Delta f(r)|^p\rangle \approx constant$, is not robust with respect to adding to the average the four other LWC datasets obtained from the same instrument during FIRE. A better estimate (using all five flights) is found to be ≈ 20 km and, viewing R as a statistic, we are simply in presence of a violation of ergodicity (another one is described in Figure 9). Assuming it is robust, the special scale R is known as the "integral" scale of the process (see Appendix). In Figure 5b we have plotted $\zeta(p)$ and highlighted two remarkable values: $\zeta(1) = H_1 \approx 0.28$, not unlike turbulent velocity, and $\zeta(2)$ which is simply related to the spectral exponent β, as described in Equation (14b). We notice the strong evidence of non-stationarity ($\zeta(p)$ is non-vanishing) in marine StCu and their multifractality (in the sense of multi-affinity, $\zeta(p)$ is nonlinear).

Multi-affine modeling is the synthetic counterpart of structure function analysis, a relatively new field judging by the following literature survey that we believe to be exhaustive at the time of publication. Schertzer and Lovejoy [52] suggested the idea of low-pass power-law filtering in Fourier space (or "fractionally" integrating) singular multiplicative cascade models; the latter are described in some detail below. Cahalan *et al.* [12] describe a class of "bounded" cascade models, shown by Marshak *et al.* [39] to have $\zeta(p) = \min[pH, 1]$ where the smoothing parameter H goes from 0 (singular cascades) to ∞ (Heaviside step functions). Viscek and Barabási [57] describe a variant of the mid-point displacement algorithm for generating fBm that yields processes with multiscaling structure functions and compares their theoretical $\zeta(p)$'s to those empirically determined for turbulent

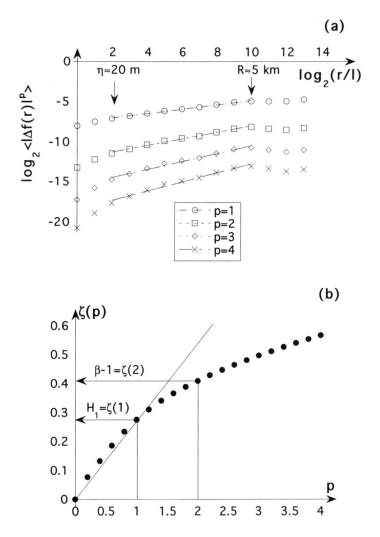

Figure 5. *Structure function analysis of cloud LWC.* (a) The quantities $\log_2 \langle |\Delta f(r)|^p \rangle$ are plotted for $\log_2(r/l) = 0, \ldots, 13$ and $p = 1, 2, 3, 4$. We have indicated the scaling range $[\eta, R]$ and the slopes are the exponents $\zeta(p)$. (b) The corresponding $\zeta(p)$ function, demonstrating the multi-affinity of LWC and highlighting the remarkable values at $p = 1, 2$ discussed in connection with Equations (14a,b).

velocity. Arnéodo *et al.* [5] and Muzy *et al.* [44] use recursively generated fields (akin to cascades but with negative weights as well as positive ones) smoothed by fractional integration. Finally, Benzi *et al.* [11] have recently proposed a wavelet-based technique for generating fields with any prescribed $\zeta(p)$ function.

We now turn to an alternative (and currently more popular) multifractal approach to statistical data characterization, singularity analysis, and another challenging issue in nonlinear processes, namely intermittency.

§4. Quantifying and Qualifying Intermittency with Singular Measures

In this section we turn our interest to random measures $\varepsilon(x)$ that are stationary ($\beta_\varepsilon < 1$); "measures" are non-negative processes defined in principle only through integrals. Geophysical data generally does not come in this (stationary, non-negative) format, a notable exception being rain rate [29]. However, a stationary measure can always be derived from a signal (say) by finite differencing and taking absolute values (cf. Figures 1a,b). Other options are possible. For instance, working with one-dimensional turbulent velocity signals, Meneveau and Sreenivasan [41] take squares rather than absolute values while Schmitt *et al.* [54] use "fractional" differentiation (a high-pass power-law filtering). Working on two-dimensional imagery, Tessier *et al.* [56] experiment with gradients and Laplacians and, working with cascade models, Lavallée *et al.* [36] demonstrate numerically this makes little difference in the ensuing singularity analysis anyway.

We will assume given some stationary random measure $\varepsilon(x) \geq 0$, $0 \leq x \leq L$, either on a grid (as for data) or in the continuum limit (as in the case of a theoretical model). Here again, wavelet-based techniques in singularity analysis were first investigated by Arnéodo *et al.* ([3], [4]). Their basic ideas are surveyed in the first subsection before reviewing the consequences of scaling; finally, we illustrate the whole approach with turbulence results from the literature and with our own LWC data, relating the new exponents to the dynamical concept of intermittency.

4.1. Wavelets and "scaling functions" for singularity analysis

The basic idea in singularity analysis is to degrade the resolution with which the measure $\varepsilon(x)$ is "observed" (mathematically speaking, via integrals over intervals of variable lengths). This is very close to the concept of wavelet analysis (viewed as a mathematical microscope with a variable magnification or, more precisely, field of view). The simplest approach is to generate a sequence of spatially degraded or coarse-grained measures $\varepsilon(r; x)$, $r/2 \leq x \leq L - r/2$ ($0 < r \leq L$), obtained by averaging $\varepsilon(x)$ over $[x - r/2, x + r/2]$. This is equivalent to a series of convolution products of

$\varepsilon(x)$ with $I_r(x)$, as defined in Equation (6a):

$$\varepsilon(r;x) = \frac{1}{r}\int_{x-r/2}^{x+r/2}\varepsilon(x')\,dx' = \frac{1}{r}\int\varepsilon(x')I_r(x-x')\,dx' = \frac{1}{r}[\varepsilon * I_r](x); \quad (15)$$

again eventual boundary complications are eliminated by making $\varepsilon(x)$ L-periodic. Making use of the symmetry of $I_r(x)$, we can rewrite Equation (15) as a wavelet-type transform:

$$\varepsilon(r;x) = \frac{1}{r}\int\varepsilon(x')I_1\left(\frac{x'-x}{r}\right)dx'. \quad (16)$$

The kernel $I_1(x)$ is however more akin to a "scaling function" than a wavelet since it is everywhere non-negative, hence a non-vanishing integral. In fact $I_2(x)$ is the scaling function of the Haar wavelet $\Psi_H(x)$ defined in Equation (9b). In the spirit of wavelet analysis however, there is no reason not to consider other weighting functions than $I_1(x)$ in Equation (16). *A priori* likely candidates need only to be symmetric, non-negative and to have a unit integral as well as an $O(1)$ "width" (say, at half-height). For instance, one could take a Gaussian-shaped function, $\varphi_{1/2}(x)$ in Equation (6b); a piece-wise linear function defined as the integral of $-\Psi_H(x)$ in Equation (9b), namely $\varphi_H(x) = \max\{0, \min[1 - x, 1 + x]\}$, would do just as well. Figure 6a illustrates all three possibilities.

More in step with standard wavelet analysis, Arnéodo *et al.* ([3], [4]) relax the conditions of non-negativity and having unit integrals; such well-known wavelets as Mexican and French "hats" can therefore be considered. Figure 6b shows these famous examples:

$$\Psi_M(x) = -\sqrt{2\pi}\,\varphi''(x) = \sqrt{2\pi}\,[1 - x^2]\varphi_1(x), \quad (17a)$$

and

$$\Psi_F(x) = \begin{cases} 1 & |x| \leq 1 \\ -1/2 & 1 < |x| \leq 3 \\ 0 & |x| > 3 \end{cases} \quad (17b)$$

The latter being a piece-wise constant approximation of the former, using a compact support. Here again there is no need for an orthonormal basis of functions. The common feature of the wavelets Ψ_M and Ψ_F, currently used in singular measure analysis, as well as the scaling-type functions I_1, $\varphi_{1/2}$ and φ_H is symmetry (hence orthogonality to linear functions in particular) and a degree of simplicity (i.e., less than two oscillations). In analogy with Equation (16), we consider the wavelet-type transform

$$T_\phi[\varepsilon](a, b) = \frac{1}{a}\int\varepsilon(x)\phi\left(\frac{x-b}{a}\right)dx, \quad a > 0, \; b \in \Re, \quad (18)$$

where $\phi(\cdot)$ designates any of the above analyzing functions. Notice the same non-standard normalization here as in Equation (10), ensuring that

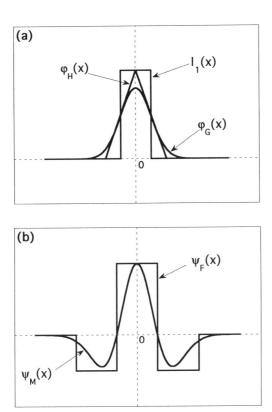

Figure 6. *(a) Weighting functions for singularity analysis.* The simplest is the indicator function of the interval of interest (all points are equally weighted), $I_1(x)$ in Equation 6a. A (non-normalized) Gaussian centered on the interval and of the same width can be used, $\varphi_G(x) = \varphi_{1/2}(x)$ in Equation (6b), or else a piece-wise linear function, $\varphi_H(x) = \max\{0, \min[1 - x, 1 + x]\}$. Notice the positive values, the symmetry and the relative simplicity (i.e., one peak at most). These kernels are not unlike the "scaling functions" of wavelet theory. *(b) Acceptable wavelets for singularity analysis.* The second derivative of $\exp(-x^2/2)$ or "Mexican hat" $\Psi_M(x)$ is illustrated as well as the simpler (piecewise-constant, compact supported) "French top hat" $\Psi_F(x)$ in Equation (17b). Notice again the symmetry and the relative simplicity (i.e., less than two complete oscillations occur).

the integral of the positive part of $\phi(x/a)$, i.e., $\theta[\phi(x/a)]$, is $\propto a$. This facilitates the following scaling comparisons; if $\varepsilon(x)$ varies little, then so will $T_\phi[\varepsilon](a, b)$ with respect to both b and a.

Arnéodo *et al.* ([3] and [4]) show that, within the general framework of scale-invariant (fractal- and multifractal) measures, the random wavelet coefficients in Equation (18) behave statistically like $\varepsilon(r; x)$ with $r \approx a$ and $x \approx b$; see also Ghez and Vaienti ([25] and [26]). In particular, the $\langle |T_\phi[\varepsilon](a, b)|^q \rangle$ will be independent of b and comparable to the 1-point statistical moment of order q for $\varepsilon(r; x)$, namely $\langle \varepsilon(r; x)^q \rangle \equiv \langle \varepsilon(r)^q \rangle$. Although a weaker statement than ergodicity, the statistical stationarity of $\varepsilon(r; x)$ justifies estimating these moments by averaging Equation (18) with respect to b. In summary, we have

$$\langle \varepsilon(r)^q \rangle \approx \frac{1}{L} \int_0^L |T_\phi[\varepsilon](r, b)|^q \, db, \qquad 0 < r \leq L, \ q \in \Re, \tag{19}$$

independently of the choice of ϕ. Some restrictions might have to be imposed on the domain of q. For instance, if $\varepsilon(r; x)$ can take null values then we must take $q > 0$. If, over and above the spatial averaging in the r.h.s. of Equation (19), *bona fide* ensemble-averages are taken then non-trivial statistical effects lead to the divergence of the higher positive order moments for many interesting theoretical models of $\varepsilon(x)$ ([37], [33], [52] and [53]). The divergence of moments is also discussed by Holley and Waymire [32] and Gupta and Waymire [29] who furthermore clarify the issue of ergodicity within the theoretical framework of multiplicative cascades as models for $\varepsilon(x)$.

One might object to using coefficients for continuously overlapping wavelets in Equation (19) for the averaging, not being independent random numbers. If required, this minor problem (redundancy is often viewed as desirable) can be fixed by replacing the integral in Equation (19) by a sum over wavelets with somewhat more disjoint supports:

$$\frac{1}{L} \int_0^L \{\cdots\} \, db \to \frac{1}{[L/r]} \sum_{B_L(r)} \{\cdots\},$$

where $[\cdot]$ means integer value and $B_L(r) = \{b \in [0, L], \ b = ir \ (i = 0, \ldots, [L/r] - 1)\}$ is the relevant set of disjoint "boxes" of size r.

Before considering the scaling properties of the average measures $\langle \varepsilon(r)^q \rangle$, we must point out that Arnéodo's group does not systematically average away the localization information in $T_\phi[\varepsilon](a, b)$. On the contrary, its members are actively researching "inverse" fractal problems. This consists in retrieving from the structures they see in $T_\phi[\varepsilon](a, b)$, viewed as a field in (a, b)-space, the dynamics of the system used to generate $\varepsilon(x)$ in the first place. To this effect they have developed the "modulus maxima wavelet transform" method and have successfully applied it to a variety of systems ranging from the deterministic Cantor measure (e.g., Arnéodo *et al.*, [3] and [4]) to random diffusion limited aggregates (e.g., Arnéodo *et al.*, [6] and [8]).

4.2. Scale-invariant singular measures, generalities

For scale-invariant measures, we will seek the exponent $K(q)$ in

$$\langle \varepsilon(r)^q \rangle \propto r^{-K(q)}. \tag{20}$$

We are following the notations and conventions suggested by Schertzer and Lovejoy [52]. An alternative but more widely used approach uses sums rather averages over r-sized sets, i.e., the focus is on $p(r; x) = r\varepsilon(r; x) \approx r|T_\phi[\varepsilon](r, x)|$ in one spatial dimension. One then defines the "partition function" $Z_q(r) = \langle \sum p(r; x)^q \rangle$ and where, as suggested above, the sum carries over disjoint r-sized boxes rather than all possible x's as in Equation (19). There are precisely $[L/r]$ such boxes at scale r in one dimension. For scale-invariant measures, the scaling of $Z_q(r)$ is parameterized as $r^{\tau(q)}$. A little algebra leads to $\tau(q) = (q - 1) - K(q)$.

Some general results follow from the above definitions. For the same reasons as for structure functions, $\langle \varepsilon(r)^q \rangle$ and $K(q)$ are smooth functions of q and, in analogy with $-\zeta(p)$ in Equation (12), $K(q)$ will be convex ($K''(q) \geq 0$), although we must still require that the prefactors in Equation (20) depend only weakly on q. Proper normalization of the 1-point p.d.f.'s of the measures $\varepsilon(r; x)$ requires $K(0) = 0$ in Equation (20), just as $\zeta(0) = 0$ in Equation (12). There is also an essential difference with structure functions. Consider the case $q = 1$; the two averages in Equations (15)–(18) and (19) – inside one-dimensional "boxes" and over these boxes when using $I_1(\cdot)$ – commute and we find $\langle \varepsilon(r) \rangle \equiv constant$, hence $K(1) = 0$. Since $K(0) = K(1) = 0$, convexity implies $K(q) \leq 0$ for $0 < q < 1$ and $K(q) \geq 0$ elsewhere as well as $K'(0) \leq 0$ and $K'(1) \geq 0$.

It is important to know what to expect for weakly variable $\varepsilon(r; x)$ fields, i.e., where $\langle \varepsilon(r)^q \rangle \approx \langle \varepsilon(r) \rangle^q$. Equation (20) then yields $K(q) = qK(1) \equiv 0$, the $\langle \varepsilon(r)^q \rangle$'s are constant with respect to r; also the ='s will apply in all of the inequalities discussed in connection with Equation (20). Conversely, it is easy to show that finite $K(q)$'s ($q \neq 0, 1$) imply extremely singular measures. To see this we first adopt units of length where $L = 1$ and units for $\varepsilon(x)$ where $\langle \varepsilon(r) \rangle \equiv 1$, $0 < r \leq 1$. As soon as $r \ll 1$, we have $\langle \varepsilon(r)^q \rangle \ll 1$ for $0 < q < 1$ (due to $-K(q) > 0$) and $\langle \varepsilon(r)^q \rangle \gg 1$ for $q > 1$ ($-K(q) < 0$). Taking the q-th power of a non-negative random variable considerably reduces its range if $0 < q < 1$ so, in order to obtain $\langle \varepsilon(r)^q \rangle \ll 1$, we need a p.d.f. for $\varepsilon(r; x)$ that is concentrated on vanishingly small values. This is however incompatible with $\langle \varepsilon(r) \rangle \equiv 1$ unless the said p.d.f. is also very skewed in the positive direction. We can envision broad expanses of very small ε-values interrupted by many narrow "spikes," defined (say) by $\varepsilon \geq 1$, with a variety of heights, the larger the rarer. This is what we mean by an "intermittent" process, i.e., one characterized by sudden bursts of high-frequency activity, recalling that $\varepsilon(x)$ is computed from gradients in the first place (a high-pass filtered version of the signal).

In the above discussion it is important to bear in mind that in practical data-analysis situations a finite $K(q)$ does not automatically imply a high degree of intermittency. Indeed, at finite spatial resolution – necessarily the case with data – a weakly variable measure will always yield small values of $K(q)$ (for $q \neq 0, 1$), a "residual" or "spurious" multiscaling. An example of this behavior is discussed by Marshak et al. [39].

An extreme case of intermittency arises when $\varepsilon(x)$ is a single δ-function placed at random on $[0,1]$; our definitions lead to $K(q) = q-1$ for $q > 0$ ([17] and [39]). However, we will generally be dealing with measures which are not as singular – nor intermittent – as this; instead of being concentrated onto a single point, the measure is distributed over many smaller peaks.

Here too a hierarchy of exponents can be defined, this time a non-decreasing one:

$$C(q) = \frac{K(q)}{q-1}, \tag{21a}$$

recalling that $K(q)$ is convex and that $K(1) = 0$. It can be directly related to the well-known non-increasing hierarchy of "generalized" dimensions

$$D(q) = 1 - C(q), \tag{21b}$$

first introduced by Grassberger [28] and Hentschel and Procaccia [31] with dynamical systems and strange attractors in mind. Following Parisi and Frisch [48] who where focusing explicitly on structure functions, Halsey et al. [30] established an interpretation for $D(q)$ in terms of variable orders of singularity. (In this case, "singularity" is used in the sense of measures: how does $\alpha(x)$ in $p(r; x) = r\varepsilon(r; x) \propto r^{-\alpha(x)}$ vary with x?)

Obtaining $K(q)$ and $C(q)$ or $D(q)$ is the goal of this approach, providing us with a means to characterize intermittency, in both quantitative and qualitative terms (examples to follow).

4.3. Geometrical, spectral and dynamical implications of statistical singularity analysis

Measures with a variable $C(q)$ are known as "multifractals" (one should specify stationary). This leaves measures with non-vanishing constant $C(q)$ – hence $K(q) \propto (q-1)$ – to be "monofractals" (again stationarity being implicit). In essence, a random (deterministic) fractal is a sparse subset of space that can be described statistically (exactly) with a single exponent, its fractal dimension. Whether viewed as a stationary random measure (present section) or as a non-stationary random function (previous section), a multifractal calls for many exponents; each of these is dominated by a specific kind of event and can therefore be associated with the fractal dimension of the set where such events occur.

Following as closely as possible our discussion of structure function analysis, it is of interest to seek a simple but meaningful way of characterizing the sparseness and/or intermittency in the system, using a single exponent rather than a whole family. It is natural to consider the mean (i.e., the case $q = 1$). We already know that $\langle \varepsilon(r) \rangle$ is unit, independently of r ($K(1) = 0$), when the measure is appropriately non-dimensionalized. We would however like to know where the events that contribute most to $\langle \varepsilon(r) \rangle$ occur. Recall that an overwhelming majority of ε-values (i.e., the most probable ones) are $\ll 1$; these "typical" events fill space, leaving only a sparse (fractal) set where $\varepsilon(r;x) \geq \langle \varepsilon(r) \rangle \equiv 1$. If "average" events (i.e., $\varepsilon(r;x) \approx 1$) are already a rare occurrence, they cannot contribute enough to the mean $\langle \varepsilon(r) \rangle$ to offset the effect of the numerous small values. We must go to much higher values hence somewhat rarer events. It can be shown that the codimension of the set where these "singularities" live is

$$C_1 = C(1) = 1 - D(1) = K'(1) \geq 0, \tag{22a}$$

where $D(1)$, the fractal dimension of this set, is known as the "information" dimension ([28] and [31]). This tells us that we can set the largest value for C_1 at 1 for $D(1) = 0$, attained in particular for randomly positioned δ-functions (all the measure is concentrated into a single point). Processes with $C_1 > 1$ are called "degenerate:" almost every realization is empty and, every now-and-then, one occurs with a huge peak, perturbing the ensemble average each time. At the opposite limit, $C_1 \to 0$ (weak variability), we find information everywhere: $D(1) = 1$.

An infinite number of exponents is needed to describe multifractals completely. For instance, there is no general relation between C_1 and the spectral exponent which is however connected with the $q = 2$ moment:

$$\beta_\varepsilon = 1 - K(2) = 1 - C(2) = D(2) \leq D(0) = 1, \tag{22b}$$

a special case of the Wiener-Khintchine theorem for stationary scaling processes (cf. Appendix). We will nevertheless retain C_1 as the single most important exponent in the whole $C(q)$ hierarchy. The geometrical significance given in connection with Equation (22a) is not as important as the fact that C_1 defines as simply as possible the linear trend in $C(q)$ or $D(q)$. In other words, we have a first order estimate of the departure of the ε-field from weak variability (non-intermittency) that is readily obtained from a $K(q)$ graph (cf. Figure 8b). In summary, we view C_1 as a direct quantifier of intermittency in the system and the complete $K(q)$-, $C(q)$- or $D(q)$

function as a means to qualify this intermittency.

Unlike the concept of non-stationarity for structure functions, there is no need to argue the dynamical relevance of intermittency. The concept was imported from turbulence into statistics in the first place. We now provide some examples both theoretical and empirical.

4.4. Examples from theory, turbulence and cloud structure

In Figure 7 we show four random processes, all theoretical cascade models $\varepsilon(l; x)$ with $l = L/1024$ and log-normal multiplicative weights W, as first proposed by Kolmogorov [35] and Obukhov [47] to emulate the variability of the dissipation field in turbulence ($\varepsilon \propto \partial u^2/\partial t$). Clearly different degrees of intermittency are present. At the top, we find a flat field (corresponding to degenerate weights) which is trivially stationary and non-intermittent: $C(q) \equiv 0$. At the bottom, we find a very intermittent process: $C(q) = qC_1$ for $q < 1/C_1$ with $C_1 = 1/2\ln 2 \approx 0.72$ ($\sigma^2_{\ln W} = 1$) and divergent moments for $q > 2\ln 2 \approx 1.34$ (including $q = 2$). Between these two extremes, we find two intermediate examples of log-normality, illustrating more-and-more intermittency with $C_1 \approx 0.05$ and 0.18. However, these models all have the same "kind" of intermittency or multifractality. The same form of $C(q)$ applies throughout, there is no change in qualitative behavior (in particular, there is always a threshold for divergence of higher moments).

Within the framework of the statistical theory of turbulence alone, many other cascade-type intermittency models have been and are still being proposed (and discussed at length). Novikov and Stewart's [46] model of "pulses within pulses," Mandelbrot's [37] model of "absolute curdling" and Frisch et al.'s [24] "β-model" are all characterized by the frequent occurrence of null multiplicative weights; they are generically described by $K(q) = C_1(q - 1)$ or $C(q) \equiv C_1$ (i.e., monoscaling) with $0 < C_1 < 1$, for $q > 0$ and $K(q) = \infty$ for $q < 0$. Schertzer and Lovejoy's [51] "α-model" and Meneveau and Sreenivasan's [42] "p-model" have log-binomial weights; they are generically described by $K(q) = \log_2[pW^q + p'W'^q]$, assuming for simplicity that the cascade proceeds by divisions into two sub-intervals in one spatial dimension, with $q \in \Re$ and parameters verifying $p + p' = 1$ (from $K(0) = 0$) and $pW + p'W' = 1$ (from $K(1) = 0$), hence from Equation (22a) $C_1 = pW \log_2 W + p'W' \log_2 W'$. Benzi et al. [10] proposed a "random β-model" which also exhibits multiscaling. Schertzer and Lovejoy [52] advocate a family of "universal" cascade models with log-Levy stable weights that are generically described by $K(q) = C_1(q^\alpha - q)/(\alpha - 1)$ with $0 < \alpha < 2$, for $q > 0$ and $K(q) = \infty$ for $q < 0$, the limiting cases $\alpha \to 0^+$ and $\alpha \to 2^-$ leading back to monofractality and log-normality respectively. Certain choices of parameters in the above can lead to the divergence of higher moments, hence to an upper bound on the range of q, namely $q < q_D$ where q_D is the solution of $K(q) = q - 1$ for $q > 1$ ([33], [37], [53] and [32]).

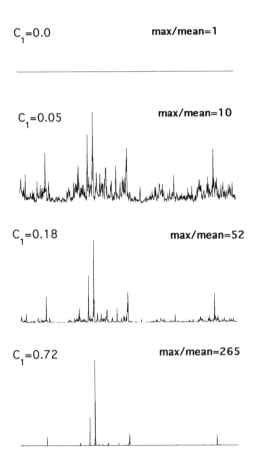

Figure 7. *Different amounts of the same kind of intermittency.* The construction of an $\varepsilon(l; x)$ field using a cascade model calls for recursively dividing subsets of $[0, L]$ into (say, twice) smaller ones and multiplying the current value of $\varepsilon(l; x)$ $(l/L = 2, 4, 8, \ldots)$ by independently chosen non-negative random numbers W with unit mean. It can be shown that, apart from the effects of divergence of higher order moments (see text), $K(q) = \log_2 \langle W^q \rangle$ [43]. We used log-normal deviates which lead to $K(q) = C_1 q(q-1)$, $q < 1/C_1$, and varied the fundamental intermittency parameter C_1 as indicated by changing the log-variance of the W's: $C_1 = \sigma_{\ln W}^2 / 2 \ln 2$ with $\sigma = 0.0, 0.25, 0.5, 1.0$. However, by enforcing throughout the log-normality, we obtain the same "flavor" of intermittency in different amounts. Other types of scale-invariant intermittency are discussed in the text.

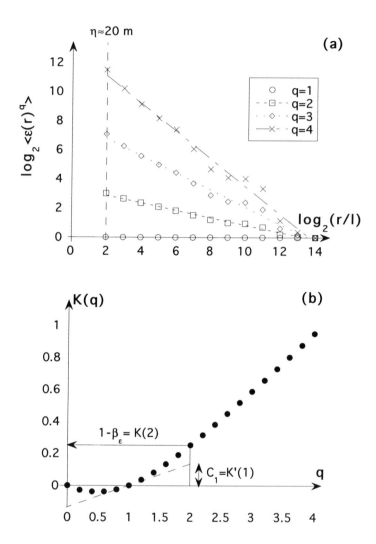

Figure 8. *Singularity analysis of cloud LWC.* (a) The quantities $\langle \varepsilon(r)^q \rangle$ are plotted for $\log_2(r/l) = 2, \ldots, 4$ and $q = 1, 2, 3, 4$. We have indicated the lower end of the scaling range η and the absolute slopes are the exponents $K(q)$. (b) The corresponding $K(q)$ function, demonstrating the multifractality of absolute gradients of the LWC field such as in Figure 1b. We have highlighted the remarkable values at $q = 1, 2$ discussed in connection with Equations (22a,b).

The most quoted singularity analysis of the dissipation field in turbulence is due to Meneveau and Sreenivasan ([41], [42]) who determined (an exponent family closely related to) $K(q)$ for wind tunnel and atmospheric boundary layer flows at high Reynolds numbers; see also [54]. Empirical values for C_1 fall in the range 0.2–0.3. Argoul *et al.* [2] show graphic evidence of the cascade process using the wavelet in Equation (17a) directly on the velocity field. In the same spirit of "inverse fractal theory" (i.e., finding the internal dynamics from the data), Arnéodo *et al.* ([3], [4]) investigate deterministic cascade models.

Figure 8a shows $\log_2[\langle|\varepsilon(r)|^q\rangle/\langle|\varepsilon(L)|^q\rangle]$ versus $\log_2(r/l)$ for $q = 1, 2, 3, 4$ and $r \geq \eta = 4l$ prepared for the time-series of LWC in Figure 1a. Specifically, we formed absolute differences over distances of 4 grid points ($\eta = 20$ m, the lower bound of the scaling regime as defined by the structure functions in Figure 5a). Four such absolute gradient fields can be obtained and one of these is illustrated in Figure 1b. The $\langle|\varepsilon(r)|^q\rangle$'s are then computed in the standard ($I_r(x)$-based) way by averaging first over space, as in Equations (15)–(19) and using disjoint r-sized boxes; then an average over all four cases is performed. *A priori*, we wish to characterize intermittency in the same range of scales $[\eta, R]$ as for the non-stationarity previously but we recall that, being caused by a single event, the break in scaling observed at $R \approx 5$ km in Figure 5a is not robust [18] and we are justified in seeking scaling behavior in $\langle|\varepsilon(r)|^q\rangle$ up to $r = L$. The absolute slopes in Figure 8a give the corresponding value of $K(q)$. In Figure 8b we have plotted $K(q)$ and highlighted two remarkable values: $K'(1) = C_1 \approx 0.1$ and $K(2)$ which is related to the spectral exponent β_ε of the ε-field, as described in the Appendix. We notice the strong evidence of intermittency in the structure of marine StCu ($K(q)$ is non-vanishing) and that it is multifractal in nature (in the sense of "multi-singular" measures, $K(q)$ is nonlinear).

§5. Summary and Bi-Multifractal Statistics

5.1. Comparative multifractal analysis from the wavelet theoretical standpoint

In the course of our survey, we have drawn many parallels between the two multifractal statistical data analysis techniques. There are also fundamental differences. To better appreciate these, we compare item-per-item definitions, results and related concepts for structure functions and singular measures.

Statistical Operation :	Structure Functions	Singular Measures
input :	$f(x)$	$\varepsilon(x)$
nature :	function	measure ($\varepsilon(x) \geq 0$)
condition $(E(k) \sim k^{-\beta})$:	non-stationary $(\beta_f > 1)$	stationary $(\beta_\varepsilon < 1)$
Define $\xi(r;x) \geq 0$:	$\|\Delta f(r;x)\| =$ $\|f(x+r) - f(x)\|$	$\varepsilon(r;x) =$ $\frac{1}{r}\int_x^{x+r} \varepsilon(y)\,dy$
$A(z)$ in $\langle \xi(r;x)^z \rangle$ $\propto r^{A(z)}$:	$\zeta(z)$	$-K(z)$
solution(s) of $A(z) = 0$:	$z = 0$	$z = 0, 1$
exponent hierarchy :	$H(z) = \zeta(z)/z$	$C(z) =$ $K(z)/(z-1)$
spectral property :	$\beta_f = \zeta(2) + 1 > 1$	$\beta_\varepsilon = 1 - K(2) < 1$
analytical properties :	continuity but non-differentiability	discontinuity and singularity
geometrical property :	roughness	sparseness
statistical/dynamical property :	non-stationarity	intermittency

Although wavelets yield only variants of existing techniques rather than radically new approaches in multifractal analysis, one does achieve a higher degree of unification. In section 3 we had "differencing" (high-pass) wavelets acting on the raw data. In section 4 we have "averaging" (low-pass) wavelets acting on the absolute differences derived from the same data (or some other measure sensitive to the high frequencies). One way or another, we are focusing on the singularities of the gradient field and this seems to imply that the two kinds of multifractality we observe (say, in clouds) are really just two facets of a deeper one, independent of how it is characterized. In this spirit, we now describe a merger of the two approaches and its potential benefits.

5.2. On $\zeta(p)$-to-$K(q)$ connections

Recalling that $\varepsilon(x)$ is derived from $f(x)$ for a given geophysical process (e.g., governing cloud structure), both $f(x)$'s structure functions and $\varepsilon(x)$'s singularity analysis can produce evidence of multifractality. It is therefore natural to ask "Is it fundamentally the same multifractality?" and, if so, "Is there a connection between $\zeta(p)$ and $K(q)$?". The answer to the former and more philosophical question is probably "yes." Davis et al. [17] address the latter question from an empirical perspective, suggesting that the fields $\|\Delta f(r;x)\|$ and $\varepsilon(r;x)$ be studied jointly. This can of course be done within the framework of wavelets by computing

$$\langle |\Delta f(r;x)|^p \varepsilon(r;x)^q \rangle \approx \frac{1}{L} \int_0^L |T_\Psi[f](r,b)|^p |T_\phi[\varepsilon](r,b)|^q \, db \propto r^{X(p,q)}, \quad (23)$$

with the appropriate ranges for p and q defined in Equations (11) and (19); concerning r, one should of course remain in the scaling regime defined in Equation (4). The new exponent function $X(p,q)$ reverts to known cases for $q = 0$ and $p = 0$ and will be concave ($\partial_{pp}^2 X \leq 0$, $\partial_{qq}^2 X \leq 0$) on its domain of definition.

Davis *et al.* [17] show how the $X(p,q)$ can be used to test the scaling hypothesis that there exists two exponents $t > 0$ and $s > 0$ such that

$$\varepsilon(r;x) = \frac{|\Delta f(r;x)|^t}{r^s}. \quad (24)$$

This generalizes Kolmogorov's [34] relation for fully developed turbulence corresponding to $s = 1$ and $t = 3$; more precisely, Kolmogorov argued Equation (24) not at every x but on average only, hence $\zeta(t = 3) = s = 1$ since we know that $\langle \varepsilon(r) \rangle$ is independent of r. If however Equation (24) applies everywhere, then it follows (by taking p/t-th powers and averaging) that

$$\zeta(p) = (\frac{s}{t})p - K(\frac{p}{t}), \quad (25)$$

as often quoted in the case of turbulence in spite of the fact that Equation (24) has not been verified beyond Kolmogorov's average at $p = t = 3$ ([55]; and references therein). So it appears that at least two new parameters are needed to connect $\zeta(\cdot)$ and $K(\cdot)$, the essential one being $s = \zeta(t)$ which allows Equation (24) to relate a stationary field (ε) to a non-stationary one (f). Indeed, for small enough C_1 the values of $K(p/t)$ in Equation (25) will be small as well (except possibly when $p \gg t$); in turbulence theory $K(p/3)$ is called an "intermittency correction" (usually at $p = 2$). This makes $(s/t)p$ reflect, to first order, the non-stationarity of $f(x)$ in Equation (25) since we then get $\zeta(p) \approx (s/t)p$ hence $H_1 \approx s/t$ (1/3 for turbulence).

5.3. The mean multifractal plane

The above ideas are important since they can in principle be used to establish, from data alone, relations such as Equation (24) that can be viewed as effective constitutive laws [17]. However in many practical situations we are not interested in dealing with the large number of exponents in $\zeta(p)$ and/or $K(q)$, let alone the larger number needed to represent $X(p,q)$. Can we restrict ourselves to a single exponent from each multifractal approach? The natural choices are H_1 and C_1 with no risk of redundancy, even if Equation (25) applies. This minimal multifractal parameterization enables us to define the "mean multifractal plane" with coordinates

$(H_1, C_1) \in [0,1] \otimes [0,1]$, the upper limit on C_1 being imposed only to exclude degenerate gradient fields. Figure 9 represents an (H_1, C_1)-plot that we have populated with empirical findings from the turbulence literature (approximate positioning) and some of our LWC studies (precise positioning).

We notice on Figure 9 that the models discussed in this paper fall primarily on either axis: additive processes (Section 3.3) have $C_1 = 0$ and $H_1 > 0$ (they are neither intermittent nor stationary) while multiplicative cascades (Section 4.3) have $H_1 = 0$ and $C_1 > 0$ (they are both stationary and intermittent). We have added Mandelbrot's [38] "Devil's staircases" which are simply integrals of cascade models, obeying Equations (24)–(25) with $s = t = 1$ according to Equation (15). They are found at $H_1 = 1$ and finite C_1 (they are almost everywhere differentiable but, having fully developed cascades as gradient fields, they are intermittent). A more familiar case is found at $H_1 = C_1 = 1$: Heaviside steps, the integrals of Dirac δ's. It seems that these standard scale-invariant models carefully avoid the locus of the data, inside the square domain. This clearly demonstrates the need for a new class of stochastic models that can access the whole multifractal domain and which are likely to have both additive and multiplicative ingredients. A list of publications where such multi-affine models are described is provided at the end of section 3; we believe this list (of five) to be exhaustive at the time of publication but there is little doubt that many more will follow. We also anticipate numerous geophysical applications for (H_1, C_1)-plots. In particular, we foresee all sorts of data-to-data and data-to-model intercomparisons for the purposes of validating numerical models or retrievals from remote sensing for instance ([16] and [18]).

§6. Conclusions

Wavelets have been extensively used to show in graphic detail how structures occur on all observable scales in many geophysical signals. A natural environment for modeling such signals is provided by stochastic processes which have no characteristic scale (sporting, e.g., power-law energy spectra) and these are best described by multiscaling or "multifractal" statistics. Since scale-invariance prevails in nature as well, we are urged to blend wavelet- and multifractal analysis techniques. The major benefit of this merger for the wavelet community is the fact that multifractal statistics have strong dynamical overtones; they can be used to characterize in quantitative and qualitative terms both intermittency and non-stationarity. The benefit for the multifractal community is a somewhat more unified view of their two main tools, singularity analysis (targeting the intermittency) and structure functions (targeting the non-stationarity). This connection is timely because many important questions are still open about the inter-

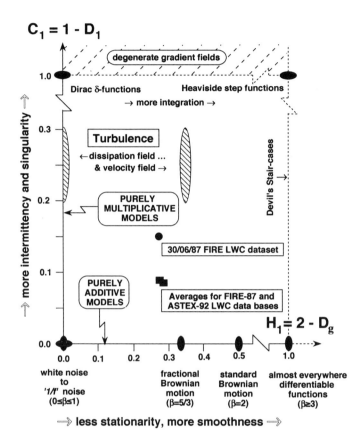

Figure 9. The "Mean Multifractal Plane" or (H_1, C_1)-plot. See text for details on the various models and the point for LWC in marine StCu (FIRE, 30/06/87). Also indicated are points mapped to the averages over whole LWC databases captured during FIRE in 1987 off the southern Californian coast and the Atlantic Stratocumulus Transition EXperiment (ASTEX) in 1992 near the Açores; see [18] and [16] respectively for details. The proximity of these two points argues for a degree of universality in the nonlinear dynamics that determines the internal distribution of liquid water in marine StCu under rather different climatological conditions. The distance between these points and the one obtained for the 30/06 flight during FIRE illustrates a dramatic violation of ergodicity in this rather typical atmospheric process, hence the need for vast databases to define the behavior of this and other atmospheric fields in statistically robust quantitative terms.

action of intermittent cascade-type fields and directly observable ones; for instance, how does the intermittency of the dissipation field affect locally the velocity field in turbulence?

The main ideas of multifractal analysis are presented from a wavelet standpoint and amply illustrated with theoretical models and empirical results drawn from the turbulence literature. New analyses of the liquid water density field are also presented in sufficient detail to serve as a tutorial. We describe how to systematically seek and how to interpret connections between structure functions and singular measures for a given type of data. Arguing that these connections are non-trivial to say the least, we define the "mean multifractal plane" by judiciously selecting a single exponent from each multiscaling approach, H_1 for structure functions and C_1 for singular measures. Finally, we use the (H_1, C_1)-plot to discuss several theoretical and empirical findings. This simple diagnostic device will be helpful in many outstanding geophysical problems of finding statistically robust and physically meaningful ways of intercomparing data of different but related origins, numerical model output included. It is used here to show the need for a newly introduced class of stochastic models known in the physics literature as "multi-affine," to demonstrate a violation of ergodicity in the distribution of liquid water in a particular marine stratocumulus cloud deck, and to argue for a degree of universality in the dynamics that determine these distributions.

A. Stationarity, Intermittency, Integral Scales and Stochastic Continuity – A Tutorial in the Framework of Scale-Invariance and Second Order Statistics

A.1. Stationarity versus ergodicity

A property of fundamental importance in the theory of stochastic processes, especially from the standpoint of data analysis, is statistical "stationarity." In essence, a stationary quantity is statistically invariant under translation. In principle one should talk about statistical "homogeneity" in the spatial domain but we prefer the term "stationarity" borrowed from the time domain; in this way we avoid any confusion with the idea of constant fields.

When analyzing data, one generally uses the spatial coordinate to perform various averaging operations, as described in the main text. This amounts to making an "ergodicity" assumption (i.e., that spatial averages will converge towards their ensemble counterparts as the amount of data increases). Stationarity is a far weaker assumption than ergodicity, but easier to work with on theoretical grounds and more reasonable in empirical situations. For example, Holley and Waymire [32] and Gupta and Waymire [29] show that multiplicative cascade models (which we will use to illustrate stationarity further on) are generally non-ergodic. The highly correlated fluctuations that characterize these models cause the (spatial) law of large numbers to fail. Schertzer and Lovejoy [53] pursue these issues in more quantitative terms, dependent on what portion of probability space (the set of all possible realizations) is sampled.

Neither ergodicity nor stationarity can be established rigorously for a dataset, no matter how long. We must rely on consistency checks and we show here how scale-invariance eases this task. The central role of the "integral scale" is also underscored. Throughout, we illustrate the concepts with additive (non-stationary and non-intermittent) and multiplicative (stationary and intermittent) scale-invariant models as well as a well-known non-scaling example (Ornstein-Uhlenbeck processes). Finally, we show that there exists a strong connection between stochastic continuity and non-stationarity for scaling processes.

A.2. Stationarity and scale-invariance in physical space, the integral scale

Let $f(x)$ be a real random process defined either continuously (for a theoretical model) or discretely (for data) on the interval $[0, L]$. A first consequence of stationarity concerns 1-point statistics, e.g., for the mean we find $\langle f(x+r) \rangle = \langle f(x) \rangle$, $0 \le r \le L$, $0 \le x \le L - r$. A 2-point statistic of obvious interest in time series analysis is $\langle [f(x+r) - \langle f(x+r) \rangle][f(x) - \langle f(x) \rangle] \rangle$; under stationary conditions, it reduces to the auto-correlation function

$$\langle f(x+r)f(x) \rangle \equiv G(r), \qquad (A.1)$$

using a linear transformation that reduces $\langle f(x) \rangle$ to zero. The identity in Equation (A.1) follows from stationarity and we notice that $G(0) = \langle f(x)^2 \rangle$, 1-point variance. A popular application of $G(r)$ is the estimation of the "integral" correlation length:

$$R = \frac{1}{G(0)} \int_0^\infty G(r)\, dr. \qquad (A.2)$$

To a first approximation, this length scale draws the line between correlated $(r < R)$ and uncorrelated $(r > R)$ values of $f(x)$ and $f(x+r)$, a sort of statistical period in the data.

For scale-invariant processes (i.e., with power-law 2-point statistics), we can anticipate

$$G(r) \propto r^{-\mu}, \qquad \mu > 0 \qquad (A.3)$$

in stationary situations. The exponent μ must be positive since we generally expect less correlation as r increases. Equation (A.3) cannot be substituted directly into Equation (A.2). Firstly, there is always an outer limit to the range of scales where Equation (A.3) can apply; for a process defined on $[0, L]$, we must require $G(r) = 0$ for $r > L$. There is also an inner scale in most situations of practical interest; for $0 \leq r < l$, we can take $f(x+r) \approx f(x)$ hence $G(r) \approx G(0)$. Between these limits (with $l \ll L$), we assume Equation (A.3) to apply; more precisely, we take $G(r) \approx G(0)[l/r]^{\mu}$. It follows that

$$\frac{R}{L} \approx (\frac{l}{L})^{\min[1,\mu]}. \tag{A.4}$$

So, for all practical purposes, the exponent μ in Equation (A.3) cannot exceed unity without leading to fields that are uncorrelated from one pixel to the next. If $\mu \to 0$ (hence $R \approx L$), there is no trend towards decorrelation; we view this as a symptom of non-stationarity and in Section A.5 will introduce a 2-point statistic better adapted to this situation than $\langle f(x+r)f(x) \rangle$.

A.3. A Fourier space criterion for stationarity

The Wiener-Khintchine theorem states that, under necessarily stationary conditions, $G(r)$ in Equation (A.1) and $E(k)$ in Equation (3) form a Fourier transform pair. So, in particular, the integral scale R can be more efficiently computed from

$$R = E(0)/ \int_0^\infty E(k)\, dk, \tag{A.5}$$

using fast Fourier transforms.

In their scaling versions Equations (A.1) and Equation (5), the Wiener-Khintchine theorem translates to

$$\beta + \mu = 1, \tag{A.6}$$

which implies, in particular ($\mu > 0$), that

$$\beta < 1. \tag{A.7}$$

This criterion for stationarity is readily applicable to any kind of data that comes on a grid and exhibits scaling behavior. Although not a proof of stationarity for data, Equation (A.7) should be verified before computing $\langle f(x + r)f(x) \rangle$ – or any other statistic requiring stationarity – via spatial averaging.

The simplest possible stochastic process is a sequence of independent random numbers; viewed as a field $f_0(x)$, $x \geq 0$, it is uncorrelated – or rather "δ-correlated" – and stationary by construction: $G_0(r) \propto \delta(r)$. The subscript "0" stands for vanishing R in Equation (A.2) or (A.5), due to the infinite denominator. Figure A1a provides a sample of such white ($\beta = 0$) noise that we will denote $f_0(x)$ for $0 \leq x \leq L = 4096$ using zero-mean unit-variance Gaussian deviates. We now describe a more interesting case of scale-invariant stationarity.

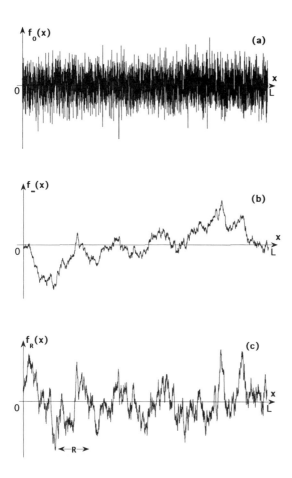

Figure A1. *Scale-Invariant Stationary and Non-Stationary Noises, Along with an Intermediate Non-Scaling Case.* (a) Unit-variance zero-mean Gaussian white noise $f_0(x)$, $0 \leq x \leq L = 4096$, which is scaling, stationary ($\beta = 0 < 1$) and discontinuous. (b) Brownian motion $f_\infty(x)$ – essentially the running integral of (a) – is scaling, continuous and non-stationary ($\beta = 2 > 1$). (c) An Ornstein-Uhlenbeck process obeying a simple stochastic ODE, $Rf_R' + f_R = f_0$ with an integral scale parameter $R = L/8$. Its spectrum is $E(k) = E(0)/[1 + (kR)^2]$. Being non-scaling, this model can be at once stationary ($\langle f_R(x + r)f_R(x) \rangle \equiv G(r) = G(0) \exp[-r/R]$) and continuous ($\langle [f_R(x + r) - f_R(x)]^2 \rangle \equiv 2[G(0) - G(r)] \rightarrow 0$ when $r \rightarrow 0$). However, the scaling cases in panels (a) and (b) are retrieved in the limits $R \rightarrow 0$ and $R \rightarrow \infty$ respectively, hence the subscripts.

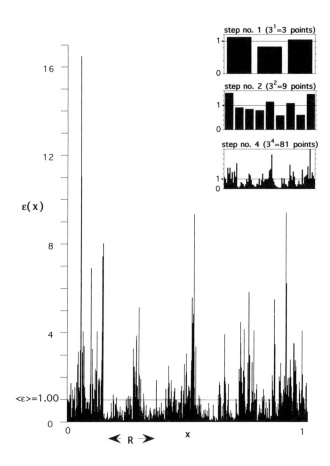

Figure A2. *Development of a Log-Normal Multiplicative Cascade.*
The inset shows the growth of the "singularities" (spikes) through the
1-st, 2-nd and 4-th cascade steps. The 6-th step is illustrated in more
detail: $\varepsilon(x)$ is represented for $0 \leq x \leq L = 1$ with a grid constant
$l = 3^{-6} = 1/729$. Because it uses unbounded multiplicative weights,
this type of cascade process is patently non-ergodic ([32], [53]). However, the spatial average of this particular realization is approximately
equal to the ensemble average (namely, unity). Below the horizontal
axis, we have indicated the integral length scale, $R \approx L/5$ in this case
(see text for details).

A.4. Multiplicative cascade models, stationarity and intermittency

As a further illustration of scale-invariant stationarity, consider a multiplicative cascade process, traditionally denoted $\varepsilon(x)$, $0 \leq x \leq L$. These are constructed by initially setting $\varepsilon(x) \equiv \varepsilon_L$ then subdividing $[0, L]$ into (say) $m = 2$ equal parts, $[0, L/2]$ and $[L/2, L]$, and multiplying $\varepsilon(x)$ in each one by W_1 and W_1', both unit-mean non-negative random variables drawn from the same distribution. This defines $\varepsilon(l; x)$ for $l = L/2$. The procedure is repeated *ad infinitum*:

$$\varepsilon(l; x) = \varepsilon_L \prod_1^n W_i, \quad l/L = m^{-n} \to 0. \tag{A.8}$$

In Figure A2 we used $m = 3$, log-normal W's with log-standard deviation $\sigma_{\ln W} = 0.5$ and log-mean $-\sigma_{\ln W}^2/2$ in order to ensure proper normalization (i.e., $\langle W \rangle = 1$); cascade steps $n = 1, 2, 4$ (inset) and 6 are illustrated. The statistical properties of

$$\varepsilon(x) = \lim_{l \to 0} \varepsilon(l; x) = \prod_1^\infty W_i \tag{A.9}$$

have been investigated primarily in the turbulence literature ([35], [47] and [37]), as a model for the intermittency of the dissipation field in very high Reynolds number flows. These processes, known as "multifractals" [48], are now widely applied in deterministic chaos [30].

Stationarity in the above sense [Equations (A.3)–(A.7)] follows from [43]

$$\langle \varepsilon(x + r)\varepsilon(x) \rangle \sim r^{-K(2)}, \quad K(2) = \log_m \langle W_i^2 \rangle \tag{A.10}$$

using notations consistent with those of section 4. Comparing with Equation (A.3), we have

$$\mu_\varepsilon = K(2) \geq 0, \tag{A.11}$$

which is known as the "intermittency" parameter. The inequality in Equation (A.11) follows directly from Schwartz's: $\langle W_i^2 \rangle \geq \langle W_i \rangle^2 = 1$. This establishes stationarity ($\beta_\varepsilon = 1 - \mu_\varepsilon < 1$) as soon as the variance of the weights is finite, $\langle W_i^2 \rangle - \langle W_i \rangle^2 > 0$. In this case "=" corresponds to degenerate multiplicative weights ($W_i \equiv 1$) hence flat fields which are trivially stationary. In Figure A2 we take $m = 3$ and the numerical values lead to $\mu_\varepsilon \approx 0.23$ and from Equation A.4 we find $R/L \approx 0.22$ for $L/l = 729$ ($n = 6$).

A.5. Non-stationary scale-invariant situations, up to the integral scale

In non-stationary cases where

$$\beta > 1, \qquad (A.12)$$

the field $f(x)$ can fluctuate wildly, given enough space, since there is relatively more energy (variance) in the large scales. Being dependent on x as well as r, the quantity in the l.h.s. of Equation (A.1) is no longer a simple statistic, dependent only on the lag r. One can however work with increments of $f(x)$: $\Delta f(r; x) = f(x + r) - f(x)$ with $0 \le r \le L$, $0 \le x \le L - r$, which will be less sensitive to the local value of $f(\cdot)$. The (2-nd order 2-point) statistical counterpart of Equation (A.1) is

$$\langle \Delta f(r; x)^2 \rangle \equiv \langle \Delta f(r)^2 \rangle \qquad (A.13)$$

where the identity assumes "stationary increments." The statistic in Equation (A.13) is known in turbulence studies as a "structure function" [43] and in geostatistics as a "variogram" [13].

In scale-invariant situations, we expect the 2-nd order structure function in Equation (A.13) to obey a power-law:

$$\langle \Delta f(r)^2 \rangle \propto r^{\zeta(2)}, \ 0 < \zeta(2) < 2, \qquad (A.14)$$

where the exponent is denoted $\zeta(2)$ for consistency with section 3 in the main text. We generally expect $\langle \Delta f(r)^2 \rangle$ to be a non-decreasing function of r, hence the sign of the exponent. The limit $\zeta(2) \to 0$ (increments independent of the scale involved) leads back to a stationary statistical regime. The opposite limit $\zeta(2) \to 2$ can be associated with the class of (almost surely everywhere) differentiable functions since $|f(x+r) - f(x)| \propto r$ will not be a rare occurrence in this case.

It is easy to see why there is always an upper bound to the non-stationary range for the scale parameter r in physical processes. If $r \to \infty$ in Equation (A.14) then $\Delta f(r; x)$, hence $f(x)$ itself, must also take arbitrarily large values whereas most geophysical fields have natural bounds. Some are necessarily non-negative (e.g., albedo, rain rate, density of an admixture in turbulence); others are limited by in absolute value, often by the input mechanism of some conserved quantity (e.g., albedo by the incident radiant energy, velocity by the kinetic energy forced into the system). This underscores the physical importance of the integral scale R; for chunks of data of size $r > R$, we will find fluctuations of comparable magnitude and, equivalently, a flatter spectrum for wavenumbers $0 \le k < 1/R$; from both stand points, we are dealing with more stationary behavior.

The Wiener-Khintchine theorem can be generalized to non-stationary situations where stationary increments prevail, leading to a Fourier connection between $\langle \Delta f(r)^2 \rangle$ and $E(k)$ [43]. For their scaling versions, Equations (A.14) and (5), this yields

$$\beta - \zeta(2) = 1, \tag{A.15}$$

in analogy with Equations (A.6) and (A.11). Notice that the spectral criterion for non-stationarity in Equation (A.12) is retrieved from (A14–A15). We must furthermore require $\beta < 3$ in order to have stationary increments by requiring the gradient field $(E_{\nabla f}(k) = k^2 E_f(k))$ to be stationary $(\beta - 2 < 1)$.

A.6. Additive scale-invariant models, non-intermittent and non-stationary

The classic example of a non-stationary stochastic process is Brownian motion (often referred to as a Wiener-Lévy process):

$$f_\infty(x) = f_\infty(0) + \int_0^x f_0(x')\,dx', \tag{A.16}$$

where we notice the additive nature of the resulting field, in contrast with the multiplicative character of $\varepsilon(l;x)$ in Equation (A.8). Brownian motion is not intermittent in the sense of Section A.4, even if the intermittency is sought in the absolute gradient field $|\nabla f_\infty(x)| = |f_0(x)|$, as customary in turbulence studies (cf. section 4). If not completely flat, $|f_0(x)|$ is very weakly variable in comparison with $\varepsilon(l;x)$ due to its close relation to Gaussian white noise. Other examples of non-intermittent non-stationary processes are discussed in the main text ("fractional" Brownian motions). An important new class of non-stationary models with intermittent gradient fields ("multi-affine" processes) is also evoked.

In Equation (A.16), we are essentially computing the position of a particle performing a random walk as a function of time $x \geq 0$, the white noise $f_0(x)$ representing the independently distributed random steps. We have $\langle f_\infty(x) \rangle = \langle f_0 \rangle x + f_\infty(0)$ where both r.h.s. terms can be made to vanish, leaving $\langle f_\infty(x) \rangle \equiv 0$, but this has no bearing on stationarity. To wit, it can be shown that, still for $\langle f_0 \rangle = f_\infty(0) = 0$, the 1-point p.d.f. of $f_\infty(x)$ is Gaussian with variance $\langle f_\infty(x)^2 \rangle$ which is known to increase linearly with $x \geq 0$. This consequence of the law of large numbers is already proof of non-stationarity. In Fourier space, we come to the same conclusion using Equation (A.12) since the integral in Equation (A.16) corresponds to a $1/k$ filtering which brings the vanishing spectral exponent of $f_0(x)$ to $\beta = 2$ for $f_\infty(x)$. (This justifies the subscript "∞" standing for a formally divergent R in Equation (A.5) due to the infinite numerator, symptom of an incipient "infra-red catastrophe.") It is easy to see that $\Delta f_\infty(r;y)$ is simply the integral of $f_0(x)$ over $[y, y+r]$ and behaves statistically like $f_\infty(x)$ in Equation (A.16) with $f_\infty(0) = 0$ and $x = r$, independently of

y. So $\langle \Delta f_\infty(r)^2 \rangle \propto r$, which proves that $\zeta(2) = 1$ in Equation (A.14) for this special case and, in turn, this confirms Equation (A.15) for Brownian motion.

A typical sample of Brownian motion is provided in Figure A1b. Like $\varepsilon(l;x)$ in turbulent cascades Equation (A.8), there is no limit to how big Brownian motion can become, starting at $f_\infty(0) = 0$, but it takes a long time to get there and, when $|f_\infty(x)|$ is large, it will of course take an equally long time for it to come back to zero, in all probability. As can be seen in Figure A1b, the "zero-crossing" set $\{x \geq 0, \; f_\infty(x) = 0\}$ where $f_\infty(x)$ is at its most probable value is in fact a very sparse set with a fractal dimension of $1/2$ [38]. This is a direct consequence of $f_\infty(x)$'s lack of stationarity. In sharp contrast, the stationary cascade models $\varepsilon(l;x)$ will return to their most probable value (which is very small when $l \ll L$) very quickly after a strong (necessarily positive) fluctuation. The typical delay will be $\approx R$ which is $\ll L$ for well-developed multiplicative cascades but $\approx L$ (formally ∞, hence the choice of subscript) in Brownian motion.

A.7. Stochastic continuity with stationarity or scale-invariance but not both

If $\langle \Delta f(r)^2 \rangle \to 0$ when $r \to 0$, then $f(\cdot)$ is said to be "stochastically" continuous and this is the case for all non-stationary scaling processes obeying Equation (A.14). Of course processes that are stationary *per se* have stationary increments (if scaling prevails, then $\beta < 1$ implies $\beta < 3$). It is easy to see that

$$\langle \Delta f(r)^2 \rangle = 2[G(0) - G(r)]. \tag{A.17}$$

So stationary processes are stochastically continuous as long as $G(r)$ is continuous at $r = 0$; in particular, this implies $G(0) < \infty$ (finite 1-point variance). This is not the case for exactly scaling processes since $G(r) \to \infty$ if $r \to 0$ in Equation (3), one of the numerous reasons why there is always a lower limit to scaling regimes in physical systems. For smaller values of the scale parameter, fields can be viewed as smoothly varying (an incipient "ultra-violet catastrophe" is thus avoided).

The log-normal multiplicative cascade process in Figure A2 obeys Equation (A.2) and therefore illustrates stochastic discontinuity, as does the white noise in Figure A1a (independent Gaussian deviates assigned to every pixel) for which $G(r) \propto \delta(r)$. The Brownian motion in Figure A1b is stochastically continuous since $\langle \Delta f(r)^2 \rangle \propto r$. Finally, in Figure A1c we illustrate an Ornstein-Uhlenbeck process denoted $f_R(x)$, $0 \leq x \leq L = 4096$, which is at once continuous and stationary but non-scaling. Apart from the variance $\sigma^2 = G_R(0)$ of its 1-point Gaussian p.d.f., the only parameter of an Ornstein-Uhlenbeck process is the integral scale R in its auto-correlation function

$$G_R(r) = G_R(0) \exp(-r/R). \qquad \text{(A.18)}$$

In Figure A1c, we took $R = L/8 = 512$. For the same process the structure function is therefore

$$\langle \Delta f_R(r)^2 \rangle = 2G_R(0)[1 - \exp(-r/R)]. \qquad \text{(A.19)}$$

Notice the cross-over from non-stationary and continuous behavior (i.e., $\langle \Delta f_R(r)^2 \rangle \propto r$) to stationary and discontinuous behavior (i.e., $\langle \Delta f_R(r)^2 \rangle \approx$ *constant*) as r goes from $\ll R$ to $\gg R$. In essence, we are dealing with Brownian motion (on scales $r \ll R$) that is forced to come back to the origin on a regular basis (at intervals of length $\approx R$). In the limit $R \to 0$ (vanishing correlation length) white noise is retrieved. For $R \to \infty$ (diverging correlation length) "unconstrained" Brownian motion is retrieved and the fact that $\langle f_R(x) f_R(x+r) \rangle \to constant > 0$ in this case is characteristic of non-stationary behavior, as anticipated in Section A.2.

A.8. Summary with analysis and modeling of geophysical data in mind

Geophysical processes are driven by complex nonlinear dynamics that typically unfold in both space and time. Can we view these as statistically periodic above some scale and treat portions of equal or greater size as statistically independent realizations? Or must we view the whole as one single irreproducible experiment and hope that space/time averages converge to a statistically meaningful value? In statistical jargon the former hypothesis is called stationarity (statistical invariance under translation), the latter as ergodicity (space/time and ensemble averages yield identical answers). In all cases, we are required to characterize these processes from data that always comes in finite quantities with some finite resolution and neither stationarity nor ergodicity can be established rigorously. We argue that stationarity, the more general concept, is also more useful in data analysis endeavors; it is also the more physically reasonable when it comes to modeling the data.

In the framework of scale-invariance, even in the widespread physical sense where a finite range of scales is involved, one can use a specific spectral criterion for stationarity: the spectral exponent (β) is less than unity. In numerous situations there exists a range of scales where non-stationary behavior $(\beta > 1)$ is observed; one can then study a stationary feature of the data such as increments (as long as $\beta < 3$). Even if a strict ergodicity assumption is physically unreasonable, we must still use spatial averaging to some degree and it is important to focus exclusively on stationary quantities. So an operational procedure for defining stationary conditions and/or features is welcome.

A given geophysical field can be stationary at large scales and non-stationary at small ones. In fact this is to be expected on physical grounds and the transition scale is known as the integral scale. Bearing this in mind, there are two classes of stationary theoretical models that we find particularly useful for tutorial purposes: Ornstein-Uhlenbeck processes and multiplicative cascade processes.

- The former are patently non-scaling but have two distinct scaling regimes: at large scales it behaves like (stationary) white noise and at small ones like (non-stationary) Brownian motion, roughly mimicking natural processes. These processes are non-differentiable but stochastically (i.e., almost everywhere) continuous and this is directly traceable to their non-stationary behavior in the small scale limit.

- The latter are currently being used in turbulence studies (to model the intermittency of the dissipation field), in hydrology (to model the spatial distribution of rain), and in deterministic chaos (to model the invariant measure supported by the strange attractor) amongst many other applications. They are characterized by small scale discontinuity (almost everywhere), itself traceable to very singular behavior on sparse fractal sets.

The concept of intermittency – sudden and intense bursts of high frequency activity – apparently conflicts with the conventional wisdom about stationarity: "Things are almost the same at all times." We argue that this is an unnecessarily restrictive outlook, excluding in particular strong deviations from mean or most probable states. A better description of stationarity that is compatible with extreme forms of variability (represented in particular by cascade models) is: "On short notice, things can quickly depart from and return to typical values." By "short notice" and "quickly" we understand one integral scale or so; to illustrate this point we show that many integral scales fit into the overall length of the process in the case of multiplicative cascades (although each one is still many pixels long, at least for a large but finite number of cascade steps).

LIST OF SYMBOLS

Scale parameters:

- general

  r temporal or spatial scale on which a statistic is to be conditioned

  k $\approx 1/r$, wavenumber (in absolute value)

- instrumental and/or computational

l inner scale, sampling scale (data), grid constant (model)
L outer scale, overall physical size of dataset or model

- physical (to be viewed as statistical properties)

η lower bound of non-stationary scaling regime (where $\beta > 1$)
R upper bound of non-stationary scaling regime, integral scale

Stochastic processes or geophysical data-streams and related random quantities:

- for general purpose

  x time or space coordinate
  $B(x)$ fractional Brownian motion
  $f(x)$ a generic random function
  $f_0(x)$ uncorrelated Gaussian fluctuations (white noise)
  $f_\infty(x)$ Levy-Wiener process (Brownian motion), integral of $f_0(x)$
  $f_R(x)$ Ornstein-Uhlenbeck process with integral scale R
  $\varepsilon(x)$ a generic random measure ($\varepsilon(x) \geq 0$)
  $\varepsilon(l; x)$ cascade process on $[0, L]$ (developed down to scale l)

- for structure functions

  $\Delta f(r; x)$ increment of $f(x)$ over scale r at x, e.g., $f(x + r) - f(x)$
  $h(x)$ local Hölder exponent (in $|\Delta f(r; x)| \propto r^{h(x)}$)

- for singularity analysis

  $\varepsilon(\eta; x)$ $= |\Delta f(\eta; x)|$, absolute small scale gradients used in singularity analysis
  $p(r; x)$ integral of $\varepsilon(x)$ or $\varepsilon(\eta; x)$ over an r-sized box at x, e.g., $[x, x + r]$ in $D = 1$
  $\alpha(x)$ local order of singularity (in $p(r; x) \propto r^{\alpha(x)}$)
  $\varepsilon(r; x)$ average of $\varepsilon(x)$ or $\varepsilon(\eta; x)$ over an r-sized box at x, i.e., $p(r; x)/r^D$

- for either

  $\xi(r; x)$ $= |\Delta f(r; x)|$, $\varepsilon(r; x)$ or $p(r; x)$
  ($\langle \xi(r; x)^z \rangle$ scales like $r^{A(z)}$)

Statistical properties:

$E(k)$ energy (or power) spectrum, spectral density
$G(r)$ $= \langle f(r + x) f(x) \rangle$, auto-correlation function (of a stationary zero-mean process)
$\langle |\Delta f(r; x)|^p \rangle$ p-th moment of stationary increments $|\Delta f(r; x)|$, structure function of order p

$\langle \varepsilon(r;x)^q \rangle$ q-th moment of stationary measures $\varepsilon(r;x)$, used in singularity analysis

σ standard deviation of a Gaussian random variable

Wavelet-related quantities:

- scaling-type functions

$I_r(x)$ $= \theta(x+r/2) - \theta(x-r/2)$, indicator function of the interval $(-r/2, +r/2)$

$\varphi_\sigma(x)$ $= \exp(-x^2/2\sigma^2)/\sigma\sqrt{2\pi}$, Gaussian function with variance σ^2

$\varphi_H(x)$ $= \max[0, \min\{1-x, 1+x\}]$, integral of Haar wavelet

- wavelet shapes

$\Psi_G(x)$ $= -\varphi'_{1/2}(x)$

$\Psi_H(x)$ Haar wavelet, see Equation (9b)

$\Psi_M(x)$ Mexican hat, see Equation (17a)

$\Psi_F(x)$ French top hat, see Equation (17b)

$-\partial I_1(x)$ $= -I'_1(x) = \delta(x-r/2) - \delta(x+r/2)$, "poor man's" wavelet [45]

- wavelet transforms of random processes

$T_\Psi[f](a;b)$ for structure functions of $f(\cdot)$: $\Psi = -\partial I_1$, Ψ_H, Ψ_G; $a = r$ and $b = x$

$T_\phi[\varepsilon](a;b)$ for singularity analysis of $\varepsilon(\cdot)$: $\phi = I_1$, φ_1, φ_H, φ_M, φ_F; $a = r$ and $b = x$

Exponents

D dimensionality of data (e.g., *number of points* $\approx (L/l)^D$)

- for second order statistics

β spectral exponent for $D \geq 1$ ($E(k) \sim 1/k^\beta$)

μ $= 1 - \beta > 0$, scaling of auto-correlation function in $D = 1$ ($G(r) \sim 1/r^\mu$)

- for structure functions ($D = 1$)

p order of statistical moment of $|\Delta f(r;x)|$ in structure function analysis

$\zeta(p)$ scaling of structure functions ($\langle |\Delta f(r;x)|^p \rangle \sim r^{\zeta(p)}$)

$H(p)$ $= \zeta(p)/p$, a non-increasing hierarchy

H_1 $= H(1) = \zeta(1)$, index of non-stationarity ($0 \leq H_1 \leq 1$)

$D_{g(f)}$ $= 2 - H_1$, fractal dimension of two-dimensional graph of data

- for singular measures ($D \geq 1$)

q order of statistical moment of $\varepsilon(r;x)$ in singularity analysis

$K(q)$ scaling of singular measures ($\langle \varepsilon(r;x)^q \rangle \sim r^{-K(q)}$)

$\tau(q)$ $= (q-1)D - K(q)$, scaling of $\sum p(r;x)^q$ (using disjoint boxes) $\sim r^{\tau(q)}$

$C(q)$ $= K(q)/(q-1)$, a non-decreasing hierarchy

$D(q)$ $= \tau(q)/(q-1) = D - C(q)$, a non-decreasing hierarchy

C_1 $= C(1) = K'(1)$, index of intermittency ($0 \leq C_1 \leq D$)

- for bi-multifractal analysis (Section 5.2)

$X(p,q)$ scaling of joint moments ($\langle |\Delta f(r;x)|^p \varepsilon(r;x)^q \rangle \sim r^{X(p,q)}$)

s,t generalized Kolmogorov exponents in Equations (24)–(25)

Miscellaneous computational quantities:

x_i discrete time or space ($x_i = il$, $i = 0, \dots, N$)

f_i f-value associated with x_i

m dividing ratio in cascade process

n number of cascade steps

N $= L/l$, number of grid/data points in $D = 1$ (also m^n in cascade models)

Mathematical functions:

$\theta(x)$ $= 0$ for $x < 0$, $= 1/2$ for $x = 0$, $= 1$ for $x > 0$, Heaviside's step function

$\delta(x)$ $= \theta'(\cdot)$, Dirac's generalized function

References

1. Anselmet, F., Y. Gagne, E. J. Hopfinger, and R. A. Antonia, High-order Velocity Structure Functions in Turbulent Shear Flows, *Journal of Fluid Mechanics*, 140, 63–80, 1984.

2. Argoul, F., A. Arnéodo, G. Grasseau, Y. Gagne, E. J. Hopfinger, and U. Frisch, Wavelet analysis of turbulence reveals the multifractal nature of the Richardson cascade, *Nature*, 338, 51–53, 1989.

3. Arnéodo, A., G. Grasseau, and M. Holschneider, Wavelet transform of multifractals, *Physical Review Letters*, 61, 2281–2284, 1988.

4. Arnéodo, A., G. Grasseau, and M. Holschneider, Wavelet transform analysis of some dynamical systems, in *Wavelets*, J. M. Combes, A. Grossmann and P. Tchamitchian, (eds.), Springer, Berlin, 1989.

5. Arnéodo, A., E. Bacry, and J. F. Muzy, Wavelet analysis of fractal signals, Direct determination of the singularity spectrum of fully de-

veloped turbulence data, in *Wavelets and Turbulence*, Springer, Berlin, 1991.

6. Arnéodo, A., A. Argoul, E. Bacry, J. F. Muzy, and M. Tabard, Golden mean arithmetic in the fractal branching of diffusion-limited aggregates, *Physical Review Letters*, 68, 3456–3459, 1991.

7. Arnéodo, A., J. F. Muzy, and E. Bacry, Wavelet analysis of fractal signals. Applications to fully developed turbulence data, in *IUTAM Symposium on Eddy Structure Identification in Free Turbulent Shear Flows*, Poitiers, October, 1992.

8. Arnéodo, A., A. Argoul, J. F. Muzy, and M. Tabard, Uncovering Fibonacci sequences in the fractal morphology of diffusion-limited aggregates, *Physical Letters A*, 171, 31–36, 1992.

9. Bacry, E., J. F. Muzy, and A. Arnéodo, Singularity spectrum of fractal signals from wavelet analysis: Exact results, *Journal of Statistical Physics*, 70, 635–674, 1993.

10. Benzi, R., G. Paladin, G. Parisi, and A. Vulpani, On the multifractal nature of fully developed turbulence and chaotic systems, *Journal of Physics A*, 17, 3521–3531, 1984.

11. Benzi, R., L. Biferale, A. Crisanti, G. Paladin, M. Vergassola, and A. Vulpani, A random process for the construction of multiaffine fields, *Physica D*, 65, 352–358, 1993.

12. Cahalan, R. F., M. Nestler, W. Ridgway, W. J. Wiscombe, and T. Bell, Marine stratocumulus spatial structure, in *4th Int. Meeting on Statistical Climatology*, Rotrura, New Zealand, 1989.

13. Chistakos, G., *Random Fields in Earth Sciences*, Academic Press, San Diego, 1992.

14. Chui, C. K., *An Introduction to Wavelets*, Academic Press, San Diego, 1991.

15. Daubechies, I., Orthonormal bases of compactly supported wavelets, *Communications in Pure and Applied Mathematics*, 41, 909–996, 1988.

16. Davis, A., A. Marshak, W. Wiscombe. and R. Cahalan, Multifractal characterizations of non-stationarity and intermittency in geophysical fields, Observed, retrieved or simulated, *Journal of Geophysical Research*, in print, 1994.

17. Davis, A., A. Marshak, and W. Wiscombe, Bi-Multifractal analysis and multi-affine modeling of non-stationary geophysical processes, Application to turbulence and clouds, *Fractals*, 3(1), 1994.

18. Davis, A., A. Marshak, W. Wiscombe. and R. Cahalan, The scale-invariant structure of marine stratocumulus deduced from observed liquid water distributions, II, Multifractal properties and model validation, *Journal of the Atmospheric Sciences*, submitted, 1994.

19. Duroure, C., and C. P. Fantodji, Wavelet analysis and Fourier analysis of stratocumulus microphysical data, report (Laboratoire de

Météorologie Physique de Clermond-Ferrand), 1993.

20. Fantodji, C. P., C. Duroure, and H. R. Larsen, The use of wavelets for the analysis of geophysical data, preprint, *Physica*, submitted, 1993.

21. Farge, M., Wavelet transforms and their applications to turbulence, *Annual Review of Fluid Mechanics*, 24, 395–457, 1992.

22. Feller, W., *An Introduction to Probability Theory and its Applications*, vol. 2, Wiley, New York, 1971.

23. Frisch, U., From global scaling, à la Kolmogorov, to local multifractal in fully developed turbulence, *Proceedings of the Royal Society of London A*, 434, 89–99, 1991.

24. Frisch, U., P. L. Sulem, and M. Nelkin, A simple dynamical model for intermittent fully developed turbulence, *Journal of Fluid Mechanics*, 87, 719–736, 1978.

25. Ghez, J. M., and S. Vaienti, On the wavelet analysis for multifractal sets, *Journal of Statistical Physics*, 57, 415–420, 1989.

26. Ghez, J. M., and S. Vaienti, Integrated wavelets on fractal sets: I, The correlation dimension, II, The generalized dimensions, *Nonlinearity*, 5, 777–804, 1992.

27. Gollmer, S. M., Harshvardan, R. F. Cahalan, and J. B. Snider, Wavelet analysis of marine stratocumulus, in *Proceedings of the 11th International Conference on Clouds and Precipitation*, Montreal (Que.), Aug. 17–22, 1992.

28. Grassberger, P., Generalized dimensions of strange attractors, *Physical Review Letters A*, 97, 227–330, 1983.

29. Gupta, V. K., and E. C. Waymire, A statistical analysis of mesoscale rainfall as a random cascade, *Journal of Applied Meteorology*, 32, 251–267, 1993.

30. Halsey, T. C., M. H. Jensen, L. P. Kadanoff, I. Procaccia, and B. I. Shraiman, Fractal measures and their singularities: The characterization of strange sets, *Physical Review A*, 33, 1141–1151, 1986.

31. Hentschel, H. G. E., and I. Procaccia, The infinite number of generalized dimensions of fractals and strange attractors, *Physica D*, 8, 435–444, 1983.

32. Holley, R., and E. C. Waymire, Multifractal dimensions and scaling exponents for strongly bounded random cascade, *The Annals of Applied Probability*, 2, 819–845, 1992.

33. Kahane, J. P., and J. Peyrière, Sur certaines martingales de Benoit Mandelbrot, *Advances in Mathematics*, 22, 131–145, 1976.

34. Kolmogorov, A. N., Local structure of turbulence in an incompressible liquid for very large Reynolds numbers, *Doklady Akademii Nauk SSSR*, 30, 299–303, 1941.

35. Kolmogorov, A. N., A refinement of previous hypothesis concerning the local structure of turbulence in viscous incompressible fluid at high

Reynolds number, *Journal of Fluid Mechanics*, 13, 82–85, 1962.

36. Lavallée, D., S. Lovejoy, D. Schertzer, and P. Ladoy, Non-linear variability, multifractal analysis and simulation of landscape topography, in *Fractals in Geophysics*, De Cola, L., and N. Lam (eds.), Kluver, 158-192, 1993. of landscape topography: Multifractal analysis and simulation. in De Cola, L., and Lam, N. (eds.), *Fractal in Geography*, Prentice Hall, 1993.

37. Mandelbrot, B. B., Intermittent turbulence in self-similar cascades: divergence of high moments and dimension of the carrier, *Journal of Fluid Mechanic*, 62, 331–358, 1974.

38. Mandelbrot, B. B., *Fractals: Form, Chance, and Dimension*, Freeman, W.H., and Co., San Francisco, 1977.

39. Marshak A., A. Davis, R. Cahalan, and W. Wiscombe, Bounded cascade models as non-stationary multifractals, *Physical Review E*, 49, 55–69, 1994.

40. Marshak A., A. Davis, W. Wiscombe, and R. Cahalan, The scale-invariant structure of marine stratocumulus deduced from observed liquid water distributions, I, Spectral properties and stationarity issues, *Journal of the Atmospheric Sciences*, submitted, 1994.

41. Meneveau, C., and K. R. Sreenivasan, The multifractal spectrum of the dissipation field in turbulent flows, *Nuclear Physics B*, 2, 49–76, 1987.

42. Meneveau, C., and K. R. Sreenivasan, Simple multifractal cascade model for fully developed turbulence, *Physical Review Letters*, 59, 1424–1427, 1987.

43. Monin, A. S., and A. M. Yaglom, *Statistical Fluid Mechanics*, vol. 2, MIT Press, Boston, Mass, 1975.

44. Muzy, J. F., E. Bacry, and A. Arnéodo, Wavelets and multifractal formalism for singular signals: Application to turbulence data, *Physical Review Letters*, 67, 3515–3518, 1991.

45. Muzy, J. F., E. Bacry, and A. Arnéodo, Multifractal formalism for fractal signals: The structure function approach versus the wavelet-transform modulus-maxima method, *Physical Review E*, 47, 875–884, 1993.

46. Novikov, E. A., and R. Stewart, Intermittency of turbulence and spectrum of fluctuations in energy dissipation, *Izvestija Akademii Nauk. SSSR, Ser. Geofiz.*, 3, 401–412, 1964.

47. Obukhov, A., Some specific features of atmospheric turbulence, *Journal of Fluid Mechanic*, 13, 77–81, 1962.

48. Parisi, G., and U. Frisch, A multifractal model of intermittency, in Ghil, M., Benzi, R., and Parisi, G. (eds.), *Turbulence and Predictability in Geophysical Fluid Dynamics*, North Holland, Amsterdam, 1985.

49. Peitgen, H.-O., and D. Saupe, (eds.), *The Science of Fractal Images*,

Springer-Verlag, New York, 1988.
50. Press, W. H., S. A. Teukolsky, W. T. Vetterling, and B. P. Flannery, *Numerical Recipes in Fortran*, second edition. Cambridge Un. Press, 1993.
51. Schertzer, D., and S. Lovejoy, On the dimension of atmospheric motions, in Tatsumi, T. (ed.) *Turbulence and Chaotic Phenomena in Fluids (IUTAM Symposium 1983)*, Elsevier North-Holland, New York, 1984.
52. Schertzer, D., and S. Lovejoy, Physical modeling and analysis of rain clouds by anisotropic scaling multiplicative processes, *Journal of Geophysical Research*, 92, 9693–9714, 1987.
53. Schertzer D., and S. Lovejoy, Hard and soft multifractal processes, *Physica A*, 185, 187–194, 1992.
54. Schmitt, F., D. Lavallée, D. Schertzer, and S. Lovejoy, Empirical determination of universal multifractal exponents in turbulent velocity fields, *Physical Review Letters*, 68, 305–308, 1992.
55. Stolovitzky G., P. Kailasnath, and K. R. Sreenivasan, Kolmogorovs refined similarity hypothesis, *Physical Review Letters*, 69, 1178–1181, 1992.
56. Tessier, Y., S. Lovejoy, and D. Schertzer, Universal multifractals: Theory and observations for rain and clouds, *Journal of Applied Meteorology*, 14, 160–164, 1993.
57. Viscek, T., and A.-L. Barabási, Multi-affine model for the velocity distribution in fully turbulent flows, *Journal of Physics A: Math. Gen.*, 24, L845–L851, 1991.

This work was supported by the Department of Energy's Atmospheric Radiation Measurement (ARM) project, grant DE-A105-90ER61069 to NASA's Goddard Space Flight Center. We acknowledge Dr. Alain Arnéodo's lectures on the elements of waveletry with many applications to scale-invariant systems. We thank Phil Austin for providing the FIRE data. We are also appreciative of the many fruitful discussions with T. Bell, R. Cahalan, C. Duroure, M. Farge, S. Gollmer, D. Lavallée, S. Lovejoy, C. Meneveau, D. Schertzer, Y. Tessier and T. Warn.

Anthony Davis
Universities Space Research Association
NASA-GSFC (Code 913)
Greenbelt
MD 20771
USA
e-mail: *davis@climate.gsfc.nasa.gov*

Alexander Marshak
Science Systems and Applications, Inc.
5900 Princess Garden Parkway
Lanham
MD 20706
USA

Warren Wiscombe
NASA Goddard Space Flight Center
Climate and Radiation Branch
Greenbelt
MD 20771
USA

Simultaneous Noise Suppression and Signal Compression Using a Library of Orthonormal Bases and the Minimum Description Length Criterion

Naoki Saito

Abstract. We describe an algorithm to estimate a discrete signal from its noisy observation, using a library of orthonormal bases (consisting of various wavelets, wavelet packets, and local trigonometric bases) and the information-theoretic criterion called minimum description length (MDL). The key to effective random noise suppression is that the signal component in the data may be represented efficiently by one or more of the bases in the library, whereas the noise component cannot be represented efficiently by any basis in the library. The MDL criterion gives the best compromise between the fidelity of the estimation result to the data (noise suppression) and the efficiency of the representation of the estimated signal (signal compression): it selects the "best" basis and the "best" number of terms to be retained out of various bases in the library in an objective manner. Because of the use of the MDL criterion, our algorithm is free from any parameter setting or subjective judgments.

This method has been applied usefully to various geophysical datasets containing many transient features.

§1. Introduction

Wavelet transforms and their relatives such as wavelet packet transforms and local trigonometric transforms are becoming increasingly popular in many fields of applied sciences. So far their most successful application area seems to be data compression; see e.g., [14], [6], [35], [30]. Meanwhile, several researchers claimed that wavelets and these transforms are also useful for reducing noise in (or denoising) signals/images [16], [7], [10], [21]. In this paper, we take advantage of both sides: we propose an algorithm for *simultaneously* suppressing random noise in data and compressing the signal, i.e., we try to "kill two birds with one stone."

Wavelets in Geophysics
Efi Foufoula-Georgiou and Praveen Kumar (eds.), pp. 299–324.

ISBN 0-12-262850-0

Throughout this paper, we consider a simple degradation model: observed data consists of a signal component and additive white Gaussian noise. Our algorithm estimates the signal component from the data using a library of orthonormal bases (including various wavelets, wavelet packets, and local trigonometric bases) and the information-theoretic criterion called the Minimum Description Length (MDL) criterion for discriminating signal from noise.

The key motivation here is that the signal component in the data can often be efficiently represented by one or more of the bases in the library whereas the noise component cannot be represented efficiently by any basis in the library.

The use of the MDL criterion frees us from any subjective parameter setting such as threshold selection. This is particularly important for real field data where the noise level is difficult to obtain or estimate *a priori*.

The organization of this paper is as follows. In Section 2, we review some of the important properties of wavelets, wavelet packets, local trigonometric transforms which constitute the "library of orthonormal bases" which will be used for efficiently representing nonstationary signals. In Section 3, we formulate our problem. We view the problem of simultaneous noise suppression and signal compression as a model selection problem out of models generated by the library of orthonormal bases. In Section 4, we review the MDL principle which plays a critical role in this paper. We also give some simple examples to help understand its concept. In Section 5, we develop an actual algorithm of simultaneous noise suppression and signal compression. We also give the computational complexity of our algorithm. Then, we extend our algorithm for higher dimensional signals (images) in Section 6. In Section 7, we apply our algorithm to several geophysical datasets, both synthetic and real, and compare the results with other competing methods. We discuss the connection of our algorithm with other approaches in Section 8, and finally, we conclude in Section 9.

§2. A Library of Orthonormal Bases

For our purpose we need to represent signals containing many transient features and edges in an efficient manner. Wavelets and their relatives, i.e., wavelet packets and local trigonometric transforms, have been found very useful for this purpose; see e.g., [14], [6], [35], [30]. As shown below, each of these transforms (or basis functions) has different characteristics. In other words, the best transform to compress a particular signal may not be good for another signal. Therefore, instead of restricting our attention to a particular basis, we consider a *library* of bases. The most suitable basis for a particular signal is selected from this collection of bases. This approach leads to a vastly more efficient representation for the signal, compared with confining ourselves to a single basis.

In this section, we briefly describe the most important properties of these transforms. Throughout this paper, we only consider real-valued discrete signals (or vectors) with finite length N ($= 2^n$). Also we limit our discussions to orthonormal transforms. Hence it suffices here to consider discrete orthonormal transforms, i.e., the orthonormal bases of $\ell^2(N)$, the N-dimensional space of vectors of finite energy.

More detailed properties of these bases can be found in the literature, most notably, in [2], [9], [13], [23], [22], [26], [33].

2.1. Wavelet bases

The wavelet transform (e.g., [13], [23]) can be considered as a smooth partition of the frequency axis. The signal is first decomposed into low and high frequency components by the convolution-subsampling operations with the pair consisting of a "lowpass" filter $\{h_k\}$ and a "highpass" filter $\{g_k\}$ directly on the discrete time domain. Let H and G be the convolution-subsampling operators using these filters and H^* and G^* be their adjoint (i.e., upsampling-anticonvolution) operations. It turns out that we can choose finite-length (L) filters and satisfy the following orthogonality (or perfect reconstruction) conditions:

$$HG^* = GH^* = 0, \quad \text{and} \quad H^*H + G^*G = I,$$

where I is the identity operator of $\ell^2(N)$. Also we have the relation $g_k = (-1)^k h_{L-1-k}$. The pair of filters $\{h_k\}_{k=0}^{L-1}$ and $\{g_k\}_{k=0}^{L-1}$ satisfying these conditions are called *quadrature mirror filters* (QMFs).

This decomposition (or expansion, or analysis) process is iterated on the low frequency components and each time the high frequency coefficients are retained intact and at the last iteration, both low and high frequency coefficients are kept. In other words, let $f = \{f_k\}_{k=0}^{N-1} \in \ell^2(N)$ be a vector to be expanded. Then, the convolution-subsampling operations transform the vector f into two subsequences Hf and Gf of lengths $N/2$. Next, the same operations are applied to the vector Hf to obtain H^2f and GHf of lengths $N/4$. If the process is iterated J ($\leq n$) times, we have the discrete wavelet coefficients $(Gf, GHf, GH^2f, \ldots, GH^Jf, H^{J+1}f)$ of length N. As a result, the wavelet transform analyzes the data by partitioning its frequency content dyadically finer and finer toward the low frequency region (i.e., coarser and coarser in the original time or space domains).

If we were to partition the frequency axis sharply using the characteristic functions (or box-car functions), then we would have ended up the so-called Shannon (or Littlewood-Paley) wavelets, i.e., the difference of two sinc functions. Clearly, however, we cannot have a finite-length filter in the time domain in this case. The other extreme is the Haar basis which partitions the frequency axis quite badly but gives the shortest filter length

($L = 2$) in the time domain.

The reconstruction (or synthesis) process is also very simple: starting from the lowest frequency components (or coarsest scale coefficients) $H^{J+1}f$ and the second lowest frequency components $GH^{J}f$, the adjoint operations are applied and added to obtain $H^{J}f = H^{*}H^{J+1}f + G^{*}GH^{J}f$. This process is iterated to reconstruct the original vector f. The computational complexity of the decomposition and reconstruction process is in both cases $O(N)$ as easily seen.

We can construct the basis vector $w_{j,k}$ at scale j and position k simply by putting $(GH^{j}f)_{l} = \delta_{l,k}$, where $\delta_{l,k}$ denotes the Kronecker delta, and synthesizing $f = w_{j,k}$ by the reconstruction algorithm. Using these basis vectors, we can express the wavelet transform in a vector-matrix form as

$$\alpha = W^{T}f,$$

where $\alpha \in \mathbf{R}^{N}$ contains the wavelet coefficients and $W \in \mathbf{R}^{N \times N}$ is an orthogonal matrix consisting of column vectors $w_{j,k}$. This basis vector has the following important properties:

- *vanishing moments:* $\sum_{l=0}^{N-1} l^{m}w_{j,k}(l) = 0$ for $m = 0, 1, \ldots, M - 1$.

The higher the degrees of vanishing moments the basis has, the better it compresses the smooth part of the signal. In the original construction of Daubechies [12], it turns out that $L = 2M$. There are several other possibilities. One of them is a family of the so-called "coiflets" with $L = 3M$ which are less asymmetric than the original wavelets of Daubechies [13].

- *regularity:* $|w_{j,k}(l + 1) - w_{j,k}(l)| \leq c\, 2^{-j\alpha}$,

where $c > 0$ is a constant and $\alpha > 0$ is called the *regularity* of the wavelets. The larger the value of α is, the smoother the basis vector becomes. This property may be important if one requires high compression rate since the shapes of the basis vectors become "visible" in those cases and one might want to avoid fractal-like shapes in the compressed signals/images [25].

- *compact support:* $w_{j,k}(l) = 0$ for $l \notin [2^{j}k, 2^{j}k + (2^{j} - 1)(L - 1)]$.

The compact support property is important for efficient and exact numerical implementation.

2.2. Wavelet packet best-bases

For oscillating signals such as acoustic signals, the analysis by the wavelet transform is sometimes inefficient because it only partitions the frequency axis finely toward the low frequency. The wavelet packet transform (e.g., [9], [22], [33]) decomposes even the high frequency bands which are kept intact in the wavelet transform. The first level decomposition is Hf and Gf just like in the wavelet transform. The second level is H^2f, GHf, HGf, G^2f. If we repeat this process for J times, we end up having JN expansion coefficients. Clearly, we have a redundant set of expansion coefficients, in fact, there are more than $2^{2^{(J-1)}}$ possible orthonormal bases. One way of selecting an efficient basis for representing the signal or vector is to use the entropy criterion [9], [33]. We can think of the wavelet packet bases as a set of different coordinate systems of $\mathbf{R}^N$. Then a signal of length N is a point in $\mathbf{R}^N$, and we try to select the most efficient coordinate system out of the given set of coordinate systems to represent this signal. The signal in an efficient coordinate system should have large magnitudes along a few axes and small magnitudes along most axes. In particular, the wavelet packet basis function becomes a unit vector along an axis of the coordinate systems. Then, it is very natural to use the entropy as a measure of efficiency of the coordinate system. The *best-basis* is the basis or coordinate system giving the minimum entropy for its coordinate distribution. The computational complexity of computing the best-basis is $O(N \log_2 N)$ as is the reconstruction of the original vector from the best-basis coefficients.

Remark. We would like to note that given a set of signals, the Karhunen-Loève basis gives the global minimum entropy. However, it is very expensive to compute; the cost is $O(N^3)$ since it involves solving an eigenvalue problem. On the other hand, the wavelet packet best-basis can be computed cheaply and is defined even for a single signal; see [34] for a comparison of these two bases using images of human faces.

2.3. Local trigonometric best-bases

Local trigonometric transforms ([9], [22], [33], [2]) can be considered as conjugates of wavelet packet transforms: they partition the time (or space) axis smoothly. In fact, Coifman and Meyer [8] showed that it is possible to partition the real-line into any disjoint intervals smoothly and construct orthonormal bases on each interval. In the actual numerical implementation, the data is first partitioned into disjoint intervals by the smooth window function, and then on each interval the data is transformed by the discrete cosine or sine transforms (DCT/DST). Since it partitions the axis smoothly, these transforms, i.e., local cosine or sine transforms (LCT/LST), have less edge (or blocking) effects than the conventional DCT/DST. Wickerhauser [33] proposed the method of dyadically partitioning the time axis and computing the best-basis using the entropy criterion similarly to the wavelet packet best-basis construction. The computational complexity in this case is about $O(N[\log_2 N]^2)$. Local trigonometric transforms are clearly efficient for the signals with localized oscillating features such as musical notes.

§3. Problem Formulation

Let us consider a discrete degradation model

$$d = f + n,$$

where $d, f, n \in \mathbf{R}^N$ and $N = 2^n$. The vector d represents the noisy observed data and f is the unknown true signal to be estimated. The vector n is white Gaussian noise (WGN), i.e., $n \sim \mathcal{N}(0, \sigma^2 I)$. Let us assume that σ^2 is unknown.

We now consider an algorithm to estimate f from the noisy observation d. First, we prepare the library of orthonormal bases mentioned in the previous section. This library consists of the standard Euclidean basis of $\mathbf{R}^N$, the Haar-Walsh bases, various wavelet bases and wavelet packet best-bases generated by Daubechies' QMFs, their less asymmetric versions (i.e., coiflets), and local trigonometric best-bases. This collection of bases is highly adaptable and versatile for representing various transient signals [7]. For example, if the signal consists of blocky functions such as acoustic impedance profiles of subsurface structure, the Haar-Walsh bases capture those discontinuous features both accurately and efficiently. If the signal consists of piecewise polynomial functions of order p, then the Daubechies wavelets/wavelet packets with filter length $L \geq 2(p+1)$ or the coiflets with filter length $L \geq 3(p+1)$ would be efficient because of the vanishing moment property. If the signal has a sinusoidal shape or highly oscillating characteristics, the local trigonometric bases would do the job. Moreover, computational efficiency of this library is also attractive; the most expensive expansion in this library, i.e., the local trigonometric expansion, costs about $O(N[\log_2 N]^2)$ as explained in the previous section.

Let us denote this library by $\mathcal{L} = \{\mathcal{B}_1, \mathcal{B}_2, \ldots, \mathcal{B}_M\}$, where $\mathcal{B}_m$ represents one of the orthonormal bases in the library, and M (typically 5 to 20) is the number of bases in this library. If we want, we can add other orthonormal bases in this library such as the Karhunen-Loève basis [1] or the prolate spheroidal wave functions [13], [36]. However, normally, the above-mentioned multiresolution bases are more than enough, considering their versatility and computational efficiency [7].

Since the bases in the library $\mathcal{L}$ compress signals/images very well, we make a strong assumption here: we suppose the unknown signal f can be *completely* represented by k ($< N$) elements of a basis $\mathcal{B}_m$, i.e.,

$$f = W_m \alpha_m^{(k)}, \tag{1}$$

where $W_m \in \mathbf{R}^{N \times N}$ is an orthogonal matrix whose column vectors are the basis elements of $\mathcal{B}_m$, and $\alpha_m^{(k)} \in \mathbf{R}^N$ is the vector of expansion coefficients of f with only k non-zero coefficients. At this point, we do not know the actual value of k and the basis $\mathcal{B}_m$. We would like to emphasize that in reality the signal f might not be strictly represented by (1). We regard (1) as a *model at hand* rather than a rigid physical model exactly *explaining* f and we will try our best under this assumption. (This is often the case if we want to fit polynomials to some data.) Now the problem of simultaneous noise suppression and signal compression can be stated as follows: *find the "best" k and m given the library $\mathcal{L}$.* In other words, we translate the estimation problem into a model selection problem where models are the bases $\mathcal{B}_m$ and the number of terms k under the additive WGN assumption.

For the purpose of data compression, we want to have k as small as possible. At the same time, we want to minimize the distortion between the estimate and the true signal by choosing the most suitable basis $\mathcal{B}_m$, keeping in mind that the larger k normally gives smaller value of error. How can we satisfy these seemingly conflicting demands?

§4. The Minimum Description Length Principle

To satisfy the above mentioned conflicting demands, we need a model selection criterion. One of the most suitable criteria for our purpose is the so-called *Minimum Description Length* (MDL) criterion proposed by Rissanen [27], [28], [29]. The MDL principle suggests that the "best" model among the given collection of models is the one giving the shortest description of the data *and* the model itself. For each model in the collection, the length of description of the data is counted as the codelength of encoding the data using that model in binary digits (bits). The length of description of a model is the codelength of specifying that model, e.g., the number of parameters and their values if it is a parametric model.

To help understand what "code" or "encoding" means, we give some simple examples. We assume that we want to transmit data by first encoding (mapping) them into a bitstream by an encoder, then receive the bitstream by a decoder, and finally try to reconstruct the data. Let $L(x)$ denote the codelength (in bits) of a vector x of deterministic or probabilistic parameters which are either real-valued, integer-valued, or taking values in a finite alphabet.

Example 4.1. *Codelength of symbols drawn from a finite alphabet.*
Let $x = (x_1, x_2, \ldots, x_N)$ be a string of symbols drawn from a finite alphabet $\mathcal{X}$, which are independently and identically distributed (i.i.d.) with probability mass function $p(x)$, $x \in \mathcal{X}$. In this case, clearly the frequently occurring symbols should have shorter codelengths than rarely occurring symbols for efficient communication. This leads to the so-called Shannon code [11] whose codelength (if we ignore the integer requirement for the codelength) can be written as

$$L(x) = -\log p(x) \qquad \text{for } x \in \mathcal{X}.$$

(From now on, we denote the logarithm of base 2 by "log", and the natural logarithm, i.e., base e by "ln".) The Shannon code has the shortest codelength *on the average*, and satisfies the so-called Kraft inequality [11]:

$$\sum_{x \in \mathcal{X}} 2^{-L(x)} \leq 1, \tag{2}$$

which is necessary and sufficient for the existence of an instantaneously decodable code, i.e., a code such that there is no codeword which is the prefix of any other codeword in the coding system. The shortest codelength on the average for the whole sequence x becomes

$$L(x) = \sum_{i=1}^{N} L(x_i) = -\sum_{i=1}^{N} \log p(x_i).$$

Example 4.2. *Codelength of deterministic integers.*
For a deterministic parameter $j \in \mathbf{Z}_N = (0, 1, \ldots, N - 1)$ (i.e., both the encoder and decoder know N), the codelength of describing j is written as $L(j) = \log N$ since $\log N$ bits are required to index N integers. This can also be interpreted as a codelength using Shannon code for a sample drawn from the uniform distribution over $(0, 1, \ldots, N - 1)$.

Example 4.3. *Codelength of an integer (universal prior for an integer).*
Suppose we do not know how large a natural number j is. Rissanen [27] proposed that the code of such j should be the binary representation of j, preceded by the code describing its length $\log j$, preceded by the code describing the length of the code for $\log j$, and so forth. This recursive strategy leads to

$$L^*(j) = \log^* j + \log c_0 = \log j + \log \log j + \cdots + \log c_0,$$

where the sum involves only the non-negative terms and the constant $c_0 \approx 2.865064$ which was computed so that equality holds in (2), i.e., $\sum_{j=1}^{\infty} 2^{-L^*(j)} = 1$. This can be generalized for an integer j by defining

$$L^*(j) = \begin{cases} 1 & \text{if } j = 0, \\ \log^* |j| + \log 4c_0 & \text{otherwise.} \end{cases} \tag{3}$$

(We can easily see that (3) satisfies $\sum_{j=-\infty}^{\infty} 2^{-L^*(j)} = 1$.)

Example 4.4. *Codelength of a truncated real-valued parameter.*
For a deterministic real-valued parameter $v \in \mathbf{R}$, the exact code generally requires infinite length of bits. Thus, in practice, some truncation must be done for transmission. Let δ be the precision and v_δ be the truncated value, i.e., $|v - v_\delta| < \delta$. Then, the number of bits required for v_δ is the

sum of the codelength of its integer part $[v]$ and the number of fractional binary digits of the truncation precision δ, i.e.,

$$L(v_\delta) = L^*([v]) + \log(1/\delta). \tag{4}$$

Having gone through the above examples, now we can state the MDL principle more clearly. Let $\mathcal{M} = \{\boldsymbol{\theta}_m : m = 1, 2, \ldots\}$ be a class or collection of models at hand. The integer m is simply an index of a model in the list. Let $\boldsymbol{x}$ be a sequence of observed data. Assume that we do not know the true model $\boldsymbol{\theta}$ generating the data $\boldsymbol{x}$. As in [29], [24], given the index m, we can write the codelength for the whole process as

$$L(\boldsymbol{x}, \boldsymbol{\theta}_m, m) = L(m) + L(\boldsymbol{\theta}_m \mid m) + L(\boldsymbol{x} \mid \boldsymbol{\theta}_m, m). \tag{5}$$

This equation says that the codelength to rewrite the data is the sum of the codelengths to describe: (i) the index m, (ii) the model $\boldsymbol{\theta}_m$ given m, and (iii) the data $\boldsymbol{x}$ using the model $\boldsymbol{\theta}_m$. The MDL criterion suggests picking the model $\boldsymbol{\theta}_{m^*}$ which gives the minimum of the total description length (5).

The last term of the right-hand side (RHS) of (5) is the length of the Shannon code of the data assuming the model $\boldsymbol{\theta}_m$ is the true model, i.e.,

$$L(\boldsymbol{x} \mid \boldsymbol{\theta}_m, m) = -\log p(\boldsymbol{x} \mid \boldsymbol{\theta}_m, m), \tag{6}$$

and the maximum likelihood (ML) estimate $\widehat{\boldsymbol{\theta}}_m$ minimizes (6) by the definition:

$$L(\boldsymbol{x} \mid \widehat{\boldsymbol{\theta}}_m, m) = -\log p(\boldsymbol{x} \mid \widehat{\boldsymbol{\theta}}_m, m) \leq -\log p(\boldsymbol{x} \mid \boldsymbol{\theta}_m, m). \tag{7}$$

However, we should consider a further truncation of $\widehat{\boldsymbol{\theta}}_m$ as shown in Example 4.4 above to check that additional savings in the description length is possible. The finer truncation precision we use, the smaller the term (7), but the larger the term $L(\widehat{\boldsymbol{\theta}}_m \mid m)$ becomes. Suppose that the model $\boldsymbol{\theta}_m$ has k_m real-valued parameters, i.e., $\boldsymbol{\theta}_m = (\theta_{m,1}, \ldots, \theta_{m,k_m})$. Rissanen showed in [27], [29] that the optimized truncation precision (δ^*) is of order $1/\sqrt{N}$ and

$$\min_\delta L(\boldsymbol{x}, \boldsymbol{\theta}_{m,\delta}, m, \delta)$$

$$= L(m) + L(\widehat{\boldsymbol{\theta}}_{m,\delta^*} \mid m) + L(\boldsymbol{x} \mid \widehat{\boldsymbol{\theta}}_{m,\delta^*}, m) + O(k_m)$$

$$\approx L(m) + \sum_{j=1}^{k_m} L^*([\widehat{\theta}_{m,j}]) + \frac{k_m}{2}\log N + L(\boldsymbol{x} \mid \widehat{\boldsymbol{\theta}}_m, m) + O(k_m), \tag{8}$$

where $\widehat{\boldsymbol{\theta}}_m$ is the optimal non-truncated value given m, $\widehat{\boldsymbol{\theta}}_{m,\delta^*}$ is its optimally truncated version, and $L^*(\cdot)$ is defined in (4). We note that the last term $O(k_m)$ in the approximation in (8) includes the penalty codelength necessary to describe the data x using the truncated ML estimate $\widehat{\boldsymbol{\theta}}_{m,\delta^*}$ instead of the true ML estimate $\widehat{\boldsymbol{\theta}}_m$. In practice, we rarely need to obtain the optimally truncated value $\widehat{\boldsymbol{\theta}}_{m,\delta^*}$ and we should compute $\widehat{\boldsymbol{\theta}}_m$ up to the machine precision, say, 10^{-15}, and use that value as the true ML estimate in (8). For sufficiently large N, the last term may be omitted, and instead of minimizing the ideal codelength (5), Rissanen proposed to minimize

$$MDL(x,\widehat{\boldsymbol{\theta}}_m,m) = L(m) + \sum_{j=1}^{k_m} L^*([\widehat{\theta}_{m,j}]) + \frac{k_m}{2}\log N + L(x \mid \widehat{\boldsymbol{\theta}}_m, m). \quad (9)$$

The minimum of (9) gives the best compromise between the low complexity in the model and high likelihood on the data.

The first term of the RHS of (9) can be written as

$$L(m) = -\log p(m), \quad (10)$$

where $p(m)$ is the probability of selecting m. If there is prior information about m as to which m is more likely, we should reflect this in $p(m)$. Otherwise, we assume each m is equally likely, i.e., $p(m)$ is a uniform distribution.

Remark. Even though the list of models $\mathcal{M}$ does not include the true model, the MDL method achieves the best result among the available models. See Barron and Cover [4] for detailed information on the error between the MDL estimate and the true model.

We also would like to note that the MDL principle does not attempt to find the absolutely minimum description of the data. The MDL always requires an available collection of models and simply suggests picking the best model from that collection. In other words, the MDL can be considered as an "oracle" for model selection [24]. This contrasts with the algorithmic complexities such as the Kolmogorov complexity which gives the absolutely minimum description of the data, however, in general, is impossible to obtain [27].

Before deriving our simultaneous noise suppression and signal compression algorithm in the context of the MDL criterion, let us give a closely related example:

Example 4.5. *A curve fitting problem using polynomials.*
Given N points of data $(x_i, y_i) \in \mathbf{R}^2$, consider the problem of fitting a polynomial through these points. The model class we consider is a set of polynomials of orders $0, 1, \ldots, N-1$. In this case, $\boldsymbol{\theta}_m = (a_0, a_1, \ldots, a_m)$ represents the $m+1$ coefficients of a polynomial of order m. We also assume

that the data is contaminated by the additive WGN with known variance σ^2, i.e.,

$$y_i = f(x_i) + e_i,$$

where $f(\cdot)$ is an unknown function to be estimated by the polynomial models, and $e_i \sim \mathcal{N}(0, \sigma^2)$. To invoke the MDL formalism, we pose this question in the information transmission setting. First we prepare an encoder which computes the ML estimate of the coefficients of the polynomial, $(\widehat{a}_0, \ldots, \widehat{a}_m)$, of the given degree m from the data. (In the additive WGN assumption the ML estimate coincides with the least squares estimate.) This encoder transmits these m coefficients as well as the estimation errors. We also prepare a decoder which receives the coefficients of the polynomial and residual errors and reconstruct the data. (We assume that the abscissas $\{x_i\}_{i=1}^{N}$ and the noise variance σ^2 are known to both the encoder and the decoder.) Then we ask how many bits of information should be transmitted to reconstruct the data. If we used polynomials of degree $N - 1$, we could find a polynomial passing through all N points. In this case, we could describe the data extremely well. In fact, there is no error between the observed data and those reconstructed by the decoder. However, we do not gain anything in terms of data compression/transmission since we also have to encode the model which requires N coefficients of the polynomial. In some sense, we did not "learn" anything in this case. If we used the polynomial of degree 0, i.e., a constant, then it would be an extremely efficient model, but we would need many bits to describe the deviations from that constant. (Of course, if the underlying data is really a constant, then the deviation would be 0.)

Let us assume there is no prior preference on the order m. Then we can easily see that the total codelength (9) in this case becomes

$$MDL(y, \widehat{\theta}_m, m) = \log N + \sum_{j=0}^{m} L^*([\widehat{a}_j]) + \frac{m+1}{2} \log N$$

$$+ \frac{N}{2} \log 2\pi\sigma^2 + \frac{\log e}{2\sigma^2} \sum_{i=1}^{N} \left(y_i - \sum_{j=0}^{m} \widehat{a}_j x_i{}^j \right)^2.$$

The MDL criterion suggests to pick the "best" polynomial of order m^* by minimizing this approximate codelength.

The MDL criterion has been successfully used in various fields such as signal detection [32], image segmentation [19], and cluster analysis [31] where the optimal number of signals, regions, and clusters, respectively, should be determined. If one knows *a priori* the physical model to explain the observed data, that model should definitely be used, e.g., the complex sinusoids in [32]. However, in general, as a descriptor of real-life signals which are full of transients or edges, the library of wavelets, wavelet packets, and local trigonometric transforms is more flexible and efficient than the set of polynomials or sinusoids.

§5. A Simultaneous Noise Suppression and Signal Compression Algorithm

We carry on our development of the algorithm based on the information transmission setting as the polynomial curve fitting problem described in the previous section. We consider again an encoder and a decoder for our problem. Given (k, m) in (1), the encoder expands the data d in the basis B_m, then transmits the number of terms k, the specification of the basis m, and k expansion coefficients, the variance of the WGN model σ^2, and finally the estimation errors. The decoder receives this information in bits and tries to reconstruct the data d.

In this case, the total codelength to be minimized may be expressed as the sum of the codelengths of: (i) two natural numbers (k, m), (ii) $(k + 1)$ real-valued parameters $(\alpha_m^{(k)}, \sigma^2)$ given (k, m), and (iii) the deviations of the observed data d from the (estimated) signal $f = W_m \alpha_m^{(k)}$ given $(k, m, \alpha_m^{(k)}, \sigma^2)$. The approximate total description length (9) now becomes

$$
\begin{aligned}
MDL&(d, \widehat{\alpha}_m^{(k)}, \widehat{\sigma}^2, k, m) \\
&= L(k, m) + L(\widehat{\alpha}_m^{(k)}, \widehat{\sigma}^2 \mid k, m) + L(d \mid \widehat{\alpha}_m^{(k)}, \widehat{\sigma}^2, k, m), \quad (11)
\end{aligned}
$$

where $\widehat{\alpha}_m^{(k)}$ and $\widehat{\sigma}^2$ are the ML estimates of $\alpha_m^{(k)}$ and σ^2, respectively.

Let us now derive these ML estimates. Since we assumed the noise component is additive WGN, the probability of observing the data given all model parameters is

$$
P(d \mid \alpha_m^{(k)}, \sigma^2, k, m) = (2\pi\sigma^2)^{-N/2} \exp\left(-\frac{\|d - W_m \alpha_m^{(k)}\|^2}{2\sigma^2}\right), \quad (12)
$$

where $\|\cdot\|$ is the standard Euclidean norm on $\mathbf{R}^N$. For the ML estimate of σ^2, first consider the log-likelihood of (12)

$$
\ln p(d \mid \alpha_m^{(k)}, \sigma^2, k, m) = -\frac{N}{2} \ln 2\pi\sigma^2 - \frac{\|d - W_m \alpha_m^{(k)}\|^2}{2\sigma^2}. \quad (13)
$$

Taking the derivative with respect to σ^2 and setting it to zero, we easily obtain

$$
\widehat{\sigma}^2 = \frac{1}{N} \|d - W_m \alpha_m^{(k)}\|^2. \quad (14)
$$

Insert this equation back to (13) to get

$$\ln p(d \mid \boldsymbol{\alpha}_m^{(k)}, \widehat{\sigma}^2, k, m) = -\frac{N}{2} \ln \left(\frac{2\pi}{N} \|d - \boldsymbol{W}_m \boldsymbol{\alpha}_m^{(k)}\|^2 \right) - \frac{N}{2}. \tag{15}$$

Let $\widetilde{d}_m = \boldsymbol{W}_m^T d$ denote the vector of the expansion coefficients of d in the basis $\mathcal{B}_m$. Since this basis is orthonormal, i.e., $\boldsymbol{W}_m$ is orthogonal, and we use the ℓ^2 norm, we have

$$\|d - \boldsymbol{W}_m \boldsymbol{\alpha}_m^{(k)}\|^2 = \|\boldsymbol{W}_m (\boldsymbol{W}_m^T d - \boldsymbol{\alpha}_m^{(k)})\|^2 = \|\widetilde{d}_m - \boldsymbol{\alpha}_m^{(k)}\|^2. \tag{16}$$

From (15), (16), and the monotonicity of the ln function, we find that maximizing (15) is equivalent to minimizing

$$\|\widetilde{d}_m - \boldsymbol{\alpha}_m^{(k)}\|^2. \tag{17}$$

Considering that the vector $\boldsymbol{\alpha}_m^{(k)}$ only contains k nonzero elements, we can easily conclude that the minimum of (17) is achieved by taking the largest k coefficients in magnitudes of $\widetilde{d}_m$ as the ML estimate of $\boldsymbol{\alpha}_m^{(k)}$, i.e.,

$$\widehat{\boldsymbol{\alpha}}_m^{(k)} = \Theta^{(k)} \widetilde{d}_m = \Theta^{(k)} (\boldsymbol{W}_m^T d), \tag{18}$$

where $\Theta^{(k)}$ is a thresholding operation which keeps the k largest elements in absolute value intact and sets all other elements to zero. Finally, inserting (18) into (14), we obtain

$$\widehat{\sigma}^2 = \frac{1}{N} \|\boldsymbol{W}_m^T d - \Theta^{(k)} \boldsymbol{W}_m^T d\|^2 = \frac{1}{N} \|(\boldsymbol{I} - \Theta^{(k)}) \boldsymbol{W}_m^T d\|^2, \tag{19}$$

where $\boldsymbol{I}$ represents the N dimensional identity operator (matrix).

Let us further analyze (11) term by term. If we do not have any prior information on (k, m), then the cost $L(k, m)$ is the same for all cases, i.e., we can drop the first term of (11) for minimization purpose. However, if one has some prior preference about the choice of basis, knowing some prior information about the signal f, $L(k, m)$ should reflect this information. For instance, if we happen to know that the original function f consists of a linear combination of dyadic blocks, then we clearly should use the Haar basis. In this case, we may use the Dirac distribution, i.e., $p(m) = \delta_{m,m_0}$, where m_0 is the index for the Haar basis in the library $\mathcal{L}$. By (10), this leads to

$$L(k, m) = \begin{cases} L(k) & \text{if } m = m_0, \\ +\infty & \text{otherwise.} \end{cases}$$

On the other hand, if we either happen to know *a priori* or want to force the number of terms retained (k) to satisfy $k_1 \le k \le k_2$, then we may want to assume the uniform distribution for this range of k, i.e.,

$$L(k, m) = \begin{cases} L(m) + \log(k_2 - k_1 + 1) & \text{if } k_1 \le k \le k_2, \\ +\infty & \text{otherwise.} \end{cases} \tag{20}$$

As for the second term of (11), which is critical for our algorithm, we have to encode k expansion coefficients $\widehat{\alpha}_m^{(k)}$ and $\widehat{\sigma}^2$, i.e., $(k+1)$ real-valued parameters. However, in this case, by normalizing the whole sequence by $\|d\|$, we can safely assume that the magnitude of each coefficient in $\widehat{\alpha}^{(k)}$ is strictly less than one; in other words, the integer part of each coefficient is simply zero. Hence we do not need to encode the integer part as in (9) if we transmit the real-valued parameter $\|d\|$. Now the description length of $(\widehat{\alpha}_m^{(k)}, \widehat{\sigma}^2)$ given (k, m) becomes approximately $\frac{k+2}{2} \log N + L^*([\widehat{\sigma}^2]) + L^*([\|d\|])$ bits since there are $k + 2$ real-valued parameters: k nonzero coefficients, $\widehat{\sigma}^2$, and $\|d\|$. After normalizing by $\|d\|$, we clearly have $\widehat{\sigma}^2 < 1$ (see (19)), so that $L^*([\widehat{\sigma}^2]) = 1$ (see (3)). For each expansion coefficient, however, we still need to specify the index of the coefficient, i.e., where the k non-zero elements are in the vector $\widehat{\alpha}_m^{(k)}$. This requires $k \log N$ bits. As a result, we have

$$L(\widehat{\alpha}_m^{(k)}, \widehat{\sigma}^2 \mid k, m) = \frac{3}{2} k \log N + c, \tag{21}$$

where c is a constant independent of (k, m).

Since the probability of observing d given all model parameters is given by (12), we have for the last term in (11)

$$L(d \mid \widehat{\alpha}_m^{(k)}, \widehat{\sigma}^2, k, m) = \frac{N}{2} \log \|(I - \Theta^{(k)}) W_m^T d\|^2 + c', \tag{22}$$

where c' is a constant independent of (k, m).

Finally we can state our simultaneous noise suppression and signal compression algorithm. Let us assume that we do not have any prior information on (k, m) for now. Then, from (11), (21), and (22) with ignoring the constant terms c and c', our algorithm can be stated as:

Pick the index (k^*, m^*) such that

$$AMDL(k^*, m^*) = \min_{\substack{0 \le k < N \\ 1 \le m \le M}} \left(\frac{3}{2} k \log N + \frac{N}{2} \log \|(I - \Theta^{(k)}) W_m^T d\|^2 \right). \tag{23}$$

Then reconstruct the signal estimate

$$\widehat{f} = W_{m^*} \alpha_{m^*}^{(k^*)}. \tag{24}$$

Let us call the objective function to be minimized in (23), the approximate MDL (AMDL) since we ignored the constant terms. Let us now show a typical behavior of the AMDL value as a function of the number of terms retained (k) in Figure 1. (In fact, this curve is generated using Example 7.1 below.) We see that the log(residual energy) always decreases as k increases. By adding the penalty term of retaining the expansion coefficients, i.e., $(3/2)k \log N$ (which is just a straight line), we have the AMDL

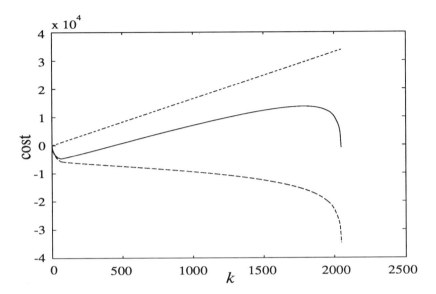

Figure 1. Graphs of AMDL versus k: AMDL [solid line] which is the sum of the $(3/2)k \log N$ term [dotted line] and the $(N/2) \log(\text{residual energy})$ term [dashed line].

curve which typically decreases for the small k, then starts increasing because of the penalty term, then finally decreases again at some large k near from $k = N$ because the residual error becomes very small. Now what we really want is the value of k achieving the minimum at the beginning of the k-axis, and we want to avoid searching for k beyond the maximum occurring for k near N. So, we can safely assume that $k_1 = 0$ and $k_2 = N/2$ in (20) to avoid searching more than necessary. (In fact, setting $k_2 > N/2$ does not make much sense in terms of data compression either.)

We briefly examine below the computational complexity of our algorithm. To obtain (k^*, m^*), we proceed as follows:

Step 1: Expand the data d into bases $\mathcal{B}_1, \ldots, \mathcal{B}_M$. Each expansion (including the best-basis selection procedure) costs $O(N)$ for wavelets, $O(N \log N)$ for wavelet packet best-bases, and $O(N[\log N]^2)$ for local trigonometric best-bases.

Step 2: Let $K(= k_2 - k_1 + 1)$ denote the length of the search range for k. For $k_1 \leq k \leq k_2$, $1 \leq m \leq M$, compute the expression in the parenthesis of the RHS in (23). This costs approximately $O(N + 3MK)$ multiplications and MK calls to the log function.

Step 3: Search the minimum entry in this table, which costs MK comparisons.

Step 4: Reconstruct the signal estimate (24), which costs $O(N)$ for wavelets, $O(N \log N)$ for wavelet packet best-bases, and $O(N[\log N]^2)$ for local trigonometric best-bases.

§6. Extension to Images

For images or multidimensional signals, we can easily extend our algorithm by using the multidimensional version of the wavelets, wavelet packets, and local trigonometric transforms. In this section, we briefly summarize the two-dimensional (2D) versions of these transforms. For the 2D wavelets, there are several different approaches. The first one, which we call the sequential method, is the tensor product of the one-dimensional (1D) wavelets, i.e., applying the wavelet expansion algorithm separately along two axes t_1 and t_2 corresponding to column (vertical) and row (horizontal) directions respectively. Let $f \in \mathbf{R}^{N_1 \times N_2}$ and H_i, G_i be the 1D convolution-subsampling operations along axis $t_i, i = 1, 2$. Then this version of the 2D wavelet transform first applies the convolution-subsampling operations along the t_1 axis to obtain $f_1 = (G_1 f, G_1 H_1 f, \ldots, G_1 H_1^{J_1} f)$, then applies the convolution-subsampling operations along the t_2 axis to get the final 2D wavelet coefficients $(G_2 f_1, G_2 H_2 f_1, \ldots, G_2 H_2^{J_2} f_1)$ of length $N_1 \times N_2$, where $J_1 (\leq \log N_1)$ and $J_2 (\leq \log N_2)$ are maximum levels of decomposition along t_1 and t_2 axes respectively. We note that one can choose different 1D wavelet bases for t_1 and t_2 axes independently. Given M different QMF pairs, there exist M^2 possible 2D wavelets using this approach.

The second approach is the basis generated from the tensor product of the multiresolution analysis. This decomposes an image f into four different sets of coefficients, $H_1 H_2 f$, $G_1 H_2 f$, $H_1 G_2 f$, and $G_1 G_2 f$, corresponding to "low-low", "high-low", "low-high", "high-high" frequency parts of the two variables, respectively. The decomposition is iterated on the "low-low" frequency part and this ends up in a "pyramid" structure of coefficients. Transforming the digital images by these wavelets to obtain the 2D wavelet coefficients are described in e.g., [20], [13].

There are also 2D wavelet bases which do not have a tensor-product structure, such as wavelets on the hexagonal grids and wavelets with matrix dilations. See e.g., [18], [17] for details.

There has been some argument as to which version of the 2D wavelet bases should be used for various applications [5], [13]. Our strategy toward this problem is this: we can put as many versions of these bases in the library as we can afford it in terms of computational time. Then minimizing the AMDL values automatically selects the most suitable one for our purpose.

As for the 2D version of the wavelet packet best-basis, the sequential method may be generalized, but it is not easily interpreted; the 1D best-bases may be different from column to column so that the resultant coefficients viewing along the row direction may not share the same frequency bands and scales unlike the 2D wavelet bases. This also makes the reconstruction algorithm complicated. Therefore, we should use the other tensor-product 2D wavelet approach for the construction of the 2D wavelet packet best-basis: we recursively decompose not only the "low-low" components but also the other three components. This process produces the "quad-tree" structure of wavelet packet coefficients instead of the "binary-tree" structure for 1D wavelet packets. Finally the 2D wavelet packet best-basis coefficients are selected using the entropy criterion [33].

The 2D version of the local trigonometric transforms can be constructed using the quad-tree structure again: the original image is smoothly folded and segmented into 4 subimages, 16 subimages, ..., and in each subimage the separable DCT/DST is applied, and then the quad-tree structure of the coefficients is constructed. Finally, the local trigonometric best-basis is selected using the entropy criterion [33].

For an image of $N = N_1 \times N_2$ pixels, the computational costs are approximately $O(N)$, $O(N \log_4 N)$, $O(N[\log_4 N]^2)$ for a 2D wavelet, a 2D wavelet packet best-basis, a 2D local trigonometric best-basis, respectively.

§7. Examples

In this section, we give several examples to show the usefulness of our algorithm.

Example 7.1. *The Synthetic Piecewise Constant Function of Donoho-Johnstone.*
We compared the performance of our algorithm in terms of the visual quality of the estimation and the relative ℓ^2 error with Donoho-Johnstone's method using the piecewise constant function used in their experiments [16]. The results are shown in Figure 2. The true signal is the piecewise constant function with $N = 2048$, and its noisy observation was created by adding the WGN sequence with $\|f\|/\|n\| = 7$. The library $\mathcal{L}$ for this example consisted of 18 different bases: the standard Euclidean basis of $\mathbf{R}^N$, the wavelet packet best-bases created with D02, D04, ..., D20, C06, C12, ..., C30, and the local cosine and sine best-bases (Dn represents the n-tap QMF of Daubechies and Cn represents the n-tap coiflet filter). In the Donoho-Johnstone method, we used the C06, i.e., 6-tap coiflet with 2 vanishing moments. We also specified the scale parameter $J = 7$, and supplied the *exact* value of σ^2. Next, we *forced* the Haar basis (D02) to

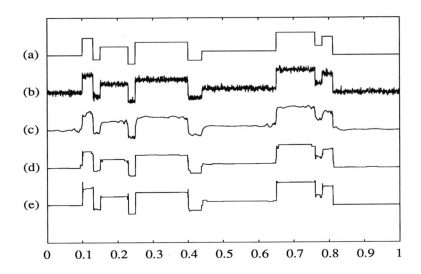

Figure 2. Results for the synthetic piecewise constant function: (a) Original piecewise constant function. (b) Noisy observation with (signal energy)/(noise energy) $= 7^2$. (c) Estimation by the Donoho-Johnstone method using coiflets C06. (d) Estimation by the Donoho-Johnstone method using Haar basis. (e) Estimation by the proposed method.

be used in their method. Finally, we applied our algorithm without specifying anything. In this case, the Haar-Walsh best-basis with $k^* = 63$ was automatically selected. The relative ℓ^2 errors are 0.116, 0.089, 0.051, respectively. Although the visual quality of our result is not too different from Donoho and Johnstone's (if we *choose* the appropriate basis for their method), our method generated the estimate with the smallest relative ℓ^2 error and slightly sharper edges. (See Section 8 for more about the Donoho-Johnstone method and its relation to our method.)

Example 7.2. *A Pure White Gaussian Noise.*
We generated a synthetic sequence of WGN with $\sigma^2 = 1.0$ and $N = 4096$. The same library as in Example 7.1 (with the best-bases adapted to this pure WGN sequence) was used. We also set the upper limit of search range $k_2 = N/2 = 2048$. Figure 3 shows the AMDL curves versus k for all bases in the library. As we can see, there is no single minimum in the graphs, and our algorithm satisfactorily decided $k^* = 0$, i.e., there is nothing to "learn" in this dataset.

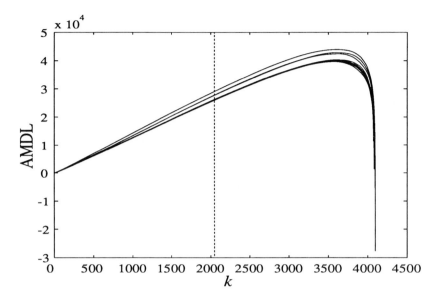

Figure 3. The AMDL curves of the White Gaussian Noise data for all bases. For each basis, $k = 0$ is the minimum value. The vertical dotted line indicates the upper limit of the search range for k.

Example 7.3. *A Natural Radioactivity Profile of Subsurface Formation.* We tested our algorithm on the actual field data which are measurement of natural radioactivity of subsurface formation obtained at an oil-producing well. The length of the data is $N = 1024$. Again, the same library was used as in the previous examples. The results are shown in Figure 4. In this case, our algorithm selected the D12 wavelet packet best-basis (Daubechies' 12-tap filter with 6 vanishing moments) with $k^* = 77$. The residual error is shown in Figure 4 (c) which consists mostly of a WGN-like high frequency component. The compression ratio is $1024/77 \approx 13.3$. However, to be able to reconstruct the signal from the surviving coefficients, we still need to record the indices of those coefficients.

Suppose we can store each index by b_i bytes of memory and the precision of the original data is b_f bytes per sample. Then the *storage reduction ratio* R_s can be computed by

$$R_s = \frac{N/r \times (b_f + b_i)}{N \times b_f} = \frac{1}{r}(1 + \frac{b_i}{b_f}), \tag{25}$$

where r is a compression ratio. The original data precision was $b_f = 8$ (bytes) in this case. Since it is enough to use $b_i = 2$ (bytes) for indices and $r = 13.3\%$, we have $R_s \approx 9.40\%$, i.e., 90.60% of the original data can be

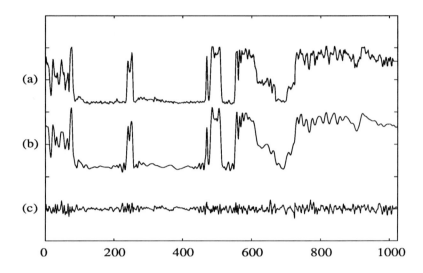

Figure 4. The estimate of the natural radioactivity profile of subsurface formation: (a) Original data which was measured in the borehole of an oil-producing well. (b) Estimation by the proposed method. (c) Residual error between (a) and (b).

discarded.

Example 7.4. *A Migrated Seismic Section.*
In this example, the data is a migrated seismic section as shown in Figure 5 (a). The data consist of 128 traces of 256 time samples. We selected six 2D wavelet packet best-bases (D02, C06, C12, C18, C24, C30) as the library. Figure 5 (b) shows the estimate by our algorithm. It automatically selected the filter C30 and the number of terms retained as $k^* = 1611$. If we were to choose a good threshold in this example, it would be fairly difficult since we do not know the accurate estimate of σ^2. The compression rate, in this case, is $(128 \times 256)/1611 \approx 20.34$. The original data precision was $b_f = 8$ as in the previous example. In this case we have to use $b_i = 3$ (1 byte for row index, 1 byte for column index, and 1 byte for scale level). If we put these and $r = 20.34\%$ into (25), we have $R_s \approx 6.76\%$, i.e., 93.24% of the original data can be discarded. Figure 5 (c) shows the residual error between the original and the estimate. We can clearly see the random noise and some strange high frequency patterns (which are considered to be numerical artifacts from the migration algorithm applied).

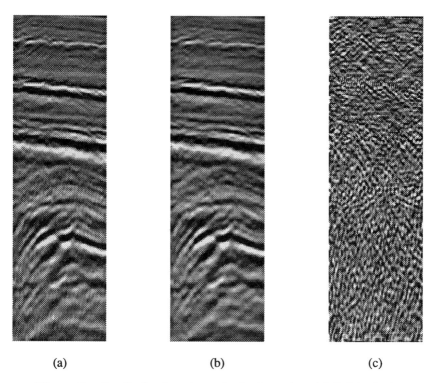

(a)   (b)   (c)

Figure 5. Results for the migrated seismic section: (a) Original seismic section with 128 traces and 256 time samples. (b) Estimation by the proposed method. (c) Residual error between (a) and (b). (Dynamic range of display (c) is different from those of (a) and (b).)

§8. Discussions

Our algorithm is intimately connected to the "denoising" algorithm of Coifman and Majid [7], [10]. Their algorithm first picks the best-basis from the collection of bases and sorts the best-basis coefficients in order of decreasing magnitude. Then they use the "theoretical compression rate" of the sorted best-basis coefficients $\{\alpha_i\}_{i=1}^{N}$ as a key criterion for separating a signal component from noise. The theoretical compression rate of a unit vector u is defined as $c(u) = 2^{H(u)}/N(u)$, where $H(u)$ is the ℓ^2-entropy of u, i.e., $H(u) = -\sum_{i=1}^{N(u)} u_i^2 \log u_i^2$, and $N(u)$ is the length of u. We note that $0 \le c(u) \le 1$ for any real unit vector u, and $c(u) = 0$ implies $u = \{\delta_{i,i_0}\}$ for some i_0 (the best possible compression), and $c(u) = 1$ implies $u = (1,\ldots,1)/\sqrt{N(u)}$ (the worst compression). Then to decide how many coefficients to keep as a signal component, they compare $c(\{\alpha_i\}_{i=k+1}^{N})$, the theoretical compression rate of the noise component (defined as the smallest $(N - k)$ coefficients), to the predetermined threshold τ. They search $k = 0, 1, \ldots$ which gives an unacceptably bad compression rate: $c(\{\alpha_i\}_{i=k+1}^N) \geq \tau$. Their algorithm critically depends on the choice of the threshold τ whereas our algorithm needs no threshold selection. On the other hand, their algorithm does not assume the WGN model we used in this paper; rather, they *defined* the noise component as a vector reconstructed from the best-basis coefficients of small magnitude.

Our algorithm can also be viewed as a simple yet flexible and efficient realization of the "complexity regularization" method for estimation of functions proposed by Barron [3]. He considered a general regression function estimation problem: given the data $(x_i, y_i)_{i=1}^N$, where $\{x_i \in \mathbf{R}^p\}$ is a sequence of the (p-dimensional) sampling coordinates (or explanatory variables) and $\{y_i \in \mathbf{R}\}$ is the observed data (or response variables), select a "best" regression function $\widehat{f}_N$ out of a list (library) $\mathcal{L}_N$ of candidate functions (models). He did not impose any assumption on the noise distribution, but assumed that the number of models in the list $\mathcal{L}_N$ depends on the number of observations N. Now the complexity regularization method of Barron is to find $\widehat{f}_N$ such that

$$R(\widehat{f}_N) = \min_{f \in \mathcal{L}_N} \left(\frac{1}{N} \sum_{i=1}^N d(y_i, f(x_i)) + \frac{\lambda}{N} L(f) \right),$$

where $d(\cdot, \cdot)$ is a measure of distortion (such as the squared error), $\lambda > 0$ is a regularization constant, and $L(f)$ is a complexity of a function f (such as the $L(m) + L(\boldsymbol{\theta}_m \mid m)$ term in (5)). He showed that various asymptotic properties of the estimator $\widehat{f}_N$ as $N \to \infty$, such as bounds on the estimation error, the rate of convergence, etc. If we restrict our attention to the finite dimensional vector space, use the library of orthonormal bases described in Section 2, adopt the length of the Shannon code (6) as a distortion measure, assume the WGN model, and finally set $\lambda = 1$, then Barron's complexity regularization method reduces to our algorithm. Our approach, although restricted in the sense of Barron, provides a computationally efficient and yet flexible realization of the complexity regularization method, especially compared to the library consisting of polynomials, splines, trigonometric series discussed in [3].

Our algorithm also has a close relationship with the denoising algorithm via "wavelet shrinkage" developed by Donoho and Johnstone [16]. (A well-written summary on the wavelet shrinkage and its applications can be found in [15].) Their algorithm first transforms the observed discrete data into a wavelet basis (specified by the user), then applies a "soft threshold" $\tau = \sigma \sqrt{\ln N}$ to the coefficients, i.e., shrinks magnitudes of all the coefficients by the amount τ toward zero. Finally the denoised data is obtained by the inverse wavelet transform. Donoho claimed informally in [15] that the reason why their method works is the ability of wavelets to compress the signal energy into a few coefficients. The main differences between our algorithm and that of Donoho and Johnstone are:

- Our method automatically selects the most suitable basis from a collection of bases whereas their method uses only a *fixed* basis specified by the user.

- Our method includes adaptive expansion by means of wavelet packets and local trigonometric bases whereas their method only uses a wavelet transform.

- Their method requires the user to set the coarsest scale parameter $J \leq n$ and a good estimate of σ^2, and the resulting quality depends on these parameters. On the other hand, our method does not require any such parameter setting.

- Their approach is based on the minimax decision theory in statistics and addresses the risk of the estimation whereas our approach uses the information-theoretic idea and combines denoising and the data compression capability of wavelets explicitly.

- Their method thresholds the coefficients *softly* whereas our method can be said to threshold *sharply*. This might cause some Gibbs-like effects in the reconstruction using our method.

Future extensions of this research are to: incorporate noise models other than Gaussian noise, extend the algorithm for highly nonstationary signals by segmenting them smoothly and adaptively, investigate the effect of sharp thresholding, and study more about the relation with the complexity regularization method of Barron as well as the wavelet shrinkage of Donoho-Johnstone.

§9. Conclusions

We have described an algorithm for simultaneously suppressing the additive WGN component and compressing the signal component in a dataset. One or more of the bases in the library, consisting of wavelets, wavelet packets, and local trigonometric bases, compress the signal component quite well, whereas the WGN component cannot be compressed efficiently by any basis in the library. Based on this observation, we have tried to estimate the "best" basis and the "best" number of terms to retain for estimating the signal component in the data using the MDL criterion. Both synthetic and real field data examples have shown the wide applicability and usefulness of this algorithm.

Acknowledgements

The author would like to thank Prof. R. Coifman and Prof. A. Barron of Yale University for fruitful discussions.

References

1. Ahmed, N. and K. R. Rao, *Orthogonal Transforms for Digital Signal Processing*, Springer-Verlag, New York, 1975.
2. Auscher, P., G. Weiss, and M. V. Wickerhauser, Local sine and cosine bases of Coifman and Meyer and the construction of smooth wavelets, in *Wavelets: A Tutorial in Theory and Applications*, C. K. Chui (ed.), Academic Press, San Diego, 237–256, 1992.
3. Barron, A. R., Complexity regularization with application to artificial neural networks, in *Proceeding NATO ASI on Nonparametric Functional Estimation*, Kluwer, 1991.
4. Barron, A. R. and T. M. Cover, Minimum complexity density estimation, *IEEE Trans. Inform. Theory*, **37** (4), 1034–1054, 1991.
5. Beylkin, G., R. Coifman, and V. Rokhlin, Fast wavelet transforms and numerical algorithms I, *Comm. Pure Appl. Math.*, **44**, 141–183, 1991.
6. Bradley, J. N. and C. M. Brislawn, Image compression by vector quantization of multiresolution decompositions, *Physica D*, **60**, 245–258, 1992.
7. Coifman, R. R. and F. Majid, Adapted waveform analysis and denoising, in *Progress in Wavelet Analysis and Applications*, Y. Meyer and S. Roques (eds.), Editions Frontieres, B.P.33, 91192 Gif-sur-Yvette Cedex, France, 63–76, 1993.
8. Coifman, R. R. and Y. Meyer. Remarques sur l'analyse de fourier à fenêtre, *Comptes Rendus Acad. Sci. Paris, Série I*, **312**, 259–261, 1991.
9. Coifman, R. R. and M. V. Wickerhauser, Entropy-based algorithms for best basis selection, *IEEE Trans. Inform. Theory*, **38** (2), 713–719, 1992.
10. Coifman, R. R. and M. V. Wickerhauser, Wavelets and adapted waveform analysis, in *Wavelets: Mathematics and Applications*, J. Benedetto and M. Frazier (eds.), chapter 10, CRC Press, Boca Raton, Florida, 1993.
11. Cover, T. M. and J. A. Thomas, *Elements of Information Theory*, Wiley Interscience, New York, 1991.
12. Daubechies, I., Orthonormal bases of compactly supported wavelets, *Comm. Pure Appl. Math.*, **41**, 909–996, 1988.
13. Daubechies, I., *Ten Lectures on Wavelets*, volume 61 of *CBMS-NSF Regional Conference Series in Applied Mathematics*, SIAM, Philadelphia, 1992.
14. DeVore, R. A., B. Jawerth, and B. J. Lucier, Image compression

through wavelet transform coding, *IEEE Trans. Inform. Theory*, **38** (2), 719–746, 1992.

15. Donoho, D. L., Wavelet shrinkage and W.V.D.: a 10-minute tour, in *Progress in Wavelet Analysis and Applications*, Y. Meyer and S. Roques (eds.), Editions Frontieres, B.P.33, 91192 Gif-sur-Yvette Cedex, France, 109–128, 1993.

16. Donoho, D. L. and I. M. Johnstone, Ideal spatial adaptation by wavelet shrinkage, preprint, Dept. of Statistics, Stanford University, Stanford, CA, Jun. 1992, revised Apr. 1993.

17. Gröchenig, K. and W. R. Madych, Multiresolution analysis, Haar bases, and self-similar tilings of R^n, *IEEE Trans. Inform. Theory*, **38** (2), 556–568, 1992.

18. Kovačević, J. and M. Vetterli, Nonseparable multidimensional perfect reconstruction filter banks and wavelet bases for R^n, *IEEE Trans. Inform. Theory*, **38** (2), 533–555, 1992.

19. Leclerc, Y. G., Constructing simple stable descriptions for image partitioning, *Intern. J. Computer Vision*, **3**, 73–102, 1989.

20. Mallat, S., A theory for multiresolution signal decomposition, *IEEE Trans. Pattern Anal. Machine Intell.*, **11** (7), 674–693, 1989.

21. Mallat, S. and Z. Zhang, Matching pursuit with time-frequency dictionaries, *IEEE Trans. Signal Processing, the special issue on wavelets and signal processing*, **41** (12), 1993, to appear.

22. Meyer, Y., *Wavelets: Algorithms and Applications*, translated and revised by R. D. Ryan, SIAM, Philadelphia, PA, 1993.

23. Meyer, Y., *Wavelets and Operators*, volume 37 of *Cambridge Studies in Advanced Mathematics*, translated by D. H. Salinger, Cambridge University Press, New York, 1993.

24. Niblack, W., *MDL Methods in Image Analysis and Computer Vision*, IEEE Conf. Comput. Vision, Pattern Recognition, Tutorial note, New York, Jun. 1993.

25. Rioul, O., Regular wavelets: a discrete-time approach, *IEEE Trans. Signal Processing, the special issue on wavelets and signal processing*, **41** (12) (1993), to appear.

26. Rioul, O. and M. Vetterli, Wavelets and signal processing, *IEEE SP Magazine*, **8** (4), 14–38, 1991.

27. Rissanen, J., A universal prior for integers and estimation by minimum description length, *Ann. Statist.*, **11** (2), 416–431, 1983.

28. Rissanen, J., Universal coding, information, prediction, and estimation, *IEEE Trans. Inform. Theory*, **30** (4), 629–636, 1984.

29. Rissanen, J., *Stochastic Complexity in Statistical Inquiry*, World Scientific, Singapore, 1989.

30. Shapiro, J. M., Image coding using the embedded zerotree wavelet algorithm, in *Proc. SPIE Conf. on Mathematical Imaging: Wavelet*

Applications in Signal and Image Processing, A. F. Laine (ed.), volume 2034, 180–193, 1993.

31. Wallace, R. S., *Finding natural clusters through entropy minimization*, PhD thesis, School of Comput. Science, Carnegie Mellon University, Pittsburgh, PA 15213, Jun. 1989.

32. Wax, M. and T. Kailath, Detection of signals by information theoretic criteria, *IEEE Trans. Acoust., Speech, Signal Processing*, **ASSP–33** (2), 387–392, 1985.

33. Wickerhauser, M. V., Lectures on wavelet packet algorithms, preprint, Dept. of Mathematics, Washington University, St. Louis, Missouri, Nov. 1991.

34. Wickerhauser, M. V., Fast approximate factor analysis, in *Proc. SPIE Conf. on Curves and Surfaces in Computer Vision and Graphics II*, volume 1610, 23–32, 1991.

35. Wickerhauser, M. V., High-resolution still picture compression, *Digital Signal Processing: A Review Journal*, **2** (4), 204–226, 1992.

36. Wilson, R., Finite prolate spheroidal sequences and their applications I: generations and properties, *IEEE Trans. Pattern Anal. Machine Intell.*, **PAMI-9** (6), 787–795, 1987.

Naoki Saito
Schlumberger-Doll Research
Old Quarry Road, Ridgefield, CT 06877
and
Department of Mathematics
Yale University
10 Hillhouse Avenue, New Haven, CT 06520
e-mail: *saito@ridgefield.sdr.slb.com*

Long-Memory Processes, the Allan Variance and Wavelets

Donald B. Percival and Peter Guttorp

Abstract. Long term memory has frequently been observed in physical time series. Statistical theory for long-memory stochastic processes is radically different from the standard time series analysis, which assumes short term memory. The Allan variance is a particular measure of variability developed for long-memory processes. This variance can be interpreted as a Haar wavelet coefficient variance, suggesting an approach towards assessing the variability of general wavelet classes. The theory is applied to a 'time' series of vertical ocean shear measurements for which some drawbacks with the Haar wavelet are observed.

§1. Introduction

In a variety of applications, time series analysts have noticed that the estimated autocovariance sequence for their data tends to decrease rather slowly, indicating that the series has 'long memory' in the sense that changes in the remote past continue to affect the present value of the series. Beran (1992) gives a good review of statistical and historical aspects of long-memory processes. Time series that are well modelled by long-memory processes have been observed, for example, by Newcomb (1895) in astronomy, by Gosset (Student, 1927) in chemistry, and by Smith (1938) in agriculture. In geophysics a famous early example is Hurst's (1951) study of the minimum annual height of the river Nile. This series has a sample auto-covariance sequence (acvs) $\hat{s}_\tau$ that is approximately proportional to $|\tau|^{-0.3}$; i.e., the sequence decays hyperbolically, thus ruling out such standard time series models as ARMA models that have exponentially decaying autoco-variance sequences (here $\hat{s}_\tau$ is an estimate of $s_\tau \equiv \text{cov}\{X_t, X_{t+\tau}\}$ when $\{X_t\}$ is a stationary process). The applicability of long-memory processes to climate data has been recently discussed by Raftery and Haslett (1989) and Smith (1992).

Wavelets in Geophysics
Efi Foufoula-Georgiou and Praveen Kumar (eds.), pp. 325–344.

The statistical properties of a long-memory process can be quite different from those of a set of independent and identically distributed (iid) observations. For example, the familiar variability properties of sample averages of iid observations are far from valid for a long-memory process. In fact, Smith (1938) observed that sample averages in agricultural uniformity trials had variances that did not decrease at the rate of the number of terms in the average, and deduced that there must be substantial spatial correlation. To see the effect of long memory on the variability of averages, consider a simple example of a long-memory process, namely, a fractional Gaussian process $\{X_t\}$ with self-similarity parameter $1/2 \leq H < 1$ (Mandelbrot and Wallis, 1969; such a process is the first difference of the fractional Brownian motion defined by Mandelbrot and Van Ness, 1968). By definition, this process has an acvs given by

$$s_\tau = \frac{s_0}{2} \left(|\tau + 1|^{2H} - 2|\tau|^{2H} + |\tau - 1|^{2H} \right), \quad \tau = \pm 1, \pm 2, \ldots,$$

where $s_0 = \text{var}\{X_t\}$ (note that the case $H = 1/2$ corresponds to the iid case because then $s_\tau = 0$ for $\tau \neq 0$, whereas $\{X_t\}$ has long-memory if $1/2 < H < 1$). Then $\text{var}\{\overline{X}\} = s_0 N^{2H-2}$, where $\overline{X}$ is the sample mean of N observations (i.e., $\sum_{t=1}^{N} X_t/N$). Note that, for values of H close to 1, the rate of decrease of variability in $\overline{X}$ is markedly different from the $1/N$ rate of the iid case. Naive application of iid statistics to the sample mean of a long-memory process can thus be very misleading. For example, if we fit a fractional Gaussian process to the Nile data, we obtain an estimate of $H = 0.85$, implying that the variance of the sample mean decreases like $1/N^{0.3}$ instead of the usual $1/N$ rate. Numerically, obtaining 100 observations from this fractional Gaussian process is equivalent to obtaining only 4 observations from an iid process!

In spite of its slow rate of decay, the sample average of a long-memory process is still a surprisingly efficient estimator of the mean level of the process (Beran and Künsch, 1985; Percival, 1985). Unfortunately, the same cannot be said for other standard statistics. In particular, the sample variance is a poor estimator of the process variance for a long-memory process because it has both severe bias and low efficiency (Beran, 1992). Due to these problems in estimation, Allan (1966) criticised the use of the sample variance as a meaningful estimator of variability for stationary processes with long memory and for nonstationary processes with infinite variance. He proposed an alternative theoretical measure of variability that is now known as the Allan variance (see Section 2 below). In terms of filtering theory, the Allan variance can be interpreted as the variance of a process after it has been subjected to an approximate band-pass filter of 'constant Q' (i.e., the ratio of the center frequency of the pass-band and the width of the pass-band is a constant). The chief advantages of the Allan variance for long-memory processes are two-fold: first, this variance can be estimated without bias and with good efficiency for such processes, and, second, estimates of the Allan variance can in turn be used to estimate the parameters of the long-memory process (as we note in Section 2, these statements also hold for certain nonstationary power-law processes). The Allan variance has been applied for over twenty-five years as the routine time domain measure of frequency stability in high-performance oscillators.

In a recent article Flandrin (1992) briefly noted that the Allan variance can be interpreted in terms of the coefficients of a Haar wavelet transform of a time series. We explore this connection in detail in Section 2. As we note in Section 3, the notion of the Allan variance can be generalized to other wavelets to define a wavelet variance. The wavelet variance is a useful way of summarizing the properties of the wavelet transform for certain processes. In particular, the parameters of long-memory processes can be deduced from the wavelet variance. We also note in Section 3 that standard estimators for the Allan variance can be easily generalized to these other wavelet variances.

In terms of computational complexity, the Allan variance is the simplest of the wavelet variances. Unfortunately, it can be misleading for certain processes of interest in geophysics. In the particular example we consider in Section 4, we find that the wavelet variance for Daubechies's 'least asymmetric' wavelet of order 8 (Daubechies, 1992) yields markedly better results (hereinafter we refer to this wavelet as the LA(8) wavelet). This example demonstrates the usefulness for data analysis of classes of wavelets beyond the simple Haar wavelet. Our analysis also demonstrates the importance of conducting a parallel spectral analysis to validate the interpretation of a wavelet variance in terms of the parameters for a long-memory process. Finally, in Section 5 we discuss some possible extensions, including some general thoughts on assessing the variability of wavelet coefficients.

§2. The Allan Variance and the Haar Wavelet

Suppose we have a time series of length N that can be regarded as one portion of one realization from the stochastic process $\{Y_t, t = 0, \pm 1, \ldots\}$ (for convenience, we assume that the sampling interval between consecutive observations is unity). Let

$$\overline{Y}_t(k) \equiv \frac{1}{k} \sum_{j=0}^{k-1} Y_{t-j} \tag{1}$$

represent the sample average of k consecutive observations, the latest one of which is Y_t. The Allan variance at scale k is denoted by $\sigma_Y^2(k)$ and is defined to be half the mean square difference between adjacent nonoverlapping

$\overline{Y}_t(k)$'s; i.e.,

$$\sigma_Y^2(k) \equiv \frac{1}{2} E\left\{[\overline{Y}_t(k) - \overline{Y}_{t-k}(k)]^2\right\}.$$

In order for $\sigma_Y^2(k)$ to be independent of the index t, we must impose a stationarity condition on the process $\{Y_t\}$, namely, that its first backward difference $Z_t \equiv Y_t - Y_{t-1}$ is a stationary process. Note that the Allan variance at scale k is a measure of how much averages of length k change from one time period of length k to the next.

To see why the Allan variance might be of interest for long-memory processes, suppose momentarily that $\{Y_t\}$ is a fractional Gaussian process with self-similarity parameter $1/2 < H < 1$. If we let $S_Y(\cdot)$ denote the spectral density function (sdf) for $\{Y_t\}$ defined for frequencies f between $-1/2$ and $1/2$, then we have

$$S_Y(f) = L_1(f)|f|^{1-2H},$$

where $L_1(\cdot)$ is a slowly varying function for $|f| \to 0$ (Beran, 1992). Thus, if we plot $\log(S_Y(f))$ versus $\log(f)$ for positive values of f close to zero, we will observe (to a good approximation) a line with a slope of $1 - 2H$. A similar result holds for the Allan variance: we have

$$\sigma_Y^2(k) = L_2(k)k^{2H-2},$$

where $L_2(\cdot)$ is a slowly varying function for $k \to \infty$ (Percival, 1983, Theorem 2.19). Thus, if we plot $\log(\sigma_Y^2(k))$ versus $\log(k)$ for large k, we will observe (to a good approximation) a line with a slope of $2H - 2$. In the analysis of frequency stability, it is common practice to produce plots of the so-called 'σ-τ curve,' which is just a plot of an estimate of $\log(\sigma_Y(k))$ versus $\log(k)$. On such a plot, a slope of β with $-1/2 \leq \beta < 0$ would be indicative of a stationary process with a self-similarity parameter $H = \beta + 1$.

As we mentioned above, the Allan variance is also well-defined when $\{Y_t\}$ is not itself a stationary process but its associated first backward difference $\{Z_t\}$ is. In such cases, we can *define* an sdf $S_Y(\cdot)$ for $\{Y_t\}$ in terms of the sdf $S_Z(\cdot)$ for $\{Z_t\}$ via the relationship

$$S_Y(f) \equiv \frac{S_Z(f)}{4\sin^2(\pi f)}$$

(this definition is motivated by the theory of linear filters). Suppose momentarily that

$$S_Y(f) = L_3(f)|f|^\alpha,$$

where $L_3(\cdot)$ is a slowly varying function for $|f| \to 0$, and $-3 < \alpha < 0$; i.e., $S_Y(\cdot)$ is a 'red' power-law process with exponent α greater than -3. Then we have

$$\sigma_Y^2(k) = L_4(k)k^{-\alpha-1},$$

where $L_4(\cdot)$ is a slowly varying function for $k \to \infty$. Thus, in principle, the Allan variance can be used to deduce the parameters for certain 'red' power-law processes.

Let us now consider the Haar wavelet transform of the time series Y_1, $\ldots$, Y_N, where we now assume that the sample size N is a power of 2 so that $N = 2^p$ for some positive integer p. By definition, this transform consists of $N - 1$ 'detail' coefficients and one 'smooth' coefficient $s_1 \equiv \sum_{t=1}^{N} Y_t/N$. The detail coefficients $d_{j,k}$ are defined for scales $k = 1, 2, 4, \ldots, N/2$ and – within the kth scale – for indices $j = 1, 2, 3, \ldots, N/2k$ as

$$d_{j,k} \equiv \frac{1}{\sqrt{2k}} \left[\sum_{l=0}^{k-1} Y_{2jk-l} - \sum_{l=0}^{k-1} Y_{2jk-k-l} \right]. \tag{2}$$

For example, we have

$$d_{j,1} \equiv [Y_{2j} - Y_{2j-1}]/\sqrt{2}, \ 1 \leq j \leq N/2$$

$$d_{j,2} \equiv [Y_{4j} + Y_{4j-1} - Y_{4j-2} - Y_{4j-3}]/2, \ 1 \leq j \leq N/4$$

$$d_{j,4} \equiv [Y_{8j} + \cdots + Y_{8j-3} - Y_{8j-4} - \cdots - Y_{8j-7}]/2\sqrt{2}, \ 1 \leq j \leq N/8$$

$$\vdots$$

$$d_{j,N/2} \equiv [Y_N + \cdots + Y_{N/2+1} - Y_{N/2} - \cdots - Y_1]/(\sqrt{2})^p.$$

We can now state the relationship between the Allan variance and the Haar wavelet coefficients $d_{j,k}$. Using Equations 1 and 2, we have

$$d_{j,k} = \left(\frac{k}{2}\right)^{1/2} [\overline{Y}_{2jk}(k) - \overline{Y}_{2jk-k}(k)].$$

If the first backward difference process $\{Z_t\}$ for $\{Y_t\}$ is stationary with zero mean, we then have

$$\text{var}\{d_{j,k}\} = E\{d_{j,k}^2\} = \frac{k}{2} E\{[\overline{Y}_{2jk}(k) - \overline{Y}_{2jk-k}(k)]^2\} = k\sigma_Y^2(k). \tag{3}$$

Thus the Allan variance at scale k is directly related to the variance of the Haar wavelet coefficient at that same scale. (Violation of the seemingly innocuous assumption that $E\{Z_t\} = 0$ can seriously impact our ability to make sense of estimates of the Allan variance – see Percival, 1983, for further discussion.)

In view of Equation 3, an obvious unbiased estimator for the Allan variance is

$$\tilde{\sigma}_Y^2(k) \equiv \frac{2}{N} \sum_{j=1}^{N/2k} d_{j,k}^2 = \frac{k}{N} \sum_{j=1}^{N/2k} [\overline{Y}_{2jk}(k) - \overline{Y}_{2jk-k}(k)]^2. \tag{4}$$

The above estimator is known in the frequency stability literature as the 'non-overlapped' estimator of the Allan variance because each value Y_t in the time series contributes to exactly one $d_{j,k}^2$ in the summation in Equation 4. The estimator $\tilde{\sigma}_Y^2(k)$ is not commonly used because of the superior statistical properties of a closely related estimator known as the maximal-overlap estimator (Greenhall, 1991). This estimator is defined as

$$\hat{\sigma}_Y^2(k) \equiv \frac{1}{2(N - 2k + 1)} \sum_{t=2k}^{N} \left(\overline{Y}_t(k) - \overline{Y}_{t-k}(k)\right)^2. \tag{5}$$

We can interpret the above estimator in terms of a filtering operation from which we can obtain the Haar wavelet coefficients $d_{j,k}$ by subsampling (also called 'downsampling'). At scale k, the filter that yields the $d_{j,k}$'s is of length $2k$ and has coefficients

$$\tilde{h}_{l,k} \equiv \begin{cases} 1/\sqrt{2k}, & l = 0, \ldots, k - 1; \\ -1/\sqrt{2k}, & l = k, \ldots, 2k - 1 \end{cases}$$

(cf. Equation 2). Note that $\sum_{l=0}^{2k-1} \tilde{h}_{l,k}^2 = 1$. If we let

$$\widetilde{W}_{t,k} \equiv \sum_{l=0}^{2k-1} \tilde{h}_{l,k} Y_{t-l}, \quad t = 2k, \ldots, N, \tag{6}$$

then we have $d_{j,k} = \widetilde{W}_{2jk,k}$; i.e., the $d_{j,k}$'s are obtained by appropriately subsampling every $2k$th value of the $\widetilde{W}_{t,k}$'s. We refer to the $\widetilde{W}_{t,k}$'s as the maximal-overlap Haar wavelet coefficients. We can now easily see the difference between the non-overlapped and maximal-overlap estimators:

$$\tilde{\sigma}_Y^2(k) = \frac{2}{N} \sum_{j=1}^{N/2k} \widetilde{W}_{2jk,k}^2, \quad \text{whereas} \quad \hat{\sigma}_Y^2(k) = \frac{1}{k(N - 2k + 1)} \sum_{t=2k}^{N} \widetilde{W}_{t,k}^2; \tag{7}$$

i.e., $\tilde{\sigma}_Y^2(k)$ makes use of just the $N/2k$ subsampled $\widetilde{W}_{t,k}$'s, whereas $\hat{\sigma}_Y^2(k)$ makes use of all $N - 2k + 1$ of the $\widetilde{W}_{t,k}$'s. Thus the maximal-overlap estimator does not use the usual decimated subseries of the discrete Haar wavelet transform, but it can be said to use a uniformly sampled version of the corresponding continuous transform. Note that use of $\hat{\sigma}_Y^2(k)$ rather than $\tilde{\sigma}_Y^2(k)$ imparts a certain degree of independence in the choice of the origin (i.e., if we form a new time series by discarding Y_1 and adding Y_{N+1}, only one of the $\widetilde{W}_{t,k}$'s in the maximal-overlap estimator changes whereas all of them change in the nonoverlapped estimator). Note also that the definition of $\hat{\sigma}_Y^2(k)$ in Equation 5 holds for all sample sizes N (i.e., N need not be a power of 2).

For computational purposes, it is more efficient to evaluate the right-hand side of Equation 5 by computing the summations needed to obtain the $\overline{Y}_t(k)$'s just once. To do so, let $X_0 \equiv 0$, and let $X_t \equiv X_{t-1} + Y_t$ for

$t = 1, \ldots, N$. Since $X_t = \sum_{u=1}^{t} Y_u$, we have

$$\overline{Y}_t(k) = \frac{1}{k} \sum_{j=0}^{k-1} Y_{t-j} = (X_t - X_{t-k})/k,$$

and hence we have

$$\overline{Y}_t(k) - \overline{Y}_{t-k}(k) = (X_t - 2X_{t-k} + X_{t-2k})/k$$

and

$$\widetilde{W}_{t,k} = (X_t - 2X_{t-k} + X_{t-2k})/\sqrt{2k}.$$

We can thus rewrite Equation 5 as

$$\hat{\sigma}_Y^2(k) = \frac{1}{2k^2(N - 2k + 1)} \sum_{t=2k}^{N} (X_t - 2X_{t-k} + X_{t-2k})^2.$$

§3. The Wavelet Variance

It is easy to extend the ideas of the previous section to wavelets other than the Haar wavelet. There are two approaches that we can take to generate the maximal-overlap wavelet coefficients $W_{t,k}$ based upon a wavelet of unit scale specified by the L_1 filter coefficients $h_{0,1}, h_{1,1}, \ldots, h_{L_1-1,1}$. In the first approach, we start by generating the appropriate filter $\{h_{l,k}\}$ for each scale k based upon the wavelet for unit scale. This can be done easily by taking the inverse discrete wavelet transform of a properly placed pulse (for details, see the subsection 'What Do Wavelets Look Like?' in Section 13.10 of Press *et al.*, 1992). Let L_k be the length of the wavelet of scale k so that, for example, $L_1 = 2$ for the Haar wavelet while $L_1 = 8$ for the LA(8) wavelet (the values of L_k for scales $k = 2, 4, \ldots$, obey the relationship $L_k = (2k - 1)(L_1 - 1) + 1$ and can also be obtained via the recursive formula $L_{2k} = 2L_k + L_1 - 2$). For each scale k we then filter our time series to obtain

$$W_{t,k} \equiv \sum_{l=0}^{L_k-1} h_{l,k} Y_{t-l}, \quad t = L_k, \ldots, N$$

(cf. Equation 6).

The second approach to obtaining the $W_{t,k}$'s is to modify the pyramid algorithm for obtaining the usual wavelet coefficients (see Section 13.10 of Press *et al.*, 1992, for a lucid description of this algorithm). The usual pyramid algorithm makes use of two filters, namely, the wavelet filter $\{h_{l,1}\}$ and the so-called scaling filter $\{g_{l,1}\}$. For the types of wavelets of interest here, the scaling filter is defined in terms of the wavelet filter via the 'quadrature mirror' relationship $g_{l,1} \equiv (-1)^{l+1} h_{L_1-l-1,1}$. Whereas the wavelet filter resembles a high-pass filter, the scaling filter resembles a low-pass filter.

The usual pyramid algorithm uses the same scaling and wavelet filters at each scale; i.e., at scale $2k$, these two filters are applied to the subsampled output from the scaling (low-pass) filter from scale k. For the maximal-overlap estimator we must eliminate all subsampling, so formally we must use different filters as we move from scale k to scale $2k$. Forming these new filters, however, is quite simple: the wavelet and scaling filters we need for scale k in the maximal-overlap pyramid algorithm are obtained by inserting $k-1$ zeros between each of the coefficients in the wavelet and scaling filters for unit scale (in effect, these zeros compensate for the elimination of subsampling). The maximal-overlap pyramid algorithm avoids all multiplications involving coefficients equal to zero by keeping track of the indices of elements of the series that need to be multiplied by the nonzero filter coefficients.

Given the wavelet filter $\{h_{l,1}\}$ and corresponding scaling filter $\{g_{l,1}\}$ for unit scale, the basic step of the maximal-overlap pyramid algorithm takes as input

- a scale k, which must be an integer power of 2, and

- a series $X_1, X_2, \ldots, X_{M_k}$ of length $M_k \equiv N - (k-1)(L_1-1)$, where N is the length of the time series (this need *not* be a power of 2);

and returns as output two series of length M_{2k}, namely,

- a low-pass series $X_1^{(l)}, X_2^{(l)}, \ldots, X_{M_{2k}}^{(l)}$ and

- a high-pass series $X_{N-M_{2k}+1}^{(h)}, X_{N-M_{2k}+2}^{(h)}, \ldots, X_N^{(h)}$.

Given its two inputs k and $\{X_t\}$, the basic step creates its two output series as follows.

1. Set $m = k(L_1 - 1) + 1$.

2. For $t = m, \ldots, M_k$, do the outer loop of 2:

   Set $X_{t-m+1}^{(l)} = g_{0,1} X_t$.
   Set $X_{N-M_k+t}^{(h)} = h_{0,1} X_t$.
   Set $u = t$.
   For $l = 1$ to $L_1 - 1$, do the inner loop of 2:

      Decrement u by k.
      Increment $X_{t-m+1}^{(l)}$ by $g_{l,1} X_u$.
      Increment $X_{N-M_k+t}^{(h)}$ by $h_{l,1} X_u$.

   End of the inner loop of 2.

   End of the outer loop of 2.

With the basic step so defined, the maximal-overlap pyramid algorithm consists of the following steps.

1. Evoke the basic step with scale $k = 1$ and $X_t = Y_t$ for $t = 1, \ldots,$ $M_1 = N$ to obtain as output the low-pass series $X_t^{(l)}$ and high-pass series $X_t^{(h)}$, both of length $M_2 = N - L_1 + 1$. Set $V_{t,1} \equiv X_t^{(l)}$ and $W_{t,1} \equiv X_t^{(h)}$; i.e., the maximal-overlap wavelet coefficients of scale 1 are just the elements of the high-pass series.

2. Evoke the basic step again, but this time with scale $k = 2$ and $X_t = V_{t,1}$ for $t = 1, \ldots, M_2$ to obtain as output the low-pass series $X_t^{(l)}$ and high-pass series $X_t^{(h)}$, both of length $M_4 = N - 3L_1 + 3$. Set $V_{t,2} \equiv X_t^{(l)}$ and $W_{t,2} \equiv X_t^{(h)}$; i.e., the maximal-overlap wavelet coefficients of scale 2 are just the elements of the high-pass series.

3. Evoke the basic step for $k = 4, 8, \ldots, 2^{\lfloor \log_2([N+L_1-2]/[L_1-1]) \rfloor - 1}$, where $\lfloor x \rfloor$ refers to the largest integer that is less than or equal to x. In particular, at the step in which the maximal-overlap wavelet coefficients for scale k are calculated, the basic step is evoked with $X_t = V_{t,k/2}$ for $t = 1, \ldots, M_k = N - (k-1)(L_1 - 1)$ to obtain as output the low-pass series $X_t^{(l)}$ and high-pass series $X_t^{(h)}$, both of length $M_{2k} = N - (2k-1)(L_1 - 1)$. Set $V_{t,k} \equiv X_t^{(l)}$ and $W_{t,k} \equiv X_t^{(h)}$; i.e., the maximal-overlap wavelet coefficients of scale k are just the elements of the high-pass series.

Note that, in contrast to the Haar wavelet, we can obtain only some of the wavelet coefficients by subsampling the $W_{t,k}$'s – the missing coefficients are those that involve circularly wrapping the time series (see the discussion in Press *et al.*, 1992). Under the same assumptions on $\{Y_t\}$ that we used to define the Allan variance, we can define the wavelet variance at scale k as

$$\nu_Y^2(k) \equiv E\{W_{t,k}^2\}/k$$

(cf. Equation 3). The equivalent of the maximal-overlap estimator for this wavelet variance would be given by

$$\hat{\nu}_Y^2(k) \equiv \frac{1}{k(N - L_k + 1)} \sum_{t=L_k}^{N} W_{t,k}^2$$

(cf. Equation 7). A plot of the square root of this quantity versus scale k on a log/log scale yields a generalization of the σ-τ curve, from which we can infer from regions of linearity that a power law process might be a good model for our data.

D. Percival and P. Guttorp

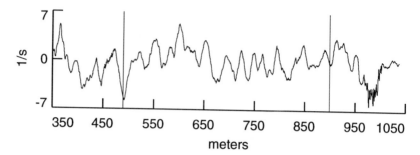

Figure 1. Plot of vertical shear measurements (in inverse seconds) versus depth (in meters). This series was collected and supplied by Mike Gregg, Applied Physics Laboratory, University of Washington. As of 1994, this series could be obtained via electronic mail by sending a message with the single line 'send lmpavw from datasets' to the Internet address statlib@lib.stat.cmu.edu – this is the address for StatLib, a statistical archive maintained by Carnegie Mellon University.

§4. An Application to Vertical Shear Measurements

Here we illustrate the ideas of the previous section by examining a 'time' series of vertical ocean shear measurements. The data were collected by an instrument that is dropped over the side of a ship and designed to then descend vertically into the ocean. As it descends, the probe collects measurements concerning the ocean as a function of depth. The ordering variable of our 'time' series is thus depth. One of the measurements is the x component of the velocity of water. This velocity is collected every 0.1 meters, first differenced over an interval of 10 meters, and then low-pass filtered to obtain a series related to vertical shear in the ocean. Vertical shear is thought to obey a power-law process over certain ranges of spatial frequency, so this series is a useful candidate for examining how well the Allan variance and other wavelet variances can deduce such a process.

Figure 1 shows the series of vertical shear measurements used in this study. The series extends from a depth of 350.0 meters down to 1037.4 meters in increments of 0.1 meters (there are 6875 data values in all). There are two thin vertical lines marked on the plot, between which there are 4096 values ranging from 489.5 meters to 899.0 meter. In what follows, we will assume that this subseries can be regarded as a portion of one realization of a process whose first backward difference is a stationary process (we need this assumption to apply meaningfully the methodology of Section 3).

The thick curves in the four rows of plots in Figure 2 show, respectively, the squared modulus of the transfer function for the Haar wavelet (left-hand

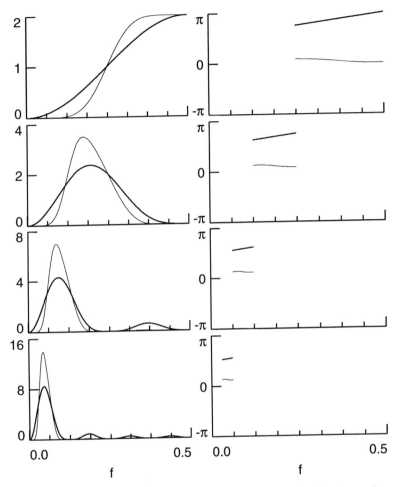

Figure 2. Squared modulus of transfer functions (left-hand column of plots) and phase functions (right-hand column) versus frequency for the Haar wavelet (thick curves) and LA(8) wavelet (thin curves) for scales 1, 2, 4 and 8 (top to bottom rows).

column) and the associated phase function over the nominal pass-band of the wavelet (right-hand column) for the four scales 1, 2, 4 and 8 (top to bottom rows). The thin curves in each plot show the corresponding functions for the LA(8) wavelet, which has the following filter coefficients:

$$
\begin{aligned}
h_{0,1} &= 0.03222310060407815 & h_{4,1} &= 0.80373875180538600 \\
h_{1,1} &= 0.01260396726226383 & h_{5,1} &= -0.49761866763256290 \\
h_{2,1} &= -0.09921954357695636 & h_{6,1} &= -0.02963552764596039 \\
h_{3,1} &= -0.29785779560560505 & h_{7,1} &= 0.07576571478935668
\end{aligned}
$$

(the source of these coefficients is the '$N = 4$' entry of Table 6.3, p. 198, Daubechies, 1992, which gives the LA(8) scaling filter coefficients normalized to sum to 2 – the above $h_{l,1}$'s were obtained via the 'quadrature mirror' relationship between scaling and wavelet filters with a renormalization so that $\sum_{l=0}^{7} h_{l,1}^2 = 1$). Note that the transfer functions for both wavelets roughly define a set of octave band filters; i.e., the transfer functions for scale k are approximately concentrated between frequencies and $1/4k$ and $1/2k$ (the spacing between the minor tick marks on the frequency axis is $1/16$). The plots show that the transfer functions for the LA(8) wavelet are a better approximation to a set of octave band filters than those for the Haar wavelet. The phase functions for the Haar wavelet are approximately linear over the nominal pass-bands, whereas those for the LA(8) wavelet are approximately constant and fairly close to zero. Thus the output from the Haar wavelet filters will be phase-shifted with respect to the input, making it difficult to line up events at various scales with the original time series. In contrast, because the filters for the LA(8) wavelet are approximately zero phase, we can more easily line up events at various scales with the original time series. Note, however, that the spans of Haar wavelet filters for scales 1, 2, 4 and 8 are, respectively, 2, 4, 8 and 16, whereas the corresponding spans for the LA(8) wavelet are 8, 22, 50 and 106. If these wavelets are used as noncircular filters as discussed in Section 3, the output from the LA(8) wavelet will become increasingly shorter compared to that of the Haar wavelet as the scale increases. In fact, for the larger scales the length of the LA(8) wavelet will exceed the length of the time series, whereas the reverse will be true for the Haar wavelet.

Figure 3 shows (from bottom to top) the renormalized outputs from the Haar wavelet filters for physical scales 0.1, 0.2, 0.4, 0.8, 1.6 and 3.2 meters (because the distance between adjacent observations is 0.1 meters, these physical scales correspond to, respectively, the unitless scales 1, 2, 4, 8, 16 and 32). In order to obtain physically meaningful units, it is necessary to renormalize the $\widetilde{W}_{t,k}$'s by dividing by $(2k)^{1/2}$. Each of these renormalized filtered series is drawn with the same vertical scale, so we see that the variability gets progressively larger as we move from shorter to longer scales (the distance between minor tick marks on the vertical scale is $1/s$). The usual Haar wavelet transform for these scales can be obtained by appropriately subsampling these filtered series. Note that all six of these filtered series appear to be approximately rescaled versions of

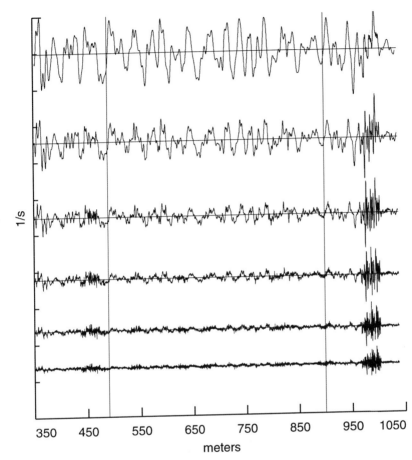

Figure 3. Renormalized outputs of Haar wavelet filters for physical scales 0.1, 0.2, 0.4, 0.8, 1.6 and 3.2 meters (bottom to top).

each other. We will see below that this is caused by a form of 'leakage' evidently due to the fact that the Haar approximation to pass-band filters is not good enough for this series and that in effect longer scale fluctuations are 'leaking' into these shorter scale fluctuations. Note that there are two bursts in the filtered series for the smaller scales, one near 450 meters and the other near 1000 meters.

Figure 4 is a repetition of Figure 3, with the LA(8) wavelet used instead of the Haar wavelet (again, we have renormalized the $W_{t,k}$'s by dividing by $(2k)^{1/2}$). The vertical scales on Figures 3 and 4 are identical, so it is evident

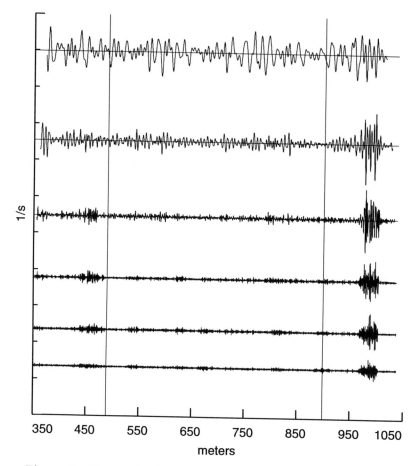

Figure 4. Renormalized outputs of LA(8) wavelet filters for physical scales 0.1, 0.2, 0.4, 0.8, 1.6 and 3.2 meters (bottom to top).

that there is much less variability at the smaller scales for the LA(8) wavelet than for the Haar wavelet. While there is some correspondence between the filtered series at different scales, the correspondence is much less marked for the LA(8) wavelet than for the Haar wavelet. Each of the filtered series in this figure is slightly shorter than the corresponding one in Figure 3 due to the longer span of the filters for the LA(8) wavelet. Note that only part of the usual LA(8) wavelet transform for these scales can be obtained by subsampling these filtered series because we are *not* filtering the data in Figure 1 as if it were circular (an assumption that would make no sense

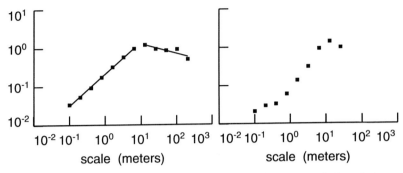

scale (meters)   scale (meters)

Figure 5. 'σ-τ' curves for the Haar wavelet (left-hand plot) and the LA(8) wavelet (right-hand).

at all for a series of ocean depth measurements!). Again we can see the bursts in the filtered series for the smaller scales near 450 and 1000 meters. An examination of the renormalized filtered series for physical scales longer than 3.2 meters shows no indication of these bursts. In the region marked by the thin vertical lines between the two bursts, the filtered series for all scales appear to be consistent with a stationarity assumption, so we have chosen this subseries of 4096 values as a candidate for analysis using the ideas discussed in the previous two sections.

The left-hand plot of Figure 5 shows the 'σ-τ' curve, a popular analysis tool in the frequency stability literature. This plot shows the square root of the Allan variance (estimated using the maximal-overlap estimator) at different physical scales plotted versus scale on a log-log plot. Theory suggests that regions of linearity correspond to a power-law process over a particular region of frequencies, with the exponent of the power-law process being related to the slope of the line (in log-log space). The first 7 values of the σ-τ curve fall on such a line almost perfectly. The line drawn through them on the plot was calculated via least squares and has a slope of 0.83. Since $\sigma_Y(k)$ (the square root of the Allan variance) varies approximately as $k^{-(\alpha+1)/2}$ for a power-law process with exponent α (see Section 2), the σ-τ curve strongly suggests the presence of a power-law process over scales of 0.1 to 6.4 meters with an exponent of $\alpha \doteq -2.66 \doteq -8/3$. The right-hand plot is the σ-τ curve corresponding to the LA(8) wavelet for the first 9 scales (recall that filtered series for the longer scales cannot be obtained due to the span of the filters for this wavelet), and it tells quite a different story. The values are not aligned in an obvious straight line as in the case of the Allan variance, and the slope that we found using the first 7 scales of the Allan variance certainly does not look reasonable for portions of the

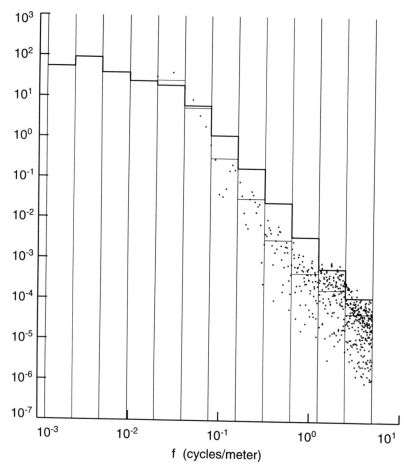

Figure 6. Comparison of spectral estimates implied by σ-τ curves with a WOSA spectral estimate.

LA(8) wavelet variance.

In Figure 6 we translate the σ-τ curves of the previous figure into estimates of the sdf over different octave bands (these have constant spacing on a log frequency scale). For example, the wavelet variance for a scale of 0.1 meters maps onto the highest octave band; the one for a scale of 0.2 meters maps onto the second highest octave band; and so forth. The thick 'staircase' on this plot corresponds to the Allan variance, while the thin staircase corresponds to the LA(8) wavelet variance. In general, the LA(8) wavelet variance yields an sdf estimate that is lower than the estimate obtained from the Haar wavelet (i.e., the Allan variance). The difference between the two sdf estimates is almost one order of magnitude in three of the octave bands. The dots in this figure show a Welch's overlapped segment averaging (WOSA) sdf estimate using a Hanning data taper and a block size of 1024 data points with segments overlapping by 50%, yielding an sdf estimate with 13.4 equivalent degrees of freedom (a comparison of the periodogram with direct sdf estimates using nontrivial data tapers indicated that the periodogram suffers from leakage and that tapering is required here; for details on the WOSA sdf estimate, see, for example, Section 6.17 of Percival and Walden, 1993). Whereas the LA(8) wavelet sdf estimate agrees fairly well with the WOSA spectral estimate, the Haar wavelet sdf estimate is too high in several of the octave bands, indicating some form of leakage evidently due to the fact that the Haar wavelet forms a fairly crude set of octave band filters.

§5. Concluding Remarks

Here we make several remarks concerning the results of the previous sections. First, a wavelet analysis of a time series that can be modelled as a power-law process can be used to deduce the properties of the underlying process, but it is dangerous to do so without a careful look at a traditional sdf estimate with good prevention against leakage. For example, if we had computed just the left-hand σ-τ curve of Figure 5, we might have been badly fooled by the degree to which the points line up as theory suggests they should in the presence of a power-law process.

Second, as others have noted, the wavelet variance is a tool that is well-adapted for studying power-law processes; however, the wavelet variance can lead us to find 'power laws' (or 'fractal behavior') in data that might be best modelled in other ways. Consider, for example, the octave band centered at 10^{-1} cycles/meter in Figure 6. The wavelet-based estimates of the power in this band are both somewhat higher than what the WOSA sdf estimate suggests is reasonable (in fact, the deficiency of power here can be attributed to the preprocessing operation in which the velocity series was first differenced over an interval of 10 meters). This octave band corresponds to a scale of 3.2 meters in the σ-τ curves of Figure 5. For both of these curves, the value at 3.2 meters can be obtained almost exactly by linearly interpolating (on a log/log scale) between the points at scales 1.6 and 6.4 meters. This suggests that the wavelet-based estimates might be biased in the sense that regions that do *not* agree with a nominal power-law behavior will tend to be 'filled in' in a manner consistent with the power-law assumption. What is vitally needed is a careful study of the bias and variance of wavelet-based estimates of the sdf for processes that deviate from a power law in at least some octave bands.

Third, for purposes of estimating the wavelet variance, results obtained by the frequency stability community using the Haar wavelet (i.e., the Allan variance) strongly favor the maximal-overlap estimator. The superiority of the maximal-overlap estimator is due to the fact that it is formed by taking the output from a filter, summing the square of each term and then dividing by the number of terms times the scale k, whereas the non-overlapped estimator takes exactly the same filter output, subsamples it, sums the square of each subsampled term and then divides by the number of subsampled terms times the scale k. It is thus intuitively reasonable that the subsampling operation yields an estimator of the Haar wavelet variance with larger variance than the nonoverlapped estimator (detailed analysis of the two estimators supports our intuition – see Greenhall, 1991). While it seems obvious that this result holds at least to some extent for wavelets other than the Haar wavelet, more work is needed to verify that the maximal-overlap estimator is indeed superior to the nonoverlapped estimator that would naturally fall out from the discrete wavelet transform (if we ignore those parts of the transform that correspond to circularly filtering a time series).

Fourth, in spite of its poor properties at small scales, the Haar wavelet seems to do quite well at long scales where in fact we cannot use the LA(8) wavelet at all (in a noncircular fashion) because the lengths of the wavelet and scaling filters are longer than our time series.

Fifth, procedures have been worked out in the frequency stability community for placing confidence limits on the Allan variance under the fairly stringent assumption that the exponent of the power law is known *a priori* (see Greenhall, 1991, for details). In principle, we can obtain similar confidence intervals for other wavelet variances. Also, by defining an estimator for the wavelet variance starting from a prewhitened direct spectral estimate, we can in fact easily obtain statistically valid confidence intervals based upon the sampling properties of direct spectral estimates. (Such an estimator would also allow us to get a handle on the covariance between estimates at different scales, a problem that has yet to be addressed even in the case of the Allan variance.)

Finally, Figure 4 points out that the real potential value of wavelets is in the area of time series with transient events. In the case of vertical shear measurements, wavelet analysis shows that these transients happen at just a few small scales. Figure 4 shows that the transient near 450 meters is confined chiefly to scales of 0.8 meters and smaller and, moreover, that the interval of time over which the transient dissipates is shorter for a scale of 0.8 meters than it is for smaller scales. These features in the data would be very hard to pick out using other analysis techniques.

References

1. Allan, D. W., Statistics of atomic frequency clocks, *Proc. IEEE*, **31**, 221–30, 1966.
2. Beran, J., Statistical methods for data with long-range dependence, *Statist. Sci.*, **7**, 404–16, 1992.
3. Beran, J. and H. R. Künsch, Location estimators for processes with long-range dependence, Research Report 40, Seminar für Statistik, ETH, Zürich, 1985.
4. Daubechies, I., Ten lectures on wavelets, *SIAM*, Philadelphia, 1992.
5. Flandrin, P., Wavelet analysis and synthesis of fractional brownian motion, *IEEE Trans. Info. Theo.*, **38**, 910–7, 1992.
6. Greenhall, C. A., Recipes for degrees of freedom of frequency stability estimators, *IEEE Trans. Instr. Meas.*, **40**, 994–9, 1991.
7. Hurst, H. E., Long-term storage capacity of reservoirs, *Trans. Amer. Soc. Civ. Eng.*, **116**, 770–9, 1951.
8. Mandelbrot, B. B. and J. W. Van Ness, Fractional Brownian motions, fractional noises and applications, *SIAM Rev.*, **10**, 422–37, 1968.
9. Mandelbrot, B. B. and J. R. Wallis, Computer experiments with fractional Gaussian noises. Part 1, averages and variances, *Water Resour. Resear.*, **5**, 228–41, 1969.
10. Newcomb, S., Astronomical constants (the elements of the four inner planets and the fundamental constants of astronomy), Supplement to the American Ephemeris and Nautical Almanac for 1897, U.S. Government Printing Office, Washington, 1895.
11. Percival, D. B., The statistics of long memory processes, unpublished Ph.D. dissertation, Department of Statistics, University of Washington, 1983.
12. Percival, D. B., On the sample mean and variance of a long memory process, Technical Report 69, Department of Statistics, University of Washington, 1985.
13. Percival, D. B. and A. T. Walden, *Spectral analysis for physical applications: multitaper and conventional univariate techniques*, Cambridge University Press, Cambridge, 1993.
14. Press, W. H., B. P. Flannery, S. A. Teukolsky and W. T. Vetterling, *Numerical recipes: the art of scientific computing (second edition)*, Cambridge University Press, Cambridge, 1992.
15. Raftery, A. E. and J. Haslett, Space-time modelling with long-memory dependence: assessing Ireland's wind power resource (with discussion), *J. Roy. Statist. Soc., Ser. C*, **38**, 1–21, 1989.
16. Smith, H. F., An empirical law describing heterogeneity in the yields of agricultural crops, *J. Agri. Sci.*, **28**, 1–23, 1938.
17. Smith, R. L., Comment [on Beran, 1992], *Statist. Sci.*, **7**, 422–5, 1992.

18. Student, Errors in Routine Analysis, *Biometrika*, **19**, 151–64, 1927.

Robert Spindel and John Harlett of the Applied Physics Laboratory, University of Washington, graciously provided discretionary funding to support Percival's work on this manuscript. Guttorp's work was partially funded by NSF grant DM S–9115756. The authors wish to thank Mike Gregg for supplying the data used in Section 4 and also to thank Chuck Greenhall and Emma McCoy for their helpful critiques.

Donald B. Percival
Applied Physics Laboratory
HN–10
University of Washington
Seattle, WA 98195
e-mail: *dbp@apl.washington.edu*

Peter Guttorp
Department of Statistics
GN–22
University of Washington
Seattle, WA 98195
e-mail: *peter@stat.washington.edu*

Bibliography

1. Akujuobi, C. M. and A. Z. Baraniecki, Wavelets and fractals: Overview of their similarities based on different application areas, *Proc. of IEEE-SP Int'l. Symposium on Time-Frequency and Time-Scale Analysis*, Victoria, 197–200, 1992.
2. Argoul, F., A. Arnéodo, J. Elezgaray, and G. Grasseau, Wavelet analysis of fractal growth processes, *Proc. of 4th EPS Liquid State Conf.*, Arcachon, France, 1988.
3. Argoul, F., A. Arnéodo, J. Elezgaray, G. Grasseau, and R. Murenzi, Wavelet transform of fractal aggregates, *Phys. Lett. A*, **135**, 327–336, 1989.
4. Argoul, F., A. Arnéodo, G. Grasseau, Y. Gagne, E. J.Hopfinger, and U. Frisch, Wavelet analysis of turbulence reveals the multifractal nature of the Richardson cascade, *Nature*, **338**, 51–53, 1989.
5. Argoul, F., A. Arnéodo, J. Elezgaray, G. Grasseau, and R. Murenzi, Wavelet analysis of the self-similarity of diffusion-limited aggregates and electrodeposition clusters, *Physical Review A*, **41(10)**, 5537–5560, 1990.
6. Arnéodo, A., F. Argoul, J. Elezgaray, and G. Grasseau, Wavelet transform analysis of fractals: Application to nonequilibrium phase transitions, in *Nonlinear Dynamics*, G. Turchetti (ed.), World Scientific, New Jersey, 130–180, 1988.
7. Arnéodo, A., G. Grasseau, and M. Holschneider, Wavelet transform of multifractals, *Phys. Rev. Lett.*, **61**, 2281–2284, 1988.
8. Arnéodo, A., G. Grasseau, and M. Holschneider, Wavelet transform analysis of some dynamical systems, in *Wavelets*, J. M. Combes, A. Grossmann, and P. Tchamitchian (eds.), Springer-Verlag, Berlin,, 315 pp., 1989.
9. Arnéodo, A., E. Bacry, and J. F. Muzy, Wavelet analysis of fractal signals - direct determination of the singularity spectrum of fully developed turbulence data, in *Wavelets and Turbulence*, Springer-Verlag, 1991.
10. Arnéodo, A., E. Bacry, J. F. Muzy, Wavelets and multifractal formalism for singular signals: Applications to turbulence data, preprint,

345

Centre de Recherche Paul Pascal, Pessac, 1991.

11. Arnéodo, A., F. Argoul, E. Bacry, J. Elezgaray, E. Freysz, G. Grasseau, J. F. Muzy, and B. Pouligny, Wavelet transform of fractals: I. From the transition to chaos to fully developed turbulence. II. Optical wavelet transform of fractal growth phenomena, in *Wavelets and Some of Their Applications*, Y. Meyer *et al.* (ed.), Springer-Verlag, 1991.

12. Arnéodo, A., F. Argoul, E. Bacry, J. F. Muzy, and M. Tabard, Golden mean arithmetic in the fractal branching of diffusion-limited aggregates, *Phys. Rev. Lett.*, **68(23)**, 3456–3459, 1992.

13. Arnéodo, A., J. F. Muzy, and E. Bacry, Wavelet analysis of fractal signals. Applications to fully developed turbulence data, *IUTAM Symp. on Eddy Structure Identification in Free Turbulent Shear Flows*, Poiters, October, 1992.

14. Arnéodo, A., F. Argoul, E. Bacry, J. Elezgaray, E. Freysz, G. Grasseau, J. F. Muzy, and B. Pouligny, 1. Wavelet transform of fractals: from the transition to chaos to fully developed turbulence. 2. Optical wavelet transform of fractal growth phenomena, in *Wavelets and Applications*, Y. Meyer (ed.), Masson, Paris and Springer, Berlin, 1992.

15. Bacry, E., A. Arneodo, U. Frisch, Y. Gagne, and E. Hopfinger, Wavelet analysis of fully developed turbulence data and measurement of scaling exponents, in *Turbulence and Coherent Structures*, O. Métais and M. Lesieur (eds.), Kluwer Academic Publishers, 203–215, 1991.

16. Bacry, E., J. F. Muzy, and A. Arneodo, Singularity spectrum of fractal signals from wavelet analysis: Exact results, *J. Stat. Phys.*, **70**, 635–674, 1993.

17. Barbaresco, F., Time-frequency analysis of a complex radar signal by a fast recursive processing of the wavelet transform for detection of moving objects in clutters, in *Proc. of IEEE-SP Int'l. Symposium on Time-Frequency and Time-Scale Analysis*, Victoria, 505–508, 1992.

18. Barrodale, I., N. R. Chapman, and C. A. Zala, Estimation of bubble pulse wavelets for deconvolution of marine seismograms, *Geophys. J. Roy. Astron. Soc.*, to appear, 1993.

19. Benzi, R. and M. Vergassola, Optimal wavelet analysis and its application to two-dimensional turbulence, *Fluid Dynamics Research*, **8**, 117–126, 1991.

20. Berger, M. A., Random affine iterated function systems - curve generation and wavelets, preprint, School of Mathematics, Georgia Inst. of Technology, Atlanta, 38 p., 1991.

21. Berger, M. A., IFS Algorithms for wavelet transforms, curves and surfaces, and image compression, preprint, School of Mathematics,

Georgia Inst. of Technology, Atlanta, 12 p.,

22. Berger, M.A. and Y. Wang, Multi-scale dilation equations and iterated function systems, preprint, School of Mathematics, Georgia Inst. of Technology, Atlanta, 30 p., 1991.

23. Bergstrom, H., and U. Hogstrom, Turbulent exchange above a pine forest II. Organized structures, *Boundary-Layer Meteorol.*, **49**, 231–263, 1989.

24. Berkhout, A. J., Least-squares inverse filtering and wavelet deconvolution, *Geophysics*, **42(7)**, 1369–1383, 1977.

25. Berkooz, G., J. Elezgaray, and P. Holmes, Coherent structures in random media and wavelets, *Physica D*, **61(1)**, 47–58, 1992.

26. Bertrand, J., P. Bertrand, and J. P. Ovarlez, Dimensionalized wavelet transform with application to radar imagery, *Proc. of IEEE Int. Conf. on Acoust., Speech and Signal Processing*, Vol. 4, Toronto, 2909–2912, 1991.

27. Bijaoui, A. and A. Fresnel, Wavelets and astronomical image analysis, in *Wavelets, Fractals, and Fourier Transforms*, M. Farge *et al.* (eds.), Clarendon Press, Oxford, 1993.

28. Bijaoui, A., E. Slezak, and G. Mars, Universe heterogeneities from a wavelet analysis, in *Wavelets, Fractals, and Fourier Transforms*, M. Farge *et al.* (eds.), Clarendon Press, Oxford, 1993.

29. Bliven, L. F. and B. Chapron, Wavelet analysis and radar scattering from water waves, *Naval Research Reviews*, **41(2)**, 11–16, 1989.

30. Bos, M. and E. Hoogendam, Wavelet transform for the evaluation of peak intensities in flow-injection analysis, *Anal. Chim. Acta*, **267(1)**, 73–80, 1992.

31. Bradshaw, G. A., Hierarchical analysis of spatial pattern and processes of Douglas-fir forests using wavelet analysis, Ph.D. thesis, Oregon State Univ., 303 p., 1991.

32. Bradshaw, G. A. and T. A. Spies, Characterizing canopy gap structure in forests using wavelet analysis, *J. Ecology*, **80**, 105–215, 1992.

33. Broecker, W. S., D. Peteet, and R. Rind, Does the ocean-atmosphere system have more than one stable mode of operation? *Nature*, **315**, 21–26, 1985.

34. Brotherton, T., T. Polland, R. Barton, and A. Krieger, Application of time-frequency and time-scale analysis to underwater acoustic transients, in *Proc. of IEEE-SP Int'l. Symposium on Time-Frequency and Time-Scale Analysis*, Victoria, 513–516, 1992.

35. Bube, K., P. Lailly, P. Sacks, F. Santosa, and W. W. Symes, Simultaneous determination of source wavelet and velocity profile using impulsive point-source reflections from a layered fluid, *Geophysical Journal*, **95(3)**, 449–462, 1988.

36. Cahalan, R. F., M. Nestler, W. Ridgway, W. J. Wiscombe, and T.

Bell, Marine stratocumulus spatial structure, Pretorna, New Zealand, paper presented at "4th International meeting on Statistical Climatology," pp. 29–25, 1989.

37. Caldwell, J., Solution of Burgers' equation by Fourier transform methods, in *Wavelets, Fractals, and Fourier Transforms*, M. Farge et al. (eds.), Clarendon Press, Oxford, 1993.

38. Chapman, N. R., I. Barrodale, and C. A. Zala, Measurement of sound-speed gradients in deep-ocean sediments using l_1-deconvolution techniques, *IEEE Journal of Oceanic Engineering*, **OE-9(1)**, 26–30, 1984.

39. Chapron, B., Application of higher order scale measurements, in *Proc. of IEEE-SP Int'l. Symposium on Time-Frequency and Time-Scale Analysis*, Victoria, 205–208, 1992.

40. Chapron, B. and F. L. Bliven, Scatterometer response and wavelet transformation analysis of water wave surface, *Traitement du Signal*, **9(1)**, 27–31, 1992.

41. Chou, Kenneth C., A stochastic modeling approach to multiscale signal processing, Ph.D. Thesis, Massachusetts Institute of Technology, May 1991.

42. Collineau, S. and Y. Brunet, Detection of turbulent coherent motions in a forest canopy. Part I: Wavelet analysis, *Boundary-Layer Meteorol.*, **65**, 357–379, 1993.

43. Combes, J. M., A. Brossmann, and Ph. Tchamitchian, Inverse problems and theoretical imaging, in *Wavelets: Time-Frequency Methods and Phase Space*, J. Combes, A. Grossmann, and Ph. Tchamitchian (eds.), Springer-Verlag, 1989.

44. Daneshvar, M. R. and C. S. Clay, Imaging of rough surfaces for impulsive and continuously radiating sources, *J. Acoust. Soc. Am.*, **82(1)**, 360–369, 1987.

45. David, P. M. and B. Chapron, Underwater acoustic signal analysis with wavelet process, *J. Acoust. Soc. Am.*, **87(5)**, 2118–2121, 1990.

46. David, P. M. and B. Chapron, Underwater acoustic, wavelets and oceanography, in *Wavelets and Applications*, Y. Meyer (ed.) Masson, Paris and Springer, Berlin, 1992.

47. Davis, A., A. Marshak, and W. Wiscombe, Bi-multifractal analysis and multi-affine modeling of non-stationary geophysical processes, application to turbulence and clouds, *Fractals*, to appear, 1993.

48. Davis, A., A. Marshak, W. Wiscombe, and R. Cahalan, Multifractal characterizations of non-stationary and intermittency in geophysical fields, observed, retrieved or simulated, submitted to *J. Geophys. Res.*, 1993.

49. Donoho, D. L., I. M. Johnstone, G. Kerkyacharian, and D. Picard, Density estimation by wavelet thresholding, Technical Report, Dept. of Statistics, Stanford, Univ., 1993.

50. Duroure, C. and C. P. Fantodiji, Wavelet analysis and Fourier analysis of stratocululus microphysical data, 1993, preprint.

51. Escalera, E. and A. Mazure, Wavelet analysis of subclustering: An illustration, Abell 754, *Astrophys. J.*, **388(1)**, 23–32, 1992.

52. Escalera, E., E. Slézak, and A. Mazure, New evidence of subclustering in the coma cluster using the wavelet analysis, *Astron. Astrophys.*, **264(2)**, 379–384, 1992.

53. Everson, R. M. and L. Sirovich, A survey of wavelet analysis applied to turbulence data, Report 89-182, Cent. Fluid Mech. Turbulence Comput., Brown Univ., Providence, RI, 1989.

54. Everson, R., L. Sirovich, K. R. Sreenivasan, Wavelet analysis of the turbulent jet, *Physical Letters A*, **146(6,7)**, 314–322, 1990.

55. Falconer, K. J., Wavelets, fractals and order-two densities, in *Wavelets, Fractals, and Fourier Transforms*, M. Farge et al. (eds.), Clarendon Press, Oxford, 1993.

56. Fantodji, C. P., C. Duroure, and H. R. Larsen, The use of wavelets for the analysis of geophysical data, submitted to *Physica*, 1993.

57. Farge, M., Transformee en ondelettes continue et application a la turbulence, *Journee Annuelle de la Societe Mathematique de France*, 1990.

58. Farge, M., Continuous wavelet transform application to turbulence, in *Wavelets and Their Applications*, G. Beylkin et al. (eds.), Jones & Bartlett, Boston, 1991.

59. Farge, M., Wavelet transforms and their applications to turbulence, *Annu. Rev. Fluid Mech*, **24**, 395–457, 1992.

60. Farge, M., The continuous wavelet transform of two-dimensional turbulent flows, in *Wavelets and Their Applications* M. B. Ruskai *et al.* (eds.), Jones & Bartlett, Boston, 275–302, 1992.

61. Farge, M., and G. Rabreau, Wavelet transform to analyze coherent structures in two-dimensional turbulent flows, *Proc. Scaling, Fractals and Nonlinear Variability in Geophysics I*, Paris, June 1988.

62. Farge, M., and G. Rabreau, Transformée en ondelettes pour détecter et analyser les structures cohérentes dans les écoulements turbulents bidimensionnels, *C. R. Acad. Sci. Paris Ser. II*, **307**, 1479–1486, 1988.

63. Farge, M. and M. Holschneider, Two-dimensional wavelet analysis of two-dimensional turbulent flows, *Proc. Sealing, Fractals and Nonlinear Variability in Geophysics II*, Barcelone, March 1989.

64. Farge, M. and G. Rabreau, Wavelet analysis of turbulent signals, *Proc. of Conf. on Chaos, Turbulence and Nonlinear Variability in Geophysics*, Barcelona, 1989.

65. Farge, M., M. Holschneider, and J. F. Colonna, Wavelet analysis of coherent structures in two-dimensional turbulent flows, in *Topological Fluid Mechanics*, H. K. Moffatt (ed.), Cambridge Univ. Press, 765–

776, 1990.

66. Farge, M., E. Goirand, and V. Wickerhauser, Wavelet packets analysis compression and filtering of two-dimensional turbulent flows, LMD, Ecole Normalé Supérieure, Paris, 1991, Preprint.

67. Farge, M., E. Goirand, Y. Meyer, F. Pascal, and M. V. Wickerhauser, Improved predictability of two-dimensional turbulent flows using wavelet packet compression, *Fluid Dyn. Res.*, **10(4-6)**, 229–250, 1992.

68. Field, D. J., Scale-invariance and self-similar 'Wavelet' transforms: an analysis of natural scenes and mammalian visual systems, in *Wavelets, Fractals, and Fourier Transforms*, M. Farge *et al.* (eds.), Clarendon Press, Oxford, 1993.

69. Flandrin, P., Time-frequency and time-scale, *Proc. of 4th Acoust., Speech and Signal Processing Workshop on Spectrum Estimation Modeling*, 77–80, 1988.

70. Flandrin, P., Fractional Brownian motion and wavelets, in *Wavelets, Fractals, and Fourier Transforms*, M. Farge *et al.* (eds.), Clarendon Press, Oxford, 1993.

71. Frick, P. and V. Zimin, Hierarchical models of turbulence, in *Wavelets, Fractals, and Fourier Transforms*, M. Farge *et al.* (eds.), Clarendon Press, Oxford, 1993.

72. Gamage, N. K. K., Modeling and analysis of geophysical turbulence: Use of optimal transforms and basis sets, Ph.D. Thesis, Oregon State University, 135 pp., 1989.

73. Gamage, N. and W. Blumen, Comparative analysis of low level cold fronts: wavelet, fourier, and empirical orthogonal function decompositions, *Monthly Weather Rev.*, **121**, 2867–2878, 1993.

74. Gamage, N. K. K. and C. Hagelberg, Detection and analysis of microfronts and associated coherent events using localized transforms, *J. Atmos. Sci.*, **50(5)**, 750–756, 1993.

75. Gambis, D., Wavelet transform analysis of the length of the day and the El-Niño/Southern Oscillation variations at intraseasonal and interannual time scales, *Ann. Geophysicae*, **10**, 331–371, 1992.

76. Genon-Catalot, V., C. Laredo, and D. Picard, Nonparametric estimation of the diffusion coefficient by wavelet methods, *Scandinavian Jour. of Statistics*, **19(4)**, 317–336, 1992.

77. Ghez, J. M. and S. Vaienti, On the wavelet analysis for multifractal sets, *J. Stat. Phys.*, **57**, 415–420, 1989.

78. Giannakis, G. B., Wavelet parameter and phase estimation using Cumulant slices, *IEEE Trans. on Geoscience and Remote Sensing*, **27(4)**, 452–455, 1989.

79. Glowinski, R., W. M. Lawton, M. Ravachol, E. Tenenbaum, Wavelet solutions of linear and nonlinear elliptic, parabolic and hyperbolic

problems in one space dimension, in *Computing Methods in Applied Sciences and Engineering*, R. Glowinski, A. Liehnewsky (eds.), SIAM Press, Philadelphia, 55–120, 1990.

80. Gollmer, S. M., R. F. Harshvardan, and J. B. Snider, Wavelet analysis of marine stratoculumus, *Proceedings 11th International Conf. on Clouds and Precipitation*, Montreal, Quebec, August, 1992.

81. Goupillaud, P. L., Three new mathematical developments on the geophysical horizon, *Geophysics: The Leading Edge of Exploration*, June, 1992.

82. Goupillaud, P., A. Grossmann, and J. Morlet, Cycle-octaves and related transforms in Seismic signal analysis, *Geoexploration*, **23**, 85–102, 1984.

83. Goursat, M., Numerical results of stochastic gradient techniques for deconvolution in seismology, *Geoexploration*, **23**, 103–119, 1984.

84. Grossmann A. and J. Morlet, Decomposition of Hardy functions into square integrable wavelets of constant shape, *SIAM J. Math. Anal.*, **15(4)**, 723–736, 1984.

85. Grossmann, A., R. Kronland-Martinet, and J. Morlet, Reading and understanding continuous wavelet transforms, in *Wavelets: Time-Frequency Methods and Phase Space*, J. Combes, A. Grossmann, and Ph. Tchamitchian (eds.), Springer-Verlag, pp. 2–20, 1989.

86. Gurbatov, S. N. and A. I. Saichev, The self-similarity of D-dimensional potential turbulence, in *Wavelets, Fractals, and Fourier Transforms*, M. Farge et al. (eds.), Clarendon Press, Oxford, 1993.

87. Hagelberg, C. R. and N. K. K. Gamage, Structure preserving wavelet decompositions of intermittent turbulence, *Boundary-Layer Meteorol.*, to appear, 1993.

88. Haynes, P. H. and W. A. Norton, Quantification of scale cascades in the stratosphere using wavelet transforms, in *Wavelets, Fractals, and Fourier Transforms*, M. Farge et al. (eds.), Clarendon Press, Oxford, 1993.

89. Heil, C. E. and D. F. Walnut, Continuous and discrete wavelet transforms, *SIAM Review*, **31(4)**, 628–666, 1989.

90. Howell, J. F. and L. Mahrt, An adaptive multiresolution data filter: application to turbulence and climatic time series, *J. Atmos. Sci.*, to appear, 1994.

91. Hudgins, L. H., Wavelet analysis of atmospheric turbulence, Ph.D. thesis, University of California, Irvine, March, 300 p., 1992.

92. Hudgins, L. H., M. E. Mayer, and C. A. Friehe, Fourier and wavelet analysis of atmospheric turbulence, in *Progress in Wavelet Analysis and Applications*, Y. Meyer and S. Roques (eds.), Editions Frontieres, 491–498, 1993.

93. Hudgins, L. H., C. A. Friehe, and M. E. Mayer, Wavelet transform

and atmospheric turbulence, *Physical Rev. Lett.*, Oct. 12, 1993.

94. Hunt, J.C.R., N.K.-R. Kevlahan, J. C. Vassilicos, and M. Farge, Wavelets, fractals and Fourier transforms: Detection and analysis of structure, in *Wavelets, Fractals, and Fourier Transforms*, M. Farge et al. (eds.), Clarendon Press, Oxford, 1993.

95. Jones, J. G., P. G. Earwicker, and G. W. Foster, Multiple-scale correlation detection, wavelet transforms and multifractal turbulence, in *Wavelets, Fractals, and Fourier Transforms*, M. Farge et al. (eds.), Clarendon Press, Oxford, 1993.

96. Kronland-Martinet, R., J. Morlet, and A. Grossmann, Analysis of sound patterns through wavelet transforms, *International J. of Pattern Recognition and Artificial Intelligence*, **1**, 273–301, 1987.

97. Kumar, P. and E. Foufoula-Georgiou, A new look at rainfall fluctuations and scaling properties of spatial rainfall using orthogonal wavelets, *J. Appl. Meteor.*, **32(2)**, 209–222, 1993.

98. Kumar, P. and E. Foufoula-Georgiou, A multicomponent decomposition of spatial rainfall fields: 1. segregation of large and small-scale features using wavelet transforms, *Water Resour. Res.*, **29(8)**, 2515–2532, 1993.

99. Kumar, P. and E. Foufoula-Georgiou, A multicomponent decomposition of spatial rainfall fields: 2. self-similarity in fluctuations, *Water Resour. Res.*, **29(8)**, 2533–2544, 1993.

100. Larsonneur, J. L. and J. Morlet, Wavelets and seismic interpretation, in *Wavelets, Time-Frequency Methods and Phase Space*, J. M. Combes, A. Grossman, P. Tchamitchian (eds.), Lecture notes on IPTI, Springer-Verlag, 126–131, 1989.

101. Levy, S. and D. W. Oldenburg, The deconvolution of phase-shifted wavelets, *Geophysics*, **47(9)**, 1285–1294, 1982.

102. Levy, S. and D. W. Oldenburg, Automatic phase correction of common-midpoint stacked data, *Geophysics*, **52(1)**, 51–59, 1987.

103. Liandrat, J. and F. Moret-Bailly, The wavelet transform: Some applications to fluid dynamics and turbulence, *European J. of Mech. B/Fluids*, **9(1)**, 1–19, 1990.

104. Lines, L. R. and T. J. Ulrych, The old and the new in seismic deconvolution and wavelet estimation, *Geophys. Prosp.*, **25**, 512–540, 1977.

105. Little, S. A., P. H. Carter, and D. K. Smith, Wavelet analysis of a bathymetric profile reveals anomalous crust, *Geophys. Res. Letts.*, **20(18)**, 1915–1918, 1993.

106. Mahrt, L., Eddy asymmetry in the sheared heated boundary layer, *Journal of the Atmospheric Sciences*, **48(3)**, 472–492, 1991.

107. Malakhov, A. and A. Yakimov, The physical models and mathematical description of l/f noise, in *Wavelets, Fractals, and Fourier Trans-*

forms, M. Farge *et al.* (eds.), Clarendon Press, Oxford, 1993.

108. Mallat, S. and S. Zhong, Characterization of signals from multiscale edges, *IEEE Trans. on Patt. Anal. and Mach. Intel.*, **14**, 710–732, 1992.

109. Mann, S. and S. Haykin, The adaptive chirplet: An adaptive generalized wavelet-like transform, *Proc. of Society of Photo-Optical Instrumentation Engineers*, **1565**, 402–413, 1991.

110. Meneveau, C., Analysis of turbulence in the orthonormal wavelet representation, Manuscript of the Center for Turbulence Research, Stanford Univ., 1990.

111. Meneveau, C., Analysis of turbulence in the orthonormal wavelet representation, *J. Fluid Mech.*, **232**, 469–520, 1991.

112. Meneveau, C., Dual spectra and mixed energy cascade of turbulence in the wavelet representation, *Physical Rev Lett.*, **11**, 1450–1453, 1991.

113. Meneveau, C. and K. R. Sreenivasan, The multifractal nature of turbulent energy dissipation, *J. Fluid Mech.*, **224**, 429–484, 1991.

114. Meneveau, C., Wavelet analysis of turbulence: the mixed energy cascade in *Wavelets, Fractals, and Fourier Transforms*, M. Farge *et al.* (eds.), Clarendon Press, Oxford, 1993.

115. Meyer, Y., *Wavelets and Applications: Proceedings of the International Conference*, Marseille, France, May 1989, Springer-Verlag, New York, 1992.

116. Meyer, Y., *Wavelets: Algorithms and Applications*, SIAM Press, Philadelphia, 1993.

117. Meyer, Y. and S. Roques, *Progress in Wavelet Analysis and Applications*, Editions Frontiéres, Gif sur Yvette, 1993.

118. Moffatt, H. K., Spiral structures in turbulent flow, in *Wavelets, Fractals, and Fourier Transforms*, M. Farge *et al.* (eds.), Clarendon Press, Oxford, 1993.

119. Morlet, J., G. Arens, I. Fourgeau, and D. Giard, Wave propagation and sampling theory, *Geophys.*, **47**, 203–236, 1982.

120. Muller, J., Morphology of disordered materials studied by multifractal analysis, in *Wavelets, Fractals, and Fourier Transforms*, M. Farge *et al.* (eds.), Clarendon Press, Oxford, 1993.

121. Muzy, J. F., E. Bacry, and A. Arnéodo, Wavelets and multifractal formalism for singular signals: Application to turbulence data, *Phys. Rev. Lett.*, **67(25)**, 3515–3518, 1992.

122. Muzy, J. F., E. Bacry, and A. Arneodo, Multifractal formalism for fractal signals: The structure function aproach versus the wavelet-transform modulus-maxima method, *Phys. Rev. E*, **47**, 875–884, 1993.

123. Meyers, S. D., B. G. Kelly, and J. J. O'Brien, An introduction to wavelet analysis in oceanography and meteorology: with application to the dispersion of Yanai waves, *Monthly Weather Review*, accepted

May 1993.

124. Otaguro, T., S. Takagi, and H. Satoy, Pattern search in a turbulent signal using wavelet analysis, *Proc. 21st Japan Symp. on Turbulence*, Tokyo, Japan, 1989.

125. Pentland, A. P., Surface interpolation using wavelets, in *Proc of 2nd European Conference on Computer Vision*, G. Sandini (ed.), Springer-Verlag, 615–619, 1992.

126. Pentland, A. P. and B. Horowitz, A practical approach to fractal-based image compression, in *Data Compression Conference*, J. A. Storer and J. H. Reif (eds.), IEEE Comput. Soc. Press, Los Alamitos, 176–185, 1991.

127. Permann, D. and I. Hamilton, Wavelet analysis of time series for the duffing oscillator: The detection of order within chaos, *Phys. Rev. Lett.*, **69(18)**, 2607–2610, 1992.

128. Persoglia, S., S. Sancin, and A. Vesnaver, Adaptive deconvolution by lattice filters: Experience on synthetic and real data, *Bollettino di Geofisica Teorica ed Applicata*, **27(107)**, 169–183, 1985.

129. Phuvan, S., T. K. Oh, N. P. Caviris, Y. Li, and H. H. Szu, Texture analysis by space-filling curves and one-dimensional Haar wavelets, *Opt. Eng.*, **31(9)**, 1899–1906, 1992.

130. Pittner, S., J. Schneid, and C. W. Ueberhuber, *Wavelet Literature Survey*, Institute for Applied and Numerical Mathematics, Technical University Vienna, Wien, Austria, 1993.

131. Pouligny, B., G. Gabriel, J. F. Muzy, A Arnéodo, F. Argoul, and E. Freysz, Optical wavelet transform and local scaling properties of fractals, *J. Appl. Cryst.*, **24(5)**, 526–530, 1991.

132. Powell, D. and C. E. Elderkin, An investigation of the application of Taylor's hypothesis to atmospheric boundary layer turbulence, *J. Atmos. Sci.*, **31**, 990–1002, 1974.

133. Ramanathan, J. and O. Zeitouni, On the wavelet transform of fractional Brownian motion, *IEEE Trans. on Inform. Theory*, **37(4)**, 1156–1158, 1991.

134. Ranchin, T. and L. Wald, The wavelet transform for the analysis of remotely sensed images, *Int. J. Remote Sensing*, **14(3)**, 615–619, 1993.

135. Rasmussen, H. O., The wavelet Gibbs phenomenon, in *Wavelets, Fractals, and Fourier Transforms*, M. Farge et al. (eds.), Clarendon Press, Oxford, 1993.

136. Redondo, J. M., Fractal models of density interfaces, in *Wavelets, Fractals, and Fourier Transforms*, M. Farge et al. (eds.), Clarendon Press, Oxford, 1993.

137. Redondo, J. M., R. M. Gonzalez, and J. L. Cano, Fractal aggregates in the atmosphere, in *Wavelets, Fractals, and Fourier Transforms*, M.

Farge *et al.* (eds.), Clarendon Press, Oxford, 1993.

138. Resnikoff, H. L. and C. S. Burrus, Relationships between the Fourier transform and the wavelet transform, *Proc. of the Society of Photo-Optical Instrumentation Engineers*, **1348**, 291–300, 1990.

139. Ricker, N., The form and laws of propagation of seismic wavelets, *Geophysics*, **18**, 10–40, 1953.

140. Rioul, O. and M. Vetterli, Wavelets and signal processing, *IEEE Signal Processing Magazine*, **8(4)**, 14–38, 1991.

141. Rioul, O. and P. Duhamel, Fast algorithms for discrete and continuous wavelet transforms, *IEEE Trans. on Inform. Theory*, **38(2)**, 569–586, 1992.

142. Rioul, O. and P. Flandrin, Time-scale energy distributions: A general class extending wavelet transforms, *IEEE Trans. on Signal Processing*, **40(7)**, 1746–1757, 1992.

143. Saether, G., K. Bendiksen, J. Muller, and E. Froland, The fractal dimension of oil-water interfaces in channel flows, in *Wavelets, Fractals, and Fourier Transforms*, M. Farge *et al.* (eds.), Clarendon Press, Oxford, 1993.

144. Saracco, G., Propagation acoustique en Régime harmonique & transitoire à travers un Milieu Inhomogène: Méthodes asymptotiques et transformation en ondelettes, Thèse de l'Univ. d'Aix-Marseille II, Faculté des Sciences de Luminy, 232 p., 1989.

145. Saracco, G., A. Grossman, and Ph. Tchamitchian, Use of wavelet transforms in the study of propagation of transient acoustic signals across a plane interface between two homogeneous media, in *Wavelets, Time-Frequency Methods and Phase Space*, J. A. Combes, A. Grossman, P. Tchamitchian (eds.), Lecture Notes on IPTI, Springer-Verlag, 139–146, 1989.

146. Saracco, G. and P. Tchamitchian, A study of acoustic transmission of transient signal in an inhomogeneous medium with the help of a wavelet transform - application to an air-water plane interface, in *Electromagnetic and Acoustic Scattering, Detection and Inverse Problems*, P.Chiapetta, B. Torréssani (eds.), World Scientific, 222–241, 1989.

147. Saracco, G., C. Gazanhes, J. Sageloli, and J.-P Sessarego, Analyse Temps-echelle de la diffusion acoustique par des coques sphériques elastiques en Régime Impulsionnel, *J. Acoustique*, **3(4)**, 381–392, 1990.

148. Saracco, G., P. Guillemain, and R. Kronland-Martinet, Characterization of elastic shells by the use of the wavelet transform, in *Proc. of IEEE Ultrasonics Symposium, Vol. 2*, Honolulu, 881–885, 1990.

149. Schiff, S. J., Resolving time-series structure with a controlled wavelet transform, *Opt Eng.*, **31(11)**, 2492–2495, 1992.

150. Schult, R. L. and H. W. Wyld, Using wavelets to solve the Burgers

equation - A comparative study, *Physical Review A*, **46(12)**, 7953–7958, 1992.

151. Shann, W.-C., Finite element methods for Maxwell's equations with stationary magnetic fields and Galerkin-wavelets methods for two-point boundary value problems, Ph.D. Thesis, Pennsylvania State Univ., 132 p., 1991.

152. Sinha, B. and K. J. Richards, The wavelet transform applied to flow around Antarctica, in *Wavelets, Fractals, and Fourier Transforms*, M. Farge et al. (eds.), Clarendon Press, Oxford, 1993.

153. Slézak, E., A. Bijaoui, and G. Mars, Identification of structures from Galaxy counts: Use of the wavelet transform, *Astron. Astrophys.*, **227(2)**, 301–316, 1990.

154. Stark, H.-G., Fractal graphs and wavelet series, *Phys. Letters A*, **143(9)**, 443–447, 1990.

155. Stark, H.-G., Continuous wavelet transform and continuous multiscale analysis, *J. Math. Anal. Appl.*, **169(1)**, 179–196, 1992.

156. Stark, J. and P. Bressloff, Iterated function systems and their applications, in *Wavelets, Fractals, and Fourier Transforms*, M. Farge et al. (eds.), Clarendon Press, Oxford, 1993.

157. Stone, D. G., Wavelet estimation, *Proc. of IEEE Int. Conf. on Acoust., Speech and Signal Processing*, **72(10)**, 1394–1402, 1984.

158. Strang, G., The optimal coefficients in Daubechies wavelets, *Physica D*, **60(1-4)**, 239–244, 1992.

159. Strichartz, R. S., Wavelet expansions of fractal measures, *J. Geom. Anal.*, **1(3)**, 269–289, 1991.

160. Tewfik, A. H. and M. Kim, Correlation structure of the discrete wavelet coefficients of fractional Brownian motion, *IEEE Trans. on Inform. Theory*, **38(2)**, 904–909, 1992.

161. Turner, B. J. and M. Y. Leclerc, Conditional sampling of coherent structures in atmospheric turbulence using the wavelet transform, *J. Atmos. and Ocean. Tech.*, to appear, 1994.

162. Vassilicos, J. C., Fractals in turbulence, in *Wavelets, Fractals, and Fourier Transforms*, M. Farge et al. (eds.), Clarendon Press, Oxford, 1993.

163. Vergassola, M. and U. Frisch, Wavelet transforms of self-similar processes, *Physica D*, **54(1,2)**, 58–64, 1991.

164. Vermeer, P. L., The multiscale and wavelet transform with applications in well log analysis, Ph.D. thesis, Techn. Univ. Delft, 200 p., 1992.

165. Vermeer, P. L. and J.A.H. Alkemade, Multiscale segmentation of well logs, in *Wavelets, Fractals, and Fourier Transforms*, M. Farge et al. (eds.), Clarendon Press, Oxford, 1993.

166. Wallace, R. S., Finding natural clusters through entropy minimiza-

tion, Ph.D. thesis, School of Comput. Science, Carnegie Mellon Univ., Pittsburg, PA 15213, Jun. 1989.

167. Wickerhauser, M. V., Lectures on wavelet packet algorithms, Technical Report. Dept. of Math., Washington Univ., 1992.

168. Wickerhauser, M. V., Acoustic signal compression with wavelet packets, in *Wavelets - A Tutorial in Theory and Applications*, C. K. Chui (ed.), Academic Press, Boston, 679–700, 1992.

169. Willis, G. E. and J. Deordorff, On the use of Taylor's translation hypothesis for diffusion in the mixed layer, *Quart. J. Roy. Meteorol. Soc.*, **102**, 817–822, 1976.

170. Wilson, R., A. D. Calway, and E.R.S. Pearson, A generalized wavelet transform for Fourier analysis: The multiresolution Fourier transform and its application to image and audio signal analysis, *IEEE Trans. on Inform. Theory*, **38(2)**, 674–690, 1992.

171. Wornell, G. W., A Karhunen-Loève-like expansion for $1/f$-processes via wavelets, *IEEE Trans. on Inform. Theory*, **36(4)**, 859–861, 1990.

172. Wornell, G. W. and A. V. Oppenheim, Estimation of fractal signals from noisy measurements using wavelets, *IEEE Trans. Sig. Proc.*, **40(3)**, 611–623, 1992.

173. Wornell, G. W. and A. V. Oppenheim, Wavelet-based representations for a class of self-similar signals with application to fractal modulation, *IEEE Trans. on Inform. Theory*, **38(2)**, 785–800, 1992.

174. Wyngaard, J. C. and S. F. Clifford, Taylor's hypothesis and high-frequency turbulence spectra, *J. Atmos. Sci.*, **34**, 922–929, 1977.

175. Xu, J.-C. and W.-C. Shann, Galerkin-wavelet methods for two-point boundary-value-problems, *Numer. Math.*, **63(1)**, 123–144, 1992.

176. Yamada, M. and K. Ohkitani, Orthonormal wavelet expansion and its application to turbulence, *Prog. Theor. Phys.*, **83(5)**, 819–823, 1990.

177. Yamada, M. and K. Ohkitani, Orthonormal wavelet analysis of turbulence, *Fluid Dynamics Res.*, **8(1-4)**, 101–115, 1991.

178. Yamada, M. and K. Ohkitani, An identification of energy cascade in turbulence by orthonormal wavelet analysis, *Prog. Theor. Phys.*, **86(4)**, 799–815, 1991.

179. Zhong, S., Edge representation from wavelet transform maxima, Ph.D. Thesis, Univ. New York, 129 p., 1991.

180. Zubair, L., Studies in turbulence using wavelet transforms for data compression and scale-separation, Thesis, Yale Univ., New Haven.

Subject Index

Detailed Table of Contents

WAVELET ANALYSIS AND ITS APPLICATIONS

Charles K. Chui, Series Editor